Point Defects in Crystals

Point Defects in Crystals

R. K. Watts

Texas Instruments, Inc.
Dallas, Texas

A Wiley-Interscience Publication

JOHN WILEY & SONS

New York / London / Sydney / Toronto

PHYSICS

Library of Congress Cataloging in Publication Data:

Watts, Roderick K   1939–
   Point defects in crystals.

   "A Wiley-Interscience publication."
   Bibliography: p.
   Includes index.
   1. Crystals—Defects. I. Title.

QD921.W24   548'.842   76-43013
ISBN 0-471-92280-3

Printed in the United States of America

10 9 8 7 6 5 4 3 2 1

To Jean and Alison

Preface

If the regular array of atoms of a crystal is interrupted by an imperfection that can be inscribed in a small sphere, the imperfection is called a point defect. A small concentration of point defects may drastically modify the electrical or optical properties of a material to make the material useful in practical applications. Familiar examples are solid-state electronic circuit elements of many types, phosphors for fluorescent lamps and television, and solid-state lasers. Academic interest in defects has also been keen for many decades. A point defect is closely surrounded by neighboring atoms, and the defect problem is a many-body problem. Constructing an economical theoretical model that describes the observed properties is challenging to the theorist, and the experimentalist tries to determine the nature of the defect by spectroscopic and other techniques.

This book is an introduction to the spectroscopic properties of point defects in crystals. It is intended for materials scientists, physicists, chemists, and students interested in crystalline materials. Much is now known about the structure of point defects, largely through spectroscopic studies. Scientists concerned with materials characterization should find convenient the collection of defect spectroscopic properties for a variety of crystal types ranging from ionic to covalent, assembled here for the first time. These are the simplest crystals with such a wide range of properties. Those interested in the tailoring of more complex or less studied materials through control of defects may also find the book useful. The reader should have the level of knowledge of basic quantum mechanics acquired in a first-year course.

The book is divided into two parts. The first six chapters deal with theoretical models of defect structure and general discussions of spectra, and the last four chapters survey the experimental situation with regard to defects in a variety of elemental and binary crystals. In contrast with a free atom that can tumble about, a defect center in a crystal has a fixed orientation and symmetry which the experimentalist tries to discover as a clue to the atomic arrangement or structure of the defect. Chapter 1 begins with a section on symmetry properties and the use of group theory to simplify calculations. Because there are several fine texts on applications of group theory, no detailed treatment or long derivations are given, but rather the most useful results are quoted and used in simple examples. The rest of this chapter is devoted to the electronic states of defects with tightly bound electrons. There are sections on rare earth and d^n transition metal ions and on a native defect, the $\mathscr{F}$ center. The second chapter treats defects in semiconductors where the electronic states are more diffuse. The effective mass theory of donor and acceptor impurities is discussed for several different types of band edge. Sections on bound excitons, isoelectronic impurities, and donor-acceptor pairs are included. Chapter 3 is concerned with vibrational properties of defects. The first part is a discussion of local mode spectra of light impurities. Vibrational coupling in electronic transitions between orbitally nondegenerate states is first discussed on the basis of the configuration coordinate model and then, briefly, in more general terms. The last part of the chapter deals with orbitally nondegenerate states and the Jahn–Teller effect. The only practical way to cover such a broad range of material so as to transmit to the reader the important results and features of the theory without becoming mired in derivations seemed to be by working out a small number of examples that illustrate the principles and show how they are applied in practice.

Chapter 4 discusses defect chemistry, with the main purpose of explaining association of simple point defects to form more complex centers. Chapter 5, on experimental methods, is not concerned with equipment but with the ways in which various spectroscopic techniques yield information on defect structure. The main subjects are magnetic resonance and optical spectroscopy. This material is supplemented by additional discussion of these topics in other chapters.

Chapter 6 contains a brief qualitative discussion of electron distributions in covalent and ionic crystals from the points of view of energy band theory and modern concepts of ionicity. This chapter is meant to give the reader a feeling for the differences and similarities among the crystals mentioned in the last four chapters. Chapter 7 reviews the experimental situation for defects in the covalent elemental crystals diamond, silicon, and germanium from column IV of the periodic table. This chapter is longer than any of those that follow, because more detail was included in this first chapter of the second part to show how deductions about defect structure follow from experimental results. Chapters 8 and 9 survey defects in III-V and II-VI compounds, and Chapter 10 is on defects in the ionic

I-VII compounds or alkali halides. The goal of the second part of the book was not an encyclopedic compilation with reference to every paper in the literature on point defects in these materials, but to include enough examples of each type of defect to give the reader a true impression of the state of present knowledge.

I should like to express my appreciation for the help of O. F. Schirmer, W. A. Sibley, and D. W. Shaw, who read and commented on portions of the unfinished manuscript.

R. K. WATTS

Dallas, Texas
June 1976

Contents

Contents

Point Defects in Crystals

Chapter 1

Electronic States of Defects with Tightly Bound Electrons

The electronic properties of many point defects are adequately described by considering a small number of electrons closely localized at a particular lattice site. This is often the case for impurity ions in crystals. Two important examples are electrons bound in partially filled f or d shells of transition ions. We examine these cases after some discussion of the ways in which symmetry properties can be exploited to simplify theory. The purpose of this chapter is to give the reader a feeling for the rather well developed theoretical approaches that have been used for such problems. References to useful tabulations of wave functions and energy level diagrams and to more detailed theoretical treatments are included. For many readers the main utility of the chapter is an introduction and guide to the use of such tables.

The last section deals with the electronic states of a much studied native defect, the $\mathcal{F}$ center in the alkali halides. The active electron is less tightly bound in this case, $\langle r \rangle$, the expected value of the radial coordinate of the electron in the ground state, being several times the value 0.5 Å which is typical for $4f$ and $3d$ electrons. It is much more localized than the electronic states of shallow donors and acceptors discussed in Chapter 2, however.

SITE SYMMETRY

In the limit of very tight binding the interaction of the electrons bound to an impurity with the host crystal can be considered a perturbation of the free ion. The most important terms in the Hamiltonian are

$$\mathcal{H} = -\frac{\hbar^2}{2m} \sum_i \nabla_i^2 - \sum_i \frac{Ze^2}{r_i} + \sum_{i>j} \frac{e^2}{r_{ij}} + \sum_i \zeta(r_i)\mathbf{l}_i \cdot \mathbf{s}_i + \mathcal{H}_c. \qquad (1.1)$$

These are, respectively, the kinetic energy, electrostatic interaction with the nucleus of charge $Z|e|$, interelectronic electrostatic repulsion, the spin-orbit interaction, and the interaction with the other lattice constituents. The sums are over the number of electrons bound to the ion. The $\mathbf{r}_i$ is a vector from the nucleus to electron i, and $r_{ij} = |\mathbf{r}_i - \mathbf{r}_j|$. Let us first set $\mathcal{H}_c$ equal to zero and review the properties of the free ion solutions.

The first two terms represent the energy of electrons moving independently in a central field. The solution of this much of $\mathcal{H}$ is a product of one electron functions ψ, the solutions of

$$\left(-\frac{\hbar^2}{2m} \nabla^2 - \frac{Ze^2}{r} \right) \psi_{nlm} = \epsilon_{nl} \psi_{nlm} \qquad (1.2)$$

These functions are

$$\psi_{nlm}(r, \theta, \varphi) = R_{nl}(r) Y_l^m(\theta, \varphi) \qquad (1.3)$$

where Y_l^m is a spherical harmonic. In this case the radial function $R_{nl}(r)$ would be proportional to an associated Laguerre polynomial. Radial functions have been calculated for most ions of interest[1,2] by the Hartree–Fock method in which some account is taken also of the term $\sum e^2/r_{ij}$. These are not expressed in closed analytical form. We are most interested in the angular functions Y_l^m, however. Some of these are listed in Table 1.1.

To solve $\mathcal{H}\psi = E\psi$ for the eigenvalues and eigenfunctions one would necessarily make many approximations. But many properties of E and ψ depend only on symmetry, and by making use of symmetry arguments we are able to derive some useful exact results.

The terms in the free ion Hamiltonian independent of spin are

$$\mathcal{H}_f = -\frac{\hbar^2}{2m} \sum_i \left(\frac{\partial^2}{\partial x_i^2} + \frac{\partial^2}{\partial y_i^2} + \frac{\partial^2}{\partial z_i^2} \right) - \sum_i \frac{Ze^2}{r_i} + \sum_{i>j} \frac{e^2}{r_{ij}} \qquad (1.4)$$

Because the energy of the system cannot depend on the coordinate system,

Table 1.1 Spherical Harmonics to $l = 3$ in Cartesian Coordinates

$$Y_0^0 = \sqrt{\frac{1}{4\pi}}$$ s

$$Y_1^1 = -\sqrt{\frac{3}{8\pi}} \, \frac{x+iy}{r}$$

$$Y_1^0 = \sqrt{\frac{3}{4\pi}} \, \frac{z}{r}$$ p

$$Y_1^{-1} = \sqrt{\frac{3}{8\pi}} \, \frac{x-iy}{r}$$

$$Y_2^2 = \sqrt{\frac{15}{32\pi}} \, \frac{(x+iy)^2}{r^2}$$

$$Y_2^1 = -\sqrt{\frac{15}{8\pi}} \, z \frac{x+iy}{r^2}$$

$$Y_2^0 = \sqrt{\frac{5}{16\pi}} \, \frac{3z^2-r^2}{r^2}$$ d

$$Y_2^{-1} = \sqrt{\frac{15}{8\pi}} \, z \frac{x-iy}{r^2}$$

$$Y_2^{-2} = \sqrt{\frac{15}{32\pi}} \, \frac{(x-iy)^2}{r^2}$$

$$Y_3^3 = -\sqrt{\frac{35}{64\pi}} \, \frac{(x+iy)^3}{r^3}$$

$$Y_3^2 = \sqrt{\frac{105}{32\pi}} \, z \frac{(x+iy)^2}{r^3}$$

$$Y_3^1 = -\sqrt{\frac{21}{64\pi}} \, (x+iy) \frac{5z^2-r^2}{r^3}$$

$$Y_3^0 = \sqrt{\frac{7}{16\pi}} \, z \frac{5z^2-3r^2}{r^3}$$ f

$$Y_3^{-1} = \sqrt{\frac{21}{64\pi}} \, (x-iy) \frac{5z^2-r^2}{r^3}$$

$$Y_3^{-2} = \sqrt{\frac{105}{32\pi}} \, z \frac{(x-iy)^2}{r^3}$$

$$Y_3^{-3} = \sqrt{\frac{35}{64\pi}} \, \frac{(x-iy)^3}{r^3}$$

the energy must have the same value if expressed in terms of a new coordinate system obtained by rotating the old.

$$x_i = R_{11}x_i' + R_{12}y_i' + R_{13}z_i'$$

$$y_i = R_{21}x_i' + R_{22}y_i' + R_{23}z_i' \qquad (1.5)$$

$$z_i = R_{31}x_i' + R_{32}y_i' + R_{33}z_i', \quad \text{for all} \quad i$$

or

$$\mathbf{r}_i = \mathbf{R}\mathbf{r}_i'$$

It can be shown that the functional form of $\mathfrak{K}_f$ will also remain the same:

$$\mathfrak{K}_f = -\frac{\hbar^2}{2m}\sum_i\left(\frac{\partial^2}{\partial x_i'^2} + \frac{\partial^2}{\partial y_i'^2} + \frac{\partial^2}{\partial z_i'^2}\right) - \sum_i \frac{Ze^2}{r_i'} + \sum_{i>j}\frac{e^2}{r_{ij}'} \qquad (1.6)$$

Any rotation, any permutation of electron labels i, and the inversion $\mathbf{r}_i \rightarrow -\mathbf{r}_i$ (all i) leave the Hamiltonian invariant in this sense and are called symmetry transformations. They do not leave an eigenfunction invariant, however, but transform it into another eigenfunction belonging to the same eigenvalue. The free ion has infinitely many symmetry transformations, but when the ion is placed in a crystal, the symmetry is much lower, and the number of symmetry transformations is finite and small.

For example, if the ion is tetrahedrally coordinated as shown in Figure 1.1, there are 24 symmetry transformations. They are

(a) A rotation of the tetrahedron by $\pm 2\pi/3$ about each $\langle 111 \rangle$ axis. There are eight of these, called C_3 rotations.

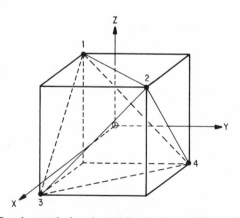

Figure 1.1   Regular tetrahedron formed by the four neighbors of a central ion.

(b) A rotation by $2\pi/2$ about each $\langle 100 \rangle$ axis (the coordinate axes). There are three of these, called C_2 rotations.
(c) A reflection in each $\{110\}$ plane. There are six of these, called σ_d reflections. The term σ_d is equivalent to an inversion in the origin, called I, followed by C_2', a rotation about a $\langle 110 \rangle$ axis: $\sigma_d = C_2'I$.
(d) An improper rotation of $\pm 2\pi/4$ about a $\langle 100 \rangle$ axis. This is a reflection in the $\{100\}$ plane, called σ_v, followed by a $\pm \pi/2$ rotation about the $\langle 100 \rangle$ axis. There are six of these, called S_4 rotations; $S_4 = C_4\sigma_v$.
(e) The identity operation, called E, which corresponds to a rotation by $0°$.

The symmetry transformations form a group, in this case the tetrahedral group T_d. A group is any collection of elements P, Q, R, ... with the following properties. There must be a way to combine any two elements PQ, and the combination must be a member of the group $PQ = R$. There must be an identity element E such that $EQ = QE = Q$ for any element Q. Every element Q must have an inverse Q^{-1}, where $QQ^{-1} = Q^{-1}Q = E$, and Q^{-1} is an element of the group. The product PQR must be uniquely defined. All these requirements are satisfied by the 24 T_d symmetry transformations. For example, from Figure 1.1 we see that a reflection in the (110) plane followed by a 180° rotation about the x axis has the effect $(1234) \rightarrow (4312)$, and the same effect is produced by an improper rotation of 90° about the z axis, or $C_2^x\sigma_d^{(110)} = S_4^{z-}$. Thus the 24 symmetry transformations are said to form a representation of the group T_d.

The most useful representations are formed by matrices expressing the behavior of wavefunctions of interest under symmetry transformations. As an example let us consider the three functions $f(r)Y_1^0$, $f(r)Y_1^1$, and $f(r)Y_1^{-1}$, or, equivalently the functions $\psi_0 = zf(r)$, $\psi_1 = -(x+iy)f(r)/\sqrt{2}$ and $\psi_{-1} = (x - iy)f(r)/\sqrt{2}$, and see what matrices are generated by letting the three symmetry transformations C_2^x, $\sigma_d^{(110)}$, and S_4^{z-} operate on them. From Figure 1.1, a clockwise rotation of coordinates corresponding to a counterclockwise rotation of the tetrahedron,

$$C_2^x\psi_0 = -z'f(r) = -\psi_0$$

$$C_2^x\psi_1 = -\frac{1}{\sqrt{2}}(x' - iy') = -\psi_{-1}$$

$$C_2^x\psi_{-1} = \frac{1}{\sqrt{2}}(x' + iy') = -\psi_1 \qquad (1.7)$$

$$\text{or} \quad C_2^x\begin{bmatrix} \psi_0 \\ \psi_1 \\ \psi_{-1} \end{bmatrix} = \begin{bmatrix} -1 & 0 & 0 \\ 0 & 0 & -1 \\ 0 & -1 & 0 \end{bmatrix}\begin{bmatrix} \psi_0 \\ \psi_1 \\ \psi_{-1} \end{bmatrix}$$

Similarly,

$$\sigma_d^{(110)}\begin{bmatrix} \psi_0 \\ \psi_1 \\ \psi_{-1} \end{bmatrix} = \begin{bmatrix} 1 & 0 & 0 \\ 0 & 0 & -i \\ 0 & i & 0 \end{bmatrix}\begin{bmatrix} \psi_0 \\ \psi_1 \\ \psi_{-1} \end{bmatrix}$$

$$S_4^{z-}\begin{bmatrix} \psi_0 \\ \psi_1 \\ \psi_{-1} \end{bmatrix} = \begin{bmatrix} -1 & 0 & 0 \\ 0 & i & 0 \\ 0 & 0 & -i \end{bmatrix}\begin{bmatrix} \psi_0 \\ \psi_1 \\ \psi_{-1} \end{bmatrix}$$

(1.8)

Matrix multiplication gives

$$C_2^x\sigma_d^{(110)} = \begin{bmatrix} -1 & 0 & 0 \\ 0 & 0 & -1 \\ 0 & -1 & 0 \end{bmatrix}\begin{bmatrix} 1 & 0 & 0 \\ 0 & 0 & -i \\ 0 & i & 0 \end{bmatrix} = \begin{bmatrix} -1 & 0 & 0 \\ 0 & i & 0 \\ 0 & 0 & -i \end{bmatrix} = S_4^{z-}$$

(1.9)

In this way a matrix, based on these functions, can be associated with every group element, and they form a representation of the group.

The trace of a matrix is called the character $\chi(R)$ of the element R. Two elements P and Q belong to the same class if $RPR^{-1} = Q$, where R is another element of the group. All elements of the same class have the same character. The group T_d has five classes, lettered (a), (b), (c), (d), (e) above. A representation is said to be reducible if all its matrices can be transformed into a form with blocks along the diagonal

$$\begin{bmatrix} m_{11} & m_{12} & \cdots & m_{1i} & & & \\ m_{21} & & & & & & \\ \vdots & & & \vdots & & 0 & \\ m_{i1} & \cdots & & m_{ii} & & & \\ & & & & m_{jj} & \cdots & m_{jk} \\ & & 0 & & \vdots & & \\ & & & & m_{kj} & \cdots & m_{kk} \end{bmatrix}$$

(1.10)

If not, the representation is irreducible. There are as many irreducible representations of a group as there are classes. (Many useful results of group theory are stated without proof. Proofs and detailed treatment can be found in references 3 and 4.) The number of times $n(\Gamma)$ that the irreducible representation Γ occurs in the reducible representation r is given by

$$n(\Gamma) = \frac{1}{h} \sum_R \chi^\Gamma(R)^* \chi^r(R) \tag{1.11}$$

h is the order of the group, the number of elements. The sum is over all elements R, and χ^Γ and χ^r are the characters of the irreducible and reducible representations, respectively.

A matrix representation is not unique. We could have based the representation in Equations 1.7, 1.8, and 1.9 on, for example, the functions $zf(r)$, $xf(r)$, and $yf(r)$, and the matrices would have been different. The characters are unique, however, because they are unchanged by a similarity transformation. This makes character tables useful. The character table for T_d is shown in Table 1.2. The five classes are written across the top, and the number of elements in a class is indicated by a number written before the symbol for the class. The one-member class containing the identity element E is conventionally written first. The labeling of irreducible representations is arbitrary and may vary from one source to another, expecially for the smaller groups representing lower symmetries.

Because E is the identity matrix, the character of E is the dimension of the irreducible representation. Conventionally, A, E, T, and G are used as labels for irreducible representations of dimension 1, 2, 3, and 4. The labels Γ_i are also in use. We use the Γ_i notation to indicate that the basis functions include the electron-spin coordinate, and the letters to indicate orbital functions. The representation of Equations 1.7 and 1.8 has characters $\chi(C_2) = -1$, $\chi(\sigma_d) = 1$, $\chi(S_4) = -1$, $\chi(E) = 3$. Indeed, this is the irreducible representation T_2. A linear combination of a set of functions $\psi_1, \psi_2, \ldots, \psi_k$ which forms a basis for an irreducible representation Γ can be found, if it exists, from

$$\psi'_\Gamma = \sum_R \chi^\Gamma(R) R \psi_j \tag{1.12}$$

Table 1.2 Character Table for T_d

		E	$8C_3$	$3C_2$	$6\sigma_d$	$6S_4$
A_1	Γ_1	1	1	1	1	1
A_2	Γ_2	1	1	1	-1	-1
E	Γ_3	2	-1	2	0	0
T_1	Γ_4	3	0	-1	-1	1
T_2	Γ_5	3	0	-1	1	-1

There are only 32 point groups—groups of transformations that leave a point (the origin) invariant—necessary to describe the point symmetry in any crystal. The point group T_d is one of these; O_h is another very important one. It is closely related to T_d and consists of the transformations which leave a regular octahedron invariant. This is the symmetry group of an ion surrounded by six equidistant neighbors at the centers of the cube faces of Figure 1.1, or by eight equidistant neighbors at the cube corners, or by 12 at the centers of the cube edges. The character table is shown in Table 1.3. The upper-left quadrant is the character table of the group O which is isomorphic with T_d; O_h has twice as many elements, because each element of O also appears multiplied by the inversion operator I. Representations that are even under inversion have subscript g, and those that are odd have subscript u. The character tables for the 32 point groups are given in reference 4. Each has an A representation whose every element leaves the basis function unchanged.

ROTATION GROUP AND DOUBLE GROUPS

Another useful group is the full rotation group—the group of all proper rotations about all axes passing through the origin. A free atom is invariant to the infinitely many elements of this group. It has infinitely many irreducible representations. The $2j+1$ eigenfunctions $|j, m_j\rangle$ of the angular momentum operators j^2 and j_z are basis functions for the irreducible representation $\mathcal{D}_j$. There is an irreducible representation corresponding to each $j = 0, 1/2, 1, 3/2, \ldots$, and all rotations by the same angle about different axes belong to the same class. It can be shown[3] that the effect of

Table 1.3 Character Table for O_h

	E	$8C_3$	$3C_2$	$6C_2'$	$6C_4$	I	$8IC_3$	$3IC_2$	$6IC_2'$	$6IC_4$
A_{1g} Γ_{1g}	1	1	1	1	1	1	1	1	1	1
A_{2g} Γ_{2g}	1	1	1	-1	-1	1	1	1	-1	-1
E_g Γ_{3g}	2	-1	2	0	0	2	-1	2	0	0
T_{1g} Γ_{4g}	3	0	-1	-1	1	3	0	-1	-1	1
T_{2g} Γ_{5g}	3	0	-1	1	-1	3	0	-1	1	-1
A_{1u} Γ_{1u}	1	1	1	1	1	-1	-1	-1	-1	-1
A_{2u} Γ_{2u}	1	1	1	-1	-1	-1	-1	-1	1	1
E_u Γ_{3u}	2	-1	2	0	0	-2	1	-2	0	0
T_{1u} Γ_{4u}	3	0	-1	-1	1	-3	0	1	1	-1
T_{2u} Γ_{5u}	3	0	-1	1	-1	-3	0	1	-1	1

the transformation $R(\varphi_0)$, a rotation by the angle φ_0 about the z axis, on $|j, m_j\rangle$ is

$$R(\varphi_0)|j, m_j\rangle = |j, m_j\rangle \exp(im_j\varphi_0) \tag{1.13}$$

This is easily seen to be the case if j is integral, because then the $|j, m_j\rangle$ are proportional to the spherical harmonics $Y_j^{m_j}$ with φ dependence $\exp(im_j\varphi)$. The trace of the matrix based on the functions $|j, m_j\rangle$ is, then,

$$\chi(\varphi_0) = \exp(ij\varphi_0) + \exp(i(j-1)\varphi_0) + \cdots + \exp(i(-j)\varphi_0)$$

$$= \frac{\sin\left(\left(j+\tfrac{1}{2}\right)\varphi_0\right)}{\sin(\varphi_0/2)} \tag{1.14}$$

If $j = 2$, the transformations C_3, C_2, C_4, and E, which correspond to rotations by $2\pi/3$, π, $2\pi/4$ and 0, have from Equation 1.14 the characters

$$\chi(C_3) = \chi(C_4) = -1, \qquad \chi(C_2) = 1, \qquad \chi(E) = 5 \tag{1.15}$$

Using these as χ^r in Equation 1.11 and the χ^Γ from Table 1.3, we find that the five d functions form bases for the E_g and T_{2g} representations of O_h. We write $\mathcal{D}_2 \rightarrow e_g + t_{2g}$. (The functions are even because l is even.) In this way the irreducible representations of the 32 point groups for which the Y_l^m form bases can be found. The basis functions, linear combinations of the Y_l^m, can be found from Equation 1.12. Sometimes it is convenient to use real basis functions, some of which are listed in Table 1.4.

Half-integral values of j occur when the electron spin is included. To indicate spin, the functions of Equation 1.3 can be written

$$\overset{\pm}{\psi}_{nlm} = \psi_{nlm}|\tfrac{1}{2}, \pm\tfrac{1}{2}\rangle \tag{1.16}$$

where $|\tfrac{1}{2}, \pm\tfrac{1}{2}\rangle$ is the spinor for $m_s = \pm\tfrac{1}{2}$. The states $|j, m_j\rangle$ are given by

$$|j, m_j\rangle = \sum_{m, m_s} \left(l\tfrac{1}{2}mm_s|jm_j\right)\psi_{lm}|\tfrac{1}{2}, m_s\rangle \tag{1.17}$$

The Wigner coefficients $(j_1 j_2 m_1 m_2|jm_j)$ are tabulated in reference 3. Equation 1.14 has the disconcerting property that for half-integral j

$$\chi(\varphi_0 + 2\pi) = -\chi(\varphi_0) \tag{1.18}$$

There are two matrices corresponding to each rotation for half-integral j; the representation is said to be unfaithful. However, because $\chi(\varphi_0 + 4\pi) =$

Table 1.4 Real Basis Functions for the Irreducible Representations of T_d and O_h formed by the Spherical Harmonics to $l = 3$

In Cartesian coordinates	In spherical harmonics	O_h label	T_d label
$\sqrt{\dfrac{1}{4\pi}}$	Y_0^0	a_{1g}	a_1
$\sqrt{\dfrac{3}{4\pi}}\dfrac{x}{r}$	$(1/\sqrt{2})(-Y_1^1 + Y_1^{-1})$	$t_{1u}x$	$t_2 x$
$\sqrt{\dfrac{3}{4\pi}}\dfrac{y}{r}$	$(i/\sqrt{2})(Y_1^1 + Y_1^{-1})$	$t_{1u}y$	$t_2 y$
$\sqrt{\dfrac{3}{4\pi}}\dfrac{z}{r}$	Y_1^0	$t_{1u}z$	$t_2 z$
$\sqrt{\dfrac{5}{16\pi}}\dfrac{3z^2 - r^2}{r^2}$	Y_2^0	$e_g u$	eu
$\sqrt{\dfrac{15}{16\pi}}\dfrac{x^2 - y^2}{r^2}$	$(1/\sqrt{2})(Y_2^2 + Y_2^{-2})$	$e_g v$	ev
$\sqrt{\dfrac{15}{4\pi}}\dfrac{yz}{r^2}$	$(i/\sqrt{2})(Y_2^1 + Y_2^{-1})$	$t_{2g}yz$	$t_2 yz$
$\sqrt{\dfrac{15}{4\pi}}\dfrac{zx}{r^2}$	$(1/\sqrt{2})(-Y_2^1 + Y_2^{-1})$	$t_{2g}zx$	$t_2 zx$
$\sqrt{\dfrac{15}{4\pi}}\dfrac{xy}{r^2}$	$(i/\sqrt{2})(-Y_2^2 + Y_2^{-2})$	$t_{2g}xy$	$t_2 xy$
$\sqrt{\dfrac{105}{4\pi}}\dfrac{xyz}{r^3}$	$(i/\sqrt{2})(-Y_3^2 + Y_3^{-2})$	a_{2u}	a_1
$\dfrac{1}{2}\sqrt{\dfrac{7}{4\pi}}\dfrac{x}{r^3}(5x^2 - 3r^2)$	$(1/4)[\sqrt{5}\,(-Y_3^3 + Y_3^{-3}) + \sqrt{3}\,(Y_3^1 - Y_3^{-1})]$	$t_{1u}x$	$t_2 x$
$\dfrac{1}{2}\sqrt{\dfrac{7}{4\pi}}\dfrac{y}{r^3}(5y^2 - 3r^2)$	$(-i/4)[\sqrt{5}\,(Y_3^3 + Y_3^{-3}) + \sqrt{3}\,(Y_3^1 + Y_3^{-1})]$	$t_{1u}y$	$t_2 y$
$\dfrac{1}{2}\sqrt{\dfrac{7}{4\pi}}\dfrac{z}{r^3}(5z^2 - 3r^2)$	Y_3^0	$t_{1u}z$	$t_2 z$
$\dfrac{1}{2}\sqrt{\dfrac{105}{4\pi}}\dfrac{x}{r^3}(y^2 - z^2)$	$(-1/4)[\sqrt{5}\,(-Y_3^1 + Y_3^{-1}) + \sqrt{3}\,(-Y_3^3 + Y_3^{-3})]$	$t_{2u}x$	$t_1 x$
$\dfrac{1}{2}\sqrt{\dfrac{105}{4\pi}}\dfrac{y}{r^3}(z^2 - x^2)$	$(i/4)[\sqrt{5}\,(Y_3^1 + Y_3^{-1}) - \sqrt{3}\,(Y_3^3 + Y_3^{-3})]$	$t_{2u}y$	$t_1 y$
$\dfrac{1}{2}\sqrt{\dfrac{105}{4\pi}}\dfrac{z}{r^3}(x^2 - y^2)$	$(1/\sqrt{2})(Y_3^2 + Y_3^{-2})$	$t_{2u}z$	$t_1 z$

$\chi(\varphi_0)$, the representation can be made faithful by augmenting each point group with the element $\mathscr{R}$, the rotation by 2π, and its product with every other element to form the "double groups."[4] The double group contains two classes for every class of the single group except for the classes containing C_2, because $\chi(3\pi) = \chi(\pi) = 0$. The characters for the double groups T_d and O are shown in Table 1.5.

Table 1.5 Character Table for Double Groups T_d and O

T_d		E	$\mathcal{R}$	$8C_3$	$8C_3\mathcal{R}$	$3C_2$ $3C_2\mathcal{R}$	$6S_4$	$6S_4\mathcal{R}$	$6\sigma_d$ $6\sigma_d\mathcal{R}$
O		E	$\mathcal{R}$	$8C_3$	$8C_3\mathcal{R}$	$3C_2$ $3C_2\mathcal{R}$	$6C_4$	$6C_4\mathcal{R}$	$6C_2'$ $6C_2'\mathcal{R}$
A_1	Γ_1	1	1	1	1	1	1	1	1
A_2	Γ_2	1	1	1	1	1	-1	-1	-1
E	Γ_3	2	2	-1	-1	2	0	0	0
T_1	Γ_4	3	3	0	0	-1	1	1	-1
T_2	Γ_5	3	3	0	0	-1	-1	-1	1
E'	Γ_6	2	-2	1	-1	0	$\sqrt{2}$	$-\sqrt{2}$	0
E''	Γ_7	2	-2	1	-1	0	$-\sqrt{2}$	$\sqrt{2}$	0
G'	Γ_8	4	-4	-1	1	0	0	0	0

Because there are now eight classes, three more irreducible representations have to be added. So that the sum of squares of dimensions of the irreducible representations remain equal to the order of the group, now 48, these must be two twofold and one fourfold representation. The spinors $|\frac{1}{2}, \pm \frac{1}{2}\rangle$ form bases for Γ_6, and the fourfold $|\frac{3}{2}, m\rangle$ form bases for Γ_8. Additional basis functions for the extra irreducible representations of the double groups are given in references 5 and 6. Functions of only the Cartesian coordinates continue to form bases only for the first five irreducible representations. When the transformation properties of only such functions are of interest, the elements containing $\mathcal{R}$ are superfluous, and the single group suffices. The Γ labels are more frequently used in the literature for these last three irreducible representations than the letters E', E'', and G'.

From Table 1.5 the meaning of the isomorphism of the two groups is apparent. They have the same character table. There is a one-to-one correspondence between the elements of the two groups.

If a free atom is placed in the lower symmetry of a crystal lattice, its symmetry group changes from the full rotation group to the symmetry group of the lattice site—T_d or O in the present example—and some degeneracy of the atomic states will be lifted. We have seen that a state characterized by l or j or $J = 2$ splits into a doublet and a triplet in cubic or tetrahedral symmetry. The thought experiment of reducing the symmetry and looking for splittings in this way is a useful device.

PRODUCT REPRESENTATIONS

Equation 1.17 is a familiar relation from atomic spectroscopy. It states that the product of two functions, each of which is part of a basis set for an irreducible representation, belongs to a basis set for a representation of the group. This is also true for the finite point groups. For example, let us form a representation of T_d from the products of the two one-electron functions from Table 1.4 $eu(1)=(1/2)(3z_1^2-r_1^2)$ and $ev(2)=(\sqrt{3}/2)(x_2^2-y_2^2)$.

$$S_4^{z-}eu = eu \qquad S_4^{z-}ev = -ev \qquad (1.19)$$

$$S_4^{z-}\begin{bmatrix} eu(1)eu(2) \\ eu(1)ev(2) \\ ev(1)eu(2) \\ ev(1)ev(2) \end{bmatrix} = \begin{bmatrix} 1 & 0 & 0 & 0 \\ 0 & -1 & 0 & 0 \\ 0 & 0 & -1 & 0 \\ 0 & 0 & 0 & 1 \end{bmatrix}\begin{bmatrix} eu(1)eu(2) \\ eu(1)ev(2) \\ ev(1)eu(2) \\ ev(1)ev(2) \end{bmatrix}$$

In the same way the other matrices can be found.

$$C_2^x \rightarrow \begin{bmatrix} 1 & 0 & 0 & 0 \\ 0 & 1 & 0 & 0 \\ 0 & 0 & 1 & 0 \\ 0 & 0 & 0 & 1 \end{bmatrix} \qquad C_3^{[111]} \rightarrow \frac{1}{4}\begin{bmatrix} 1 & -\sqrt{3} & -\sqrt{3} & 3 \\ \sqrt{3} & 1 & -3 & -\sqrt{3} \\ \sqrt{3} & -3 & 1 & -\sqrt{3} \\ 3 & \sqrt{3} & \sqrt{3} & 1 \end{bmatrix}$$

$$\qquad (1.20)$$

$$\sigma_d^{[110]} \rightarrow \begin{bmatrix} 1 & 0 & 0 & 0 \\ 0 & -1 & 0 & 0 \\ 0 & 0 & -1 & 0 \\ 0 & 0 & 0 & 1 \end{bmatrix}$$

The characters of this representation are 4, 1, 4, 0, and 0 for E, C_3, C_2, S_4, and σ_d respectively, and these are the squares of the characters of the E irreducible representation. The representation is reducible. Equation 1.11 can be used to show that it reduces to an A_1, an A_2, and an E representation. This is written $e \times e = A_1 + A_2 + E$. Such products of irreducible representations are tabulated in reference 5 for the 32 point groups. Also given are the coupling coefficients, corresponding to the Wigner coefficients of Equation 1.17, so that basis functions for the irreducible representations (A_1, A_2, and E in this case) can be written as linear combinations of products (of eu and ev here). The multiplication table for O and T_d is given in Table 1.6.

Table 1.6 Multiplication Table for O and T_d

	Γ_1	Γ_2	Γ_3	Γ_4	Γ_5	Γ_6	Γ_7	Γ_8
Γ_1	Γ_1	Γ_2	Γ_3	Γ_4	Γ_5	Γ_6	Γ_7	Γ_8
Γ_2		Γ_1	Γ_3	Γ_5	Γ_4	Γ_7	Γ_6	Γ_8
Γ_3			$\Gamma_1+\Gamma_2+\Gamma_3$	$\Gamma_4+\Gamma_5$	$\Gamma_4+\Gamma_5$	Γ_8	Γ_8	$\Gamma_6+\Gamma_7+\Gamma_8$
Γ_4				$\Gamma_1+\Gamma_3+\Gamma_4+\Gamma_5$	$\Gamma_2+\Gamma_3+\Gamma_4+\Gamma_5$	$\Gamma_6+\Gamma_8$	$\Gamma_7+\Gamma_8$	$\Gamma_6+\Gamma_7+2\Gamma_8$
Γ_5					$\Gamma_1+\Gamma_3+\Gamma_4+\Gamma_5$	$\Gamma_7+\Gamma_8$	$\Gamma_6+\Gamma_8$	$\Gamma_6+\Gamma_7+2\Gamma_8$
Γ_6						$\Gamma_1+\Gamma_4$	$\Gamma_2+\Gamma_5$	$\Gamma_3+\Gamma_4+\Gamma_5$
Γ_7							$\Gamma_1+\Gamma_4$	$\Gamma_3+\Gamma_4+\Gamma_5$
Γ_8								$\Gamma_1+\Gamma_2+\Gamma_3+2\Gamma_4+2\Gamma_5$

In most quantum mechanical calculations the matrix elements of some operator F are sought. Let F transform according to the irreducible representation Γ_F. If F has components that transform according to more than one irreducible representation, each can be treated separately. Let the wave functions ψ_n^j and ψ_m^k in $\langle \psi_n^j | F | \psi_m^k \rangle$ be the jth and kth members of the basis sets of the irreducible representations Γ_n and Γ_m. Then the product $\Gamma_F \times \Gamma_m$ can be decomposed as indicated. Unless the decomposition contains a term transforming according to Γ_n, the matrix element will be zero because of the orthogonality of basis functions. Even if there is a Γ_n term in the expansion, the matrix element will still be zero unless the term transforms as the jth basis function, because these, too, are orthogonal.

Especially important is the case in which F is the Hamiltonian. Because the Hamiltonian is invariant to all elements of the symmetry group, it transforms as Γ_1. Because $\Gamma_1 \times \Gamma_k = \Gamma_k$ for all k, then

$$\langle \psi_n^j | \mathcal{H} | \psi_m^k \rangle = E_n \delta_{jk} \delta_{nm}, \tag{1.21}$$

where E_n is independent of j and k. If we want to find energy levels corresponding to a certain Hamiltonian, we can simplify the calculation by starting with functions with the correct symmetry properties. All matrix elements between functions transforming according to different irreducible representations will be zero.

RARE EARTH IONS

The atoms of the rare earth transition series, Ce through Yb, are usually incorporated in insulating crystals as the trivalent or, less frequently, the divalent ions. These ions have electronic configurations $(4f)^n$, all other electrons being in closed shells. The orbital radius of a $4f$ electron is only about half the ionic radius. The large ionic radius, about 1 Å, is largely due to the filled $5s$ and $5p$ shells. Because of the large radius, neighboring ions in the crystal are rather distant and do not influence the $4f$ electrons very much. The gross optical properties are almost independent of the host crystal. In this sense the $4f$ electrons are the most tightly bound with which we shall deal.

The energy splittings caused by the interaction of the $4f$ electrons with the surrounding crystal are small, a few hundred to one thousand cm^{-1}. The spin-orbit splittings vary from 600 cm^{-1} to 3000 cm^{-1}, and the interelectronic electrostatic interaction energies are slightly larger. Let us look more closely at the spectroscopic properties of a particularly simple ion, Yb^{3+} in an environment with O_h symmetry, which results when Yb^{3+} replaces Ca^{2+} in CaF_2 without local charge compensation, for example. The spin-orbit splitting for Yb^{3+} is $\sim 10^4$ cm^{-1}, at least an order of magnitude larger than the splittings due to interaction with the crystal. Although the Yb^{3+} configuration is $(4f)^{13}$, this can be treated as a single particle—a hole in the $4f$ shell. The Hamiltonian of interest is

$$\mathcal{H} = -\zeta(r)\mathbf{l}\cdot\mathbf{s} + \mathcal{H}_c \qquad (1.22)$$

In dealing with a hole instead of electrons, we put a negative sign before the spin orbit term, because ζ is defined for electrons as

$$\zeta(r) = -\frac{\hbar^2|e|}{2m^2c^2r}\frac{d}{dr}V(r) \qquad (1.23)$$

$V(r)$ is the electrostatic potential due to the central field, and $d/(dr)V(r)$ is negative.

Because the spin-orbit term is much larger than $\mathcal{H}_c$, it is diagonalized first. There are 14 states $|3^{\pm},m\rangle = R_{4f}(r)Y_3^m|1/2,m_s\rangle$ and they can be combined according to Equation 1.17 to form eight states $|J=7/2, M_{7/2}\rangle$ and six states $|J=5/2, M_{5/2}\rangle$. Since

$$(\mathbf{l}\cdot\mathbf{s})|J,M_j\rangle = \tfrac{1}{2}\big[J(J+1) - l(l+1) - s(s+1)\big]|J,M_j\rangle \qquad (1.24)$$

the energies of the two multiplets are $E(7/2) = -(3/2)\zeta_{4f}$ and $E(5/2) =$

$2\zeta_{4f}$; the $^2F_{7/2}$ multiplet lies lower. The ζ_{4f} is defined by

$$\zeta_{4f} = \int_0^\infty \left[R_{4f}(r) \right]^2 \zeta(r) r^2 \, dr \qquad (1.25)$$

The energy difference between the two multiplets $(7/2)\zeta_{4f}$ is so much larger than the splittings due to $\mathcal{H}_c$ that to a good approximation in the case of Yb^{3+} we can diagonalize $\mathcal{H}_c$ within each multiplet separately. Let us first consider the $^2F_{5/2}$ upper multiplet. From Equation 1.17 and a table of Wigner coefficients

$$|5/2, 5/2\rangle = \sqrt{6/7}\, |\overset{-}{3}\rangle - \sqrt{1/7}\, |\overset{+}{2}\rangle$$

$$|5/2, 3/2\rangle = \sqrt{5/7}\, |\overset{-}{2}\rangle - \sqrt{2/7}\, |\overset{+}{1}\rangle$$

$$\qquad (1.26)$$

$$\vdots$$

$$|5/2, -5/2\rangle = \sqrt{1/7}\, |-\overset{-}{2}\rangle - \sqrt{6/7}\, |-\overset{+}{3}\rangle$$

Because l is always 3, it was not written in the kets on the right. The states $|3, m\rangle$ can be written in terms of the real functions of Table 1.4:

$$|3\rangle = \frac{1}{4}\left(i\sqrt{5}\, |t_{1u}y\rangle + i\sqrt{3}\, |t_{2u}y\rangle - \sqrt{5}\, |t_{1u}x\rangle + \sqrt{3}\, |t_{2u}x\rangle \right)$$

$$|2\rangle = (1/\sqrt{2})(i|a_{2u}\rangle + |t_{2u}z\rangle)$$

$$\qquad (1.27)$$

$$\vdots$$

$$|-3\rangle = \frac{1}{4}\left(i\sqrt{5}\, |t_{1u}y\rangle + i\sqrt{3}\, |t_{2u}y\rangle + \sqrt{5}\, |t_{1u}x\rangle - \sqrt{3}\, |t_{2u}x\rangle \right)$$

The matrix elements of $\mathcal{H}_c$ can now be evaluated for $^2F_{5/2}$ by combining Equations 1.26 and 1.27. $\mathcal{H}_c$ is assumed to be independent of the electron spin. The matrix is

	5/2	3/2	1/2	$-1/2$	$-3/2$	$-5/2$
5/2	$\frac{1}{28}(2\delta_1+3\delta_2)$	0	0	0	$\frac{\sqrt{5}}{28}(2\delta_1+3\delta_2)$	0
3/2	0	$-\frac{3}{28}(2\delta_1+3\delta_2)$	0	0	0	$\frac{\sqrt{5}}{28}(2\delta_1+3\delta_2)$
1/2	0	0	$\frac{1}{14}(2\delta_1+3\delta_2)$	0	0	0
$-1/2$	0	0	0	$\frac{1}{14}(2\delta_1+3\delta_2)$	0	0
$-3/2$	$\frac{\sqrt{5}}{28}(2\delta_1+3\delta_2)$	0	0	0	$-\frac{3}{28}(2\delta_1+3\delta_2)$	0
$-5/2$	0	$\frac{\sqrt{5}}{28}(2\delta_1+3\delta_2)$	0	0	0	$\frac{1}{28}(2\delta_1+3\delta_2).$

Matrix of $\mathcal{H}_c$ for $^2F_{5/2}$

$$(1.28)$$

Equation 1.21 was used to define the quantities δ_1 and δ_2:

$$\langle t_{1u}|\mathcal{H}_c|t_{1u}\rangle = \epsilon(t_{1u}) \qquad \delta_1 = \epsilon(t_{1u}) - \epsilon(t_{2u})$$

$$\langle t_{2u}|\mathcal{H}_c|t_{2u}\rangle = \epsilon(t_{2u}) \qquad \delta_2 = \epsilon(t_{2u}) - \epsilon(a_{2u}) \qquad (1.29)$$

$$\langle a_{2u}|\mathcal{H}_c|a_{2u}\rangle = \epsilon(a_{2u}) \qquad 3\epsilon(t_{1u}) + 3\epsilon(t_{2u}) + \epsilon(a_{2u}) = 0$$

In the same way the 8×8 matrix of $\mathcal{H}_c$ among the $^2F_{7/2}$ states can be found. One of the elements is

$$\langle 7/2, 7/2|\mathcal{H}_c|7/2, 7/2\rangle = \frac{1}{56}(8\delta_1 + 11\delta_2) \qquad (1.30)$$

The spin-orbit energies can be added to the diagonal elements of the two matrices and the energy levels and wavefunctions can be found as functions of ζ_{4f}, δ_1, and δ_2. (The wavefunctions are independent of ζ_{4f} in the approximation used; that is, elements $\langle 7/2, M_{7/2}|\mathcal{H}_c|5/2, M_{5/2}\rangle$ can be neglected.) From Equations 1.11 and 1.14 and Table 1.5,

$$\mathcal{D}_{7/2} \rightarrow \Gamma_{6u} + \Gamma_{7u} + \Gamma_{8u}, \qquad \mathcal{D}_{5/2} \rightarrow \Gamma_{7u} + \Gamma_{8u} \qquad (1.31)$$

The $^2F_{5/2}$ multiplet is split by $\mathcal{H}_c$ into a doublet and a quartet, and $^2F_{7/2}$ splits into two doublets and a quartet.

Using only symmetry arguments we have seen that the effect of $\mathcal{H}_c$ is to split $^2F_{5/2}$ into two levels and $^2F_{7/2}$ into three levels, and the splitting can be described by only two parameters, δ_1 and δ_2. We have made no assumption about the form of $\mathcal{H}_c$ except that it is independent of spin. This result is an example of the power of an approach based on symmetry. The representations used are convenient because in this case ζ_{4f} is much larger than δ_1 or δ_2. The problem could also be solved by beginning with the real functions of Table 1.4 which diagonalize $\mathcal{H}_c$, forming product representations of these with the spinors $|1/2, \pm 1/2\rangle$, and diagonalizing $\mathcal{H}$ among these. This would be more convenient for large $\mathcal{H}_c$, as in the case of the actinides with $(5f)^n$ configurations.[7] Both approaches lead to the same result, of course.

Symmetry arguments do not yield an estimation of the magnitudes and signs of δ_1 and δ_2, however. For this a definite model for $\mathcal{H}_c$ is necessary. The model usually assumed is that of the crystal field, in which the neighboring ions of the crystal are considered point charges. The finite extent of the charge distributions of these ions is neglected, and the wavefunctions of the central ion are supposed not to overlap these charges. This model is most appropriate for the rare earths, although even for these ions it leads only to crude, approximate results. For the case of Yb^{3+} in O_h

symmetry we expand the potential due to the surrounding point charges q_i at $\mathbf{R}_i$ in powers of x/R_i, y/R_i, and z/R_i and multiply by the charge of the hole $|e|$ to obtain $\mathcal{H}_c$.

$$\mathcal{H}_c = |e| \sum_i q_i |\mathbf{R}_i - \mathbf{r}|^{-1} \tag{1.32}$$

Now $\mathcal{H}_c$ must transform as A_{1g}. The only function from Table 1.4 that transforms as A_{1g} is the constant. However, if we were to extend Table 1.4 to functions of higher powers of the coordinates,[8] we should find that the fourth-degree polynomial $(x^4 + y^4 + z^4 - 3r^4/5)$ and the sixth-degree function

$$x^6 + y^6 + z^6 + (15/4)(x^2y^4 + x^2z^4 + y^2x^4 + z^4y^2 + z^2x^4 + z^2y^4) - (15/14)r^6$$

also transform as A_{1g}. We need consider no higher powers as long as we are interested only in matrix elements between two $4f$ functions. Therefore, the expansion has the form

$$\mathcal{H}_c = C_0 + C_4 f_4(x,y,z) + C_6 f_6(x,y,z) \tag{1.33}$$

where f_4 and f_6 stand for the two polynomials. Although the constant term C_0 is the largest it will just change the zero level of the energy and can be dropped. The contributions to C_4 and C_6 from the first shell of eight nearest neighbors in eightfold cubic coordination are

$$C_4^{(1)} = -\frac{70q|e|}{9R_1^5}, \quad C_6^{(1)} = -\frac{224q|e|}{9R_1^7} \tag{1.34}$$

The matrix elements of $\mathcal{H}_c$ in Equation 1.33 can be calculated by writing f_4 and f_6 in terms of spherical harmonics and using a well-known formula for the integral of the product of three spherical harmonics.

$$f_4(x,y,z) = \frac{4\sqrt{\pi}}{15} r^4 \left\{ Y_4^0(\theta,\varphi) + \sqrt{\frac{5}{14}} \left[Y_4^4(\theta,\varphi) + Y_4^{-4}(\theta,\varphi) \right] \right\}$$

$$f_6(x,y,z) = -\frac{1}{7}\sqrt{\frac{\pi}{13}} r^6 \left\{ Y_6^0(\theta,\varphi) - \sqrt{\frac{7}{2}} \left[Y_6^4(\theta,\varphi) + Y_6^{-4}(\theta,\varphi) \right] \right\}$$

$$\tag{1.35}$$

Then from Equation 1.26 the matrix element $\langle 5/2, 5/2 | \mathcal{H}_c | 5/2, 5/2 \rangle$ is given by

$$\left\langle \frac{5}{2}, \frac{5}{2} \middle| \mathcal{H}_c \middle| \frac{5}{2}, \frac{5}{2} \right\rangle = \frac{6}{7} \langle 3 | \mathcal{H}_c | 3 \rangle + \frac{1}{7} \langle 2 | \mathcal{H}_c | 2 \rangle \tag{1.36}$$

The matrix elements $\langle 3|Y_L^M|3\rangle$ and $\langle 2|Y_L^M|2\rangle$ can be evaluated from the expression

$$\langle lm'|Y_L^M|lm\rangle = (-1)^{-m'}\left[\frac{(2l+1)^2(2L+1)}{4\pi}\right]^{1/2}\begin{pmatrix} l & L & l \\ -m' & Mm \end{pmatrix}\begin{pmatrix} lLl \\ 000 \end{pmatrix} \quad (1.37)$$

with $l=3$. The $3j$ symbols ( ) are tabulated in reference 9. They are proportional to the Wigner coefficients (Equation 1.17) and are often more convenient to use because of their more symmetrical form. The relationship is

$$\begin{pmatrix} j_1 & j_2 & j_3 \\ m_1 & m_2 & m_3 \end{pmatrix} = \frac{(-1)^{j_1-j_2-m_3}}{\sqrt{2j_3+1}}(j_1j_2m_1m_2|j_3-m_3)$$

The two matrices of $\mathcal{H}_c$, one based on the six states of $^2F_{5/2}$ and another on the eight states of $^2F_{7/2}$ can be found in this way. The result for $^2F_{5/2}$ is

	5/2	3/2	1/2	−1/2	−3/2	−5/2
5/2	$\frac{2}{105}C_4\langle r^4\rangle$	0	0	0	$\frac{2\sqrt5}{105}C_4\langle r^4\rangle$	0
3/2	0	$-\frac{2}{35}C_4\langle r^4\rangle$	0	0	0	$\frac{2\sqrt5}{105}C_4\langle r^4\rangle$
1/2	0	0	$\frac{4}{105}C_4\langle r^4\rangle$	0	0	0
−1/2	0	0	0	$\frac{4}{105}C_4\langle r^4\rangle$	0	0
−3/2	$\frac{2\sqrt5}{105}C_4\langle r^4\rangle$	0	0	0	$-\frac{2}{35}C_4\langle r^4\rangle$	0
−5/2	0	$\frac{2\sqrt5}{105}C_4\langle r^4\rangle$	0	0	0	$\frac{2}{105}C_4\langle r^4\rangle$

$$(1.38)$$

$\langle r^n\rangle$ is the expected value of r^n for a $4f$ orbital. From comparison with the matrix (1.28) it can be seen that

$$C_4\langle r^4\rangle = \frac{15}{8}(2\delta_1+3\delta_2) \quad (1.39)$$

From this result and equating matrix elements in the same way for $^2F_{7/2}$,

$$C_6\langle r^6\rangle = \frac{39}{20}(4\delta_1-5\delta_2) \quad (1.40)$$

The two sets of parameters δ_1, δ_2 and $C_4\langle r^4\rangle$, $C_6\langle r^6\rangle$ are two equivalent ways of describing the splitting caused by $\mathcal{H}_c$. From the point-charge model, C_4 and C_6 can be estimated. Usually, however, these are determined empirically from optical data.

Figure 1.2 shows the energy levels of Tm^{2+} in CaF_2 measured by Kiss.[10] Tm^{2+} is isoelectronic with Yb^{3+}. Because Yb^{3+} requires charge compensation, usually several different symmetries are observed for Yb^{3+} in CaF_2 corresponding to different types of association with a local charge compensating defect which lowers the symmetry at the Yb^{3+} site. The Γ_6 level of $^2F_{7/2}$ was not observed. The Zeeman effect in the luminescence spectrum is used to identify the levels.[11] Let us see what values of our parameters these measured energy levels imply. The splitting of $^2F_{5/2}$ depends only on $C_4\langle r^4\rangle$.

$$E\left(^2F_{5/2}\Gamma_8\right) - E\left(^2F_{5/2}\Gamma_7\right) = \frac{12}{105}C_4\langle r^4\rangle = 410.4 \text{ cm}^{-1} \quad (1.41)$$

$$C_4\langle r^4\rangle = 3.6 \times 10^3 \text{ cm}^{-1}$$

The splitting of $^2F_{7/2}$ depends also on $C_6\langle r^6\rangle$.

$$E\left(^2F_{7/2}\Gamma_8\right) - E\left(^2F_{7/2}\Gamma_7\right) = \frac{8}{77}C_4\langle r^4\rangle + \frac{10}{429}C_6\langle r^6\rangle = 555.8 \text{ cm}^{-1} \quad (1.42)$$

$$C_6\langle r^6\rangle = 7.8 \times 10^3 \text{ cm}^{-1}$$

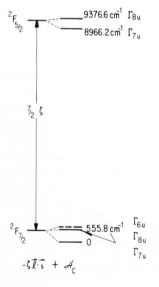

Figure 1.2 Energy levels of the f^{13} configuration in a cubic crystal field. The measured level positions for CaF_2: Tm^{2+} are indicated.

The eight nearest neighbor fluorines are 2.36 Å from the Ca^{2+} site. Freeman and Watson's[12] calculated Yb^{3+} wavefunctions imply $\langle r^4 \rangle = 0.96\, a_0^4$, and $\langle r^6 \rangle = 3.10\, a_0^6$, where a_0 is the Bohr radius; the $\langle r^n \rangle$ values for Tm^{2+} would be very slightly larger. From Equation 1.34 the point-charge model leads to

$$C_4\langle r^4 \rangle = 0.9 \times 10^3 \text{ cm}^{-1}, \qquad C_6\langle r^6 \rangle = 0.5 \times 10^3 \text{ cm}^{-1} \qquad (1.43)$$

Including contributions from more distant shells of neighbors changes these values only slightly because of the strong dependence on R_i. There are 12 Ca^{2+} next nearest neighbor ions, 24 F^- third nearest neighbors, and so on.[13] But from Equation 1.34 the contribution of an ion at distance R_i to C_4 is proportional to R_i^{-5} and the contribution to C_6 is proportional to R_i^{-7}. Our fairly typical result is that the point-charge model predicts the correct signs for $C_4\langle r^4 \rangle$ and $C_6\langle r^6 \rangle$ but not the correct magnitudes, the discrepancy for $C_6\langle r^6 \rangle$ being the larger.

Equations 1.42, 1.39, 1.40, and 1.41 can be combined to yield the relative positions of the a_{2u}, t_{1u}, and t_{2u} orbitals. The result is that the a_{2u} level lies lowest, the t_{1u} level is higher by 0.99×10^3 cm^{-1}, and t_{2u} is ~ 17 cm^{-1} above t_{1u}. Figure 1.3 shows that in eightfold coordination the a_{2u} orbital would be expected to lie lowest, because it has lobes of hole density directed toward the negative anions.

The preceding calculations have illustrated certain points. The method is practical because we are dealing with a single particle—a single hole in the otherwise filled f shell, $(4f)^{13}$. The case of a single electron $(4f)^1$ would be almost identical. For other configurations the decomposition of the states $|\psi J M_J\rangle$ into single particle states (Equation 1.26) is cumbersome, however. (Here ψ represents other quantum numbers necessary to specify the state, such as S, L, and any other labels which may help to avoid ambiguity.) If we are interested only in the first-order effect of $\mathcal{H}_c$, that is, in matrix elements of $\mathcal{H}_c$ between states $|\psi J M_J\rangle$ of the same multiplet ψJ, then the simplest way to calculate splittings due to $\mathcal{H}_c$ is to write $\mathcal{H}_c$ not in terms of Cartesian coordinates (Equation 1.33) or spherical harmonics (Equation 1.35), but in terms of polynomials in J_x, J_y, J_z which transform in the same way under rotations as spherical harmonics. Then $\langle \psi J M_J' | \mathcal{H}_c | \psi J M_J \rangle$ can be obtained directly. Writing $\mathcal{H}_c$ in terms of these operators rather than Cartesian coordinates is allowed because the matrix elements of $\mathcal{H}_c$ in the two forms are proportional to each other. Expressions for such operator equivalents of, for example, $f_4(x,y,z)$ and $f_6(x,y,z)$ of (1.33) and many matrix elements of these operators can be found in reference 14.

When a more complex configuration is involved, then, the first step is to decide from symmetry considerations what form $\mathcal{H}_c$ must have and how

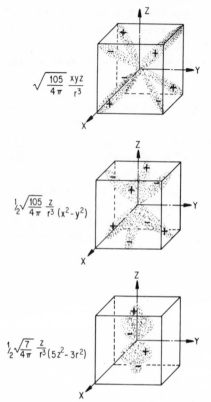

$$\sqrt{\frac{105}{4\pi}}\,\frac{xyz}{r^3}$$

$$\tfrac{1}{2}\sqrt{\frac{105}{4\pi}}\,\frac{z}{r^3}(x^2-y^2)$$

$$\tfrac{1}{2}\sqrt{\frac{7}{4\pi}}\,\frac{z}{r^3}(5z^2-3r^2)$$

Figure 1.3   Contours of the $f\,a_{2u}$, $t_{2u}z$, and $t_{1u}z$ orbitals of Table 1.4.

many independent parameters C_n there will be. For f electrons ($l=3$) $n \leqslant 2 \cdot 3 = 6$. Then $\mathcal{H}_c$ can be written in terms of the operators J_i, a matrix written for each multiplet ψJ of interest, and the splittings calculated in terms of the $C_n \langle r^n \rangle$. These crystal field parameters can then be varied to obtain a best fit to the frequencies of the observed spectral lines. Ideally, there are very many more observed transitions than parameters in contrast to the case of Tm^{2+} in a cubic field treated here.

With the $^2F_{5/2} - {}^2F_{7/2}$ energy separation $\sim 10^4\ cm^{-1}$ there is the possibility of optical transitions in the wavelength region near 1 μm. Electric dipole transitions are strictly forbidden in this case. The electric dipole operator $\mathbf{M}^{E1} = e\mathbf{r}$ is odd under the inversion, its components transforming as Γ_{4u}, and so does not connect two states of the same parity. In this case, the site having inversion symmetry, there is no mechanism for admixing states of other parity into the 4f functions. If the site had T_d symmetry, however, we see from Table 1.4 that the function xyz transforms as Γ_1 of T_d. There would then be a term $C_3 xyz$ in $\mathcal{H}_c$. This term, being of odd parity, would not affect the energy level calculations because $\langle f|xyz|f\rangle = 0$, but could mix even parity highly excited states into the wavefunctions, since, for example, $\langle f|xyz|d\rangle \neq 0$. Electric dipole transitions that become allowed by this mechanism are usually dominant in the spectra of rare earth ions in sites lacking inversion symmetry and are called forced electric dipole transitions. They can occur if $\Delta J \leqslant 6$.

The magnetic dipole operator $-\beta(\mathbf{L} + g_0 \mathbf{S})$ transforms as Γ_{4g} and does connect states of the same parity. Magnetic dipole transitions are allowed in this case because $\Delta J = 7/2 - 5/2 = 1$. The selection rules in the crystal can be found by the method outlined in the discussion that led to Equation 1.21. From Table 1.6

$$\Gamma_{6u} \times \Gamma_{4g} = \Gamma_{6u} + \Gamma_{8u}$$

$$\Gamma_{7u} \times \Gamma_{4g} = \Gamma_{7u} + \Gamma_{8u} \qquad (1.44)$$

$$\Gamma_{8u} \times \Gamma_{4g} = \Gamma_{6u} + \Gamma_{7u} + 2\Gamma_{8u}$$

Therefore, in a luminescence experiment five lines might be observed: $^2F_{5/2}\Gamma_{7u} \rightarrow {}^2F_{7/2}\Gamma_{7u}$, Γ_{8u}; $^2F_{5/2}\Gamma_{8u} \rightarrow {}^2F_{7/2}\Gamma_{6u}$, Γ_{7u}, Γ_{8u}. Electric quadrupole transitions are also allowed, but should be weak compared with the magnetic dipole transitions; they have not been observed.

The $4f$ electrons are weakly coupled to the lattice, as evidenced by the small crystal field splittings. Weak vibronic lines are seen in the spectra, however, which correspond to excitation or de-excitation of a lattice phonon simultaneously with the electronic transition. Such transistions are usually most prominent in the spectra of Ce^{3+} and Yb^{3+}, which also have relatively larger crystal field splittings. In O_h symmetry odd vibrations may force electric dipolar vibronic transitions.

Usually the positions of the components of excited rare earth multiplets ($^2F_{5/2}$ in the case of Yb^{3+} or Tm^{2+}) are found from low-temperature absorption spectra. At sufficiently low temperature only the lowest component of the ground multiplet is populated, and transitions from this level only are seen. The positions of the components of the ground multiplet can be found from low-temperature emission spectra in which only transitions from the lowest component of an excited multiplet to the components of the ground multiplet appear. It is not practical to find these from absorption measurements because the transition wavelengths lie in the far infrared. Transitions between levels belonging to the same multiplet cannot be seen in luminescence spectra. These transitions are always nonradiative, the total splitting of a multiplet being comparable with the maximum phonon frequency. The $4f \rightarrow 4f$ transition spectra consist of sharp lines of width a few cm^{-1} or less in crystals.

Electronic Raman scattering has also been useful in finding the positions and symmetries of the components of rare earth ground multiplets. By this technique these far-infrared transitions can be observed with visible light. An incident photon of frequency ν_1 (in the visible) polarized in the i direction is inelastically scattered. The frequency of the scattered photon is $\nu_2 = \nu_1 - \nu_{kl}$, and is polarized along the j direction; $i, j = x, y, z$. The ν_{kl} is the (infrared) frequency of the transition between the ground state l and an excited state k; $h\nu_{kl} = E_k - E_l$. The transition probability for this process is proportional to the square of the matrix element of a scattering tensor α,

$$\langle k | \alpha_{ij} | l \rangle \propto \sum_n \frac{\langle k | M_i | n \rangle \langle n | M_j | l \rangle}{E_n - E_k - h\nu_1} + \frac{\langle k | M_j | n \rangle \langle n | M_i | l \rangle}{E_n - E_k - h\nu_2} \qquad (1.45)$$

$|n\rangle$ is a highly excited state of opposite parity so that the transition $l \rightarrow n$ would lie in the ultraviolet. (For divalent rare earths these excited states can lie much lower, however. For Tm^{2+} in CaF_2 the $4f^{12}5d$ absorption bands extend as low as 15,000 cm^{-1}.[10]) It is often a good approximation to replace the energy denominators by an average value if they do not change rapidly with n and to compress the sums by means of the closure condition,[15]

$$\sum_n \langle k | M_i | n \rangle \langle n | M_j | l \rangle = \langle k | M_i M_j | l \rangle \qquad (1.46)$$

M stands for M^{E1} which transforms as Γ_{4u}. From Table 1.6,

$$\Gamma_{4u} \times \Gamma_{4u} = \Gamma_{1g} + \Gamma_{3g} + \Gamma_{4g} + \Gamma_{5g} \qquad (1.47)$$

In our example Γ_k is Γ_{7u}, and

$$\Gamma_{7u} \times (\Gamma_{1g} + \Gamma_{3g} + \Gamma_{4g} + \Gamma_{5g}) = \Gamma_{6u} + 2\Gamma_{7u} + 3\Gamma_{8u} \qquad (1.48)$$

Therefore, there are combinations of incident and scattered polarizations for which the Γ_6 and Γ_8 components of $^2F_{7/2}$ could be observed.

Every energy level of an atom with an odd number of electrons is at least doubly degenerate in the absence of a magnetic field. In cubic symmetry such an ion has doubly and fourfold degenerate levels whose wavefunctions are Kramers pairs ψ, ψ^* or pairs of Kramers pairs. In a magnetic field $\mathbf{H}$ this degeneracy is removed. The Zeeman effect is described by the addition to the Hamiltonian

$$\mathcal{H}_Z = g_J \beta \mathbf{H} \cdot \mathbf{J} \qquad (1.49)$$

β is the Bohr magneton, and g_J is the Landé g factor, given for the case of Russell Saunders coupling by

$$g_J = 1 + \left[\frac{J(J+1) - L(L+1) + S(S+1)}{2J(J+1)} \right](g_0 - 1) \qquad (1.50)$$

$$g_0 = 2.0023$$

Let us calculate the splitting of the $^2F_{7/2}\Gamma_{7u}$ ground level in a magnetic field $\mathbf{H} = H(m\mathbf{i} + n\mathbf{j} + p\mathbf{k})$. The wavefunctions of the doublet are

$$|^2F_{7/2}\Gamma_{7u} \pm \rangle = \frac{\sqrt{3}}{2}|7/2, \pm 5/2\rangle - \frac{1}{2}|7/2, \mp 3/2\rangle \qquad (1.51)$$

With

$$\mathbf{H} \cdot \mathbf{J} = \frac{1}{2}H\left[(m - in)J_+ + (m + in)J_- + 2pJ_z \right] \qquad (1.52)$$

the eigenvalues of (1.41) are

$$E_\pm = \pm \frac{3}{2}g_J\beta H(m^2 + n^2 + p^2)^{1/2}$$

$$= \pm \frac{3}{2}g_J\beta H$$

$$= \pm \langle \Gamma_{7u} + |J_z|\Gamma_{7u} + \rangle g_J\beta H \qquad (1.53)$$

A similar result is obtained for any Kramers doublet in cubic symmetry. The splitting is isotropic, that is, independent of the direction of $\mathbf{H}$. This is not the case for a Γ_8 state. In this case the splitting is given by[16]

$$E_{1,2} = \pm g_j\beta H$$

$$\times \left\{ \frac{1}{2}\left[P^2 + Q^2 \pm \left(P^2 + Q^2 - 4P^2Q^2 - 4r(n^2m^2 + n^2p^2 + p^2m^2) \right)^{1/2} \right] \right\}^{1/2}$$

$$E_{3,4} = \mp g_J\beta H \left\{ \frac{1}{2}\left[P^2 + Q^2 \pm (P^2 + Q^2 - \cdots)^{1/2} \right] \right\}^{1/2} \qquad (1.54)$$

$$P = \langle \Gamma_8 a + |J_z|\Gamma_8 a + \rangle, \qquad Q = \langle \Gamma_8 b + |J_z|\Gamma_8 b + \rangle$$

$$r = \frac{3}{16}(P - 3Q)(3P - Q)(P^2 + Q^2)^2$$

The four components of the Γ_8 state are $|\Gamma_8 a\pm\rangle$ and $|\Gamma_8 b\pm\rangle$, where $|\Gamma_8 a\pm\rangle$ transform in the same way as $|3/2, \pm 3/2\rangle$, and $|\Gamma_8 b\pm\rangle$ transform as $|3/2, \pm 1/2\rangle$.

In an electron paramagnetic resonance experiment the hyperfine splitting due to the interaction of the nuclear spin $\mathbf{I}$ and the electronic angular momentum can also be observed. A term $a\mathbf{I}\cdot\mathbf{J}$ can be added to $\mathcal{H}_Z$ to represent this interaction. Each line is then split into $2I+1$ components with relative intensities corresponding to the isotopic abundances. This hyperfine splitting is a "signature" of the ion and facilitates identification of the resonance.

Except for Ce^{3+} and Yb^{3+} the energy level structures for the trivalent ions are quite complicated because of the large degeneracies; the $(4f)^7$ configuration has a degeneracy of 3432. However, the energy levels of the free trivalent ions are well known now through comparison of calculated energy levels with experiment.[17] Linear combinations of states characterized by m_l and m_s for each electron are formed to give states characterized by L, S, M_L, and M_S. These states are split apart by the term $\Sigma_{i>j}e^2 r_{ij}^{-1}$, states with same L and S being mixed. The spin-orbit interaction term splits each of these states further into a number of levels $^{2S+1}L_J$. The levels are always labeled in this Russell-Saunders notation, although for all but some of the lowest lying levels the coupling is really intermediate; that is, the spin-orbit term is not small compared with $\Sigma e^2 r_{ij}^{-1}$, and states with the same J but different L and S are mixed. The notation is such that a $^{2S+1}L_J$ state, really a mixture of states of different L, S combinations, approaches $^{2S+1}L_J$ as the spin-orbit interaction goes to zero. Calculated wavefunctions are available for all the free rare earth ions. Reference 18 contains a guide to this literature.

When the $^{2S+1}L_J$ multiplets lie so close together that their separation is not large compared with the crystal field splittings, then mixing of these states by $\mathcal{H}_c$ must be taken into account. Similarly, when the crystal is in a magnetic field $\mathcal{H}_Z$ may mix different crystal-field-split components of a multiplet when these are not widely separated compared to Zeeman energies.

Extension of the calculation of crystal field splittings to lower symmetries is straightforward. The lower the symmetry, the more independent parameters $C_n^m\langle r^n\rangle$ there are corresponding to terms $C_n^m\langle r^n\rangle Y_n^m$ in $\mathcal{H}_c$. Many experimental crystal spectra have been fitted to expressions of this kind, and it has been found that the crystal field parameters are very nearly independent of $^{2S+1}L_J$ level; that is, the crystal field splitting of all the $^{2S+1}L_J$ levels can be described by the same set of parameters. This implies that the correlations among $4f$ electrons are modified very little by the crystal. The point-charge model almost always fails to predict quantitatively the crystal field parameters. Quantitative agreement with experimental values has been obtained in calculations, taking into account the overlap of the $4f$ orbitals with the orbitals of the nearest-neighbor ions—especially important for the $C_6^m\langle r^6\rangle$ parameters—and shielding by the $5s$ and $5p$ shells.[19]

Figure 1.4 shows energy levels coming from the $4f$ configuration of several trivalent rare earth ions. The positions of the $^{2S+1}L_J$ multiplets are nearly independent of crystal host. The small crystal field splittings are not shown. An ion with configuration $(4f)^{14-n}$ has the same $^{2S+1}L_J$ multiplets as an ion with congifuration $(4f)^n$. Er^{3+} and Nd^{3+} are such a pair of ions with $n=3$. The ^4I multiplets are oppositely ordered for the two. If the $(4f)^{11}$ configuration is considered a configuration of three holes rather than eleven electrons, it is easy to see that the same multiplets should occur as for $(4f)^3$ and that the spin-orbit splitting should be opposite in sign. As the number of $4f$ electrons increases, the spin-orbit splitting becomes larger, and the coupling departs more from the Russel-Saunders limit. In Figure 1.5 a luminescence spectrum of Nd^{3+} in $Y_3Al_5O_{12}$ is shown. The Nd^{3+} substitutes for Y^{3+}. In the low symmetry of the site all degeneracy except Kramers degeneracy is lifted. We note the typical rare earth spectrum consisting of groups of sharp lines. The crystal field in $Y_3Al_5O_{12}$ is relatively large, leading to larger separations of lines in a group than in most other crystals.

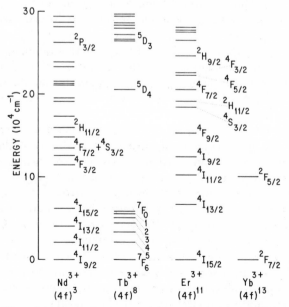

Figure 1.4 Energy levels of several free trivalent rare earth ions.

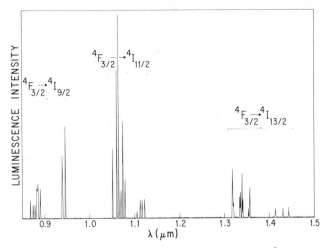

Figure 1.5 Room temperature luminescence spectrum of Nd^{3+} in $Y_3Al_5O_{12}$.

d^n TRANSITION METAL IONS

Ions of the three transition metal series Sc-Cu, Y-Ag, and Hf-Au have also been studied extensively as impurities in crystals. The interesting optical and magnetic properties are due to the electrons of the incomplete $3d$, $4d$, or $5d$ shell, the other electrons being to a first approximation in closed shells. The energy levels are much more sensitive to the crystalline environment than those of the rare earths, splittings due to $\mathcal{H}_c$ varying from 3×10^3 cm^{-1} to 3×10^4 cm^{-1}. Because much more experimental work has been done on the iron group ions, the discussion of this section concentrates on them. The same theory also applies to the $4d$ and $5d$ ions, except that the spin-orbit coupling is larger and cannot always be treated as a perturbation.

For the $3d$ ions the term $\mathcal{H}_c$ of the Hamiltonian may be smaller or larger than the interelectron electrostatic interactions; both are much larger than the spin-orbit interaction. There are two starting points for a theoretical treatment. The $3d$ electrons can be coupled together first, as in a treatment of the free ion, to form various terms ^{2S+1}L among which $\Sigma e^2 r_{ij}^{-1}$ is diagonal and the matrix elements of $\mathcal{H}_c$ found between these states (weak field scheme). The n electrons can be placed in the configurations $(t_2)^q (e)^{n-q}$ (see Table 1.4), which diagonalize $\mathcal{H}_c$, and terms $^{2S+1}\Gamma$ can be formed from these between which matrix elements of $\Sigma e^2 r_{ij}^{-1}$ are calculated (strong field scheme). Because the second method is the more different from that appropriate to the rare earths, we use it. As an example we treat the case of two d electrons d^2 in T_d symmetry.

Three strong field configurations are possible, e^2, t_2e, and t_2^2. From Table 1.6 the terms that can be formed from these configurations are

$$e \times e = A_1 + A_2 + E, \qquad t_2 \times t_2 = A_1 + E + T_1 + T_2$$

$$e \times t_2 = T_1 + T_2 \tag{1.55}$$

Consider e^2 first. If the electrons have parallel spin, they must occupy different orbitals. Therefore, an A state with $M_s = 1$ is

$$| \overset{+}{u} \ \overset{+}{v} | = \frac{1}{\sqrt{2}} \left[\overset{+}{u}(1) \overset{+}{v}(2) - \overset{+}{u}(2) \overset{+}{v}(1) \right] \tag{1.56}$$

From Equation 1.19 we see that $S_4^{z-} | \overset{+}{u} \ \overset{+}{v} | = - | \overset{+}{u} \ \overset{+}{v} |$. This, then, is the A_2 state, written in the notation $| \gamma^n \ ^{2S+1}\Gamma M_s \rangle$,

$$|e^2 \ ^3A_2 1\rangle = | \overset{+}{u} \ \overset{+}{v} | \tag{1.57}$$

The other $S = 1$ states are found by the group theoretical methods already described, or they can be found in the tables of references 20 and 21. The $^3T_2$ states are

$$|t_2e \ ^3T_2 xy 1\rangle = | \left(\overset{+}{xy} \right) \overset{+}{v} |$$

$$|t_2e \ ^3T_2 yz 1\rangle = - \frac{1}{2} | \left(\overset{+}{yz} \right) \overset{+}{u} | + \frac{\sqrt{3}}{2} | \left(\overset{+}{yz} \right) \overset{+}{v} | \tag{1.58}$$

$$|t_2e \ ^3T_2 zx 1\rangle = - \frac{1}{2} | \left(\overset{+}{zx} \right) \overset{+}{u} | - \frac{\sqrt{3}}{2} | \left(\overset{+}{zx} \right) \overset{+}{v} |$$

Thus the e^2 configuration has terms $^3A_2$, $^1A_1$, ^1E; the t_2e configuration terms are $^3T_1$, $^3T_2$, $^1T_1$, and $^1T_2$; and t_2^2 has terms $^3T_1$, ^1E, $^1T_2$, and $^1A_1$. In a crystal field of O_h or T_d symmetry the e and t_2 orbitals have different energies, and we call this energy difference Δ:

$$\epsilon(t_2) = \frac{2}{5} \Delta, \qquad \epsilon(e) = - \frac{3}{5} \Delta \tag{1.59}$$

For fourfold (T_d) coordination and eightfold (O_h) coordination $\Delta > 0$. For sixfold (O_h) coordination $\Delta < 0$. (The lobes of electron density point toward the centers of the cube edges for t_2 orbitals and toward the face centers for e orbitals.) Thus the crystal field energies of the configurations e^2, t_2e, and t_2^2 are $-(6/5)\Delta$, $-(1/5)\Delta$, and $(4/5)\Delta$, respectively.

The matrix elements of the interelectron electrostatic repulsion $\Sigma e^2 r_{ij}^{-1}$,

$$\left\langle ab \left| \frac{e^2}{r_{12}} \right| cd \right\rangle = \int \int a(1)b(2) \frac{e^2}{r_{12}} c(1)d(2) \, d\tau_1 \, d\tau_2 \tag{1.60}$$

are tabulated by Griffith[20] for the case in which the orbitals a, b, c, d are the strong field orbitals u, v, xy, zx, or yz in terms of the positive Racah parameters[21] A, B, and C. Because $\Sigma e^2 r_{ij}^{-1}$ is independent of spin and transforms as A_1, it mixes only states $^{2S+1}\Gamma$ of the same S and Γ. The energies of the $S = 1$ states are

$$E\left(^3A_2\right) = A - 8B - \frac{6}{5}\Delta$$

$$E\left(^3T_2\right) = A - 8B - \frac{1}{5}\Delta \qquad\qquad (1.61)$$

$$E_{\pm}\left(^3T_1\right) = A - \frac{1}{2}B + \frac{3}{10}\Delta \pm \left[225B^2 - 18B\Delta + \Delta^2\right]^{1/2}$$

Similar expressions are found for the $S = 0$ states; their energies depend on C as well as A, B, and Δ. Because the energies of all states are equal to A plus expressions containing only B, C, and Δ, all energy differences depend only on B, C, and Δ. For positive Δ the ground state is $^3A_2$. The energy differences $E\left(^{2S+1}\Gamma\right) - E\left(^3A_2\right)$ are plotted as a function of Δ/B in Figure 1.6. The ratio C/B was set equal to 4.5, a typical value. Similar figures are found in reference 20 for the other d^n configurations. In T_d symmetry forced electric dipole transitions could lead to the absorption transitions $^3A_2 \rightarrow ^3T_1$, because $A_2 \times T_2 = T_1$.

The spin-orbit coupling energies are relatively small. To first order each T_1 or T_2 state is split by $\zeta\Sigma \mathbf{l}_i \cdot \mathbf{s}_i$ into three components as an atomic P state splits; E and A states do not split. In general, there will be nonzero matrix elements of $\zeta\Sigma \mathbf{l}_i \cdot \mathbf{s}_i$ between two terms $^{2S+1}\Gamma$ and $^{2S'+1}\Gamma'$ if $\Gamma \times \Gamma'$ contains Γ_4 and $S - S' = 0$ or ± 1.

The three ground-state functions transform as bases for Γ_4 because they are products of an orbital A_2 function and $S = 1$ spin functions, which transform as T_2 or Γ_5. $A_2 \times T_2 = T_1$ or Γ_4. They are

$$|e^2 \ ^3A_2\Gamma_4 1\rangle = |\overset{+}{u}\ \overset{+}{v}| = |uv|\,|1,1\rangle$$

$$|e^2 \ ^3A_2\Gamma_4 0\rangle = \frac{1}{\sqrt{2}}|\overset{+}{u}\ \overset{-}{v}| + \frac{1}{\sqrt{2}}|\overset{-}{u}\ \overset{+}{v}| = |uv|\,|1,0\rangle \qquad (1.62)$$

$$|e^2 \ ^3A_2\Gamma_4 -1\rangle = |\overset{-}{u}\ \overset{-}{v}| = |uv|\,|1,-1\rangle$$

The kets on the right are the spin functions $|S, M_s\rangle$. The Zeeman splitting

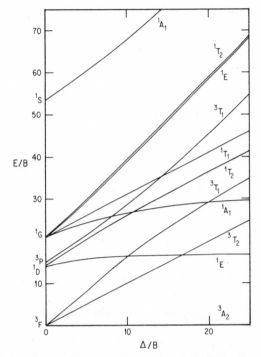

Figure 1.6 Energy levels of the d^2 configuration in a tetrahedral field. Energy divided by the electrostatic Racah parameter B is plotted versus Δ/B. C is set equal to $4.5B$.

is isotropic and

$$\mathcal{H}_z = \beta H \sum_{i=1}^{2} (l_{zi} + 2s_{zi}) = \beta H (L_z + 2S_z) \qquad (1.63)$$

All matrix elements of L_z between the states (1.62) are zero. The Zeeman energies are given by

$$E(\Gamma_4 1) = 2\beta H, \qquad E(\Gamma_4 0) = 0, \qquad E(\Gamma_4 - 1) = -2\beta H \qquad (1.64)$$

The ground state splits in a magnetic field as three spin states $|S=1, M_s\rangle$ with g factor 2. Actually, this is not quite right. The spin orbit coupling mixes a little of two excited states, $|t_2 e^3 T_2 \Gamma_4\rangle$ and $|t_2 e^1 T_2 \Gamma_4\rangle$, into $|e^2 \, ^3A_2 \Gamma_4\rangle$. Moreover the g factor of the electron spin is $g_0 = 2.0023$, not 2. This difference is observable in an EPR measurement. The corrected Zeeman energies are

$$\frac{E(\Gamma_4 1)}{\beta H} = g_0 + \left(\frac{2i\zeta_{3d}}{\Delta}\right)\langle v|l_z|xy\rangle$$

$$= g_0 - \frac{4\zeta_{3d}}{\Delta}$$

$$\frac{E(\Gamma_4 0)}{\beta H} = 0 \qquad\qquad (1.65)$$

$$\frac{E(\Gamma_4 -1)}{\beta H} = -g_0 - \left(\frac{2i\zeta_{3d}}{\Delta}\right)\langle v|l_z|xy\rangle$$

$$= -g_0 + \frac{4\zeta_{3d}}{\Delta}$$

For example, Ti^{2+} has the $(3d)^2$ configuration, substitutes for Cd in CdTe, and is coordinated by four Te's.[22] The ground state g factor is 1.915. The optical absorption spectrum consists of two bands at 1.93 and 1.09 μm, which are assigned to the transitions $^3A_2 \rightarrow {}^3T_1(F)$ and $^3A_2 \rightarrow {}^3T_1(P)$, respectively. From these optical data we obtain $\Delta = 3.12 \times 10^3$ cm^{-1} and $B = 332$ cm^{-1}. These are typical values for an ion in fourfold coordination. For the free Ti^{2+} ion B is about twice as large. The $|\Delta|$ is larger in sixfold or eightfold coordination.

Suppose that the tetrahedron formed by the four coordinating anions were slightly distorted by a small inward or outward relaxation along a $\langle 100 \rangle$ axis so that the symmetry were lowered to D_{2d}. Because the basis functions for Γ_4 of T_d form bases for the onefold Γ_2 and twofold Γ_5 irreducible representations of D_{2d}, the ground state, for example, could split into a doublet and a singlet.[5]

MOLECULAR ORBITAL CLUSTER MODEL

It is generally found from interpretation of the optical spectra of ions with d^n configurations that the magnitude of Δ depends on the identity of the coordinating anions, being larger for coordination by fluorine than by chlorine, for example; Δ is larger the more positive the central metal ion; and the parameters B and C are smaller than the free-ion values and depend on the coordinating anions. There are also strong high-energy absorption bands that cannot be explained by the single-ion theory. In the electron paramagnetic resonance spectra hyperfine interactions are often seen with the nuclear magnetic moments of the coordinating ions, and the tensors characterizing the interactions cannot be explained by a simple dipolar interaction. In addition, the hyperfine interaction with the nuclear moment of the central metal ion depends on the host lattice. All these difficulties can be somewhat alleviated at the expense of increased complexity by a molecular orbital approach.

In the molecular orbital cluster model the nearest coordinating anions or ligands are no longer treated only as point charges, but the extent of their outermost electrons is taken into account. These electrons may transfer partially to the empty $3d$ orbitals of the metal ion, and the $3d$ electrons may sometimes be found in the partially emptied ligand orbitals. The larger the charge on the metal ion, the greater its attraction for the ligand electrons and the larger these effects should be. Let us see how the treatment of a tetrahedral cluster is modified in the molecular orbital approach.

The orbitals of interest are linear combinations of d and ligand orbitals whose energies are not very different. The ligand orbitals are assumed to be s and p. Under the operations of the T_d symmetry group the four s orbitals are transformed into each other. The four p orbitals whose lobes point along the lines connecting the metal ion to each ligand ($\langle 111 \rangle$ directions) also transform among themselves. These orbitals are symmetric about the $0-1$, $0-2$, $0-3$, and $0-4$ lines of Figure 1.1 and are called σ orbitals. The other eight p orbitals have lobes that are perpendicular to these lines. They transform among themselves and are called π orbitals. The σ orbitals are written $\sigma_i = \sqrt{1-\alpha^2}\, s_i + \alpha p_{zi}$, where s_i is an s orbital on ligand i, and p_{zi} is the p orbital on ligand i with positive lobe directed toward the metal ion. The characters of the representation for which these four σ_i form a basis are found from consideration of Figure 1.1. They are

$$\begin{array}{ccccc} E & C_3 & C_2 & S_4 & \sigma_d \\ 4 & 1 & 0 & 0 & 2 \end{array} \tag{1.66}$$

By inspection of the T_d character table (Table 1.2) it can be seen that this representation reduces to a_1 and t_2: $\sigma = a_1 + t_2$. The a_1 basis function is obviously,

$$\Phi(a_1) = \frac{1}{2}(\sigma_1 + \sigma_2 + \sigma_3 + \sigma_4) \tag{1.67}$$

The three linear combinations of σ_i forming a basis for t_2 are found with the help of Equation 1.12, Figure 1.1, and Table 1.2.

$$\psi_1(t_2) = \sum_R \chi^{t_2}(R)R\sigma_1 = 6\sigma_1 - 3(\sigma_2 + \sigma_3 + \sigma_4)$$

$$\psi_2(t_2) = \sum \chi^{t_2}(R)R\sigma_2 = 6\sigma_2 - 3(\sigma_1 + \sigma_3 + \sigma_4) \qquad (1.68)$$

$$\psi_3(t_2) = \sum \chi^{t_2}(R)R\sigma_3 = 6\sigma_3 - 3(\sigma_1 + \sigma_2 + \sigma_4)$$

The linear combinations of ψ_1, ψ_2, and ψ_3 which transform as x, y, and z can be found by operating on them with three group elements. Let $\psi_x = a\psi_1 + b\psi_2 + c\psi_3$. Because $C_2^x x = x$, then we must have

$$C_2^x(a\psi_1 + b\psi_2 + c\psi_3) = a\psi_1 + b\psi_2 + c\psi_3 \qquad (1.69)$$

Because

$$C_2^x \begin{bmatrix} \sigma_1 \\ \sigma_2 \\ \sigma_3 \\ \sigma_4 \end{bmatrix} = \begin{bmatrix} \sigma_4 \\ \sigma_3 \\ \sigma_2 \\ \sigma_1 \end{bmatrix} \qquad (1.70)$$

we find $a=0$, $b=c$, and

$$\psi_x = \frac{1}{2}(\sigma_2 - \sigma_1 + \sigma_3 - \sigma_4) \qquad (1.71)$$

Overlap between σ orbitals is neglected in normalization throughout. Similarly,

$$\psi_y = \sigma_d^{(110)}\psi_x = \frac{1}{2}(\sigma_2 - \sigma_1 + \sigma_4 - \sigma_3)$$

$$\psi_z = \sigma_d^{(011)}\psi_y = \frac{1}{2}(\sigma_2 - \sigma_4 + \sigma_1 - \sigma_3) \qquad (1.72)$$

$\sigma_d^{(110)}$ is the reflection in the (110) plane. In the same way it can be shown that the π orbitals from bases for e, t_1, and t_2. The σ orbitals can combine with the metal t_2 orbitals and the π orbitals with the metal e and t_2 orbitals. Because this treatment is meant to be only an illustrative example of a method, the π orbitals are ignored as a simplification. The new orbitals are

$$|t_2 i\rangle = a_1|dt_2 i\rangle + a_2|\psi_i\rangle, \qquad i = x,y,z$$

$$|dt_2 x\rangle = |yz\rangle \qquad (1.73)$$

An overlap integral S is defined by

$$S = \langle dt_2 | \psi \rangle = 2 \langle yz | \sigma_2 \rangle > 0 \qquad (1.74)$$

The e orbitals $|du\rangle$ and $|dv\rangle$ are unchanged because no σ orbitals mix with them.

The Hamiltonian operator is

$$\mathcal{H} = -\left(\frac{\hbar^2}{2m}\right)\nabla^2 - \frac{Z_m e^2}{r} - \sum_{l=1}^{4} \frac{Z_l e^2}{|\mathbf{r} - \mathbf{R}_l|} \qquad (1.75)$$

$Z_m e$ and $Z_l e$ are the effective charges on the metal and ligand ion cores. The mixing coefficients a_1 and a_2 are found by substituting Equation 1.73 into $\mathcal{H}|t_2\rangle = E|t_2\rangle$ and multiplying from the left by the bras $\langle dt_2|$ or $\langle \psi|$. The index i is dropped, because the three orbitals must remain degenerate. The resulting equations are

$$a_1(H_{dd} - E) + a_2(H_{\sigma d} - SE) = 0$$
$$a_2(H_{\sigma d} - SE) + a_2(H_{\sigma\sigma} - E) = 0 \qquad (1.76)$$

$$H_{dd} = \langle dt_2 | \mathcal{H} | dt_2 \rangle, \qquad H_{\sigma\sigma} = \langle \psi | \mathcal{H} | \psi \rangle$$
$$H_{\sigma d} = \langle \psi | \mathcal{H} | dt_2 \rangle \qquad (1.77)$$

These equations have a nontrivial solution if the determinant of coefficients is zero; that is, if

$$(E - H_{dd})(E - H_{\sigma\sigma}) - (ES - H_{\sigma d})^2 = 0 \qquad (1.78)$$

The two solutions are E_a and E_b, where $E_b < E_a$. If $H_{\sigma d}(E_a - E_b)^{-1}$ and S are much less than 1, then the corresponding states, written in a slightly different form, are

$$|t_2 b\rangle = N_b(\eta | dt_2 \rangle + |\psi\rangle)$$
$$|t_2 a\rangle = N_a(|dt_2\rangle - \mu|\psi\rangle) \qquad (1.79)$$

$$\mu = \frac{\eta + S}{1 + \eta S} \approx \eta + S \qquad \mu, \eta < 1$$

The single-particle energy levels are shown schematically in Figure 1.7. The bonding orbital $t_2 b$ has electron density concentrated largely on the ligands; the antibonding orbital $t_2 a$ is largely localized on the metal. If the metal has two d electrons, these and the eight σ electrons from the ligands

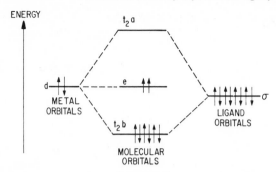

Figure 1.7 Schematic representation of the combination of d metal orbitals with the σ ligand orbitals of four neighbors to form t_2a antibonding, e nonbonding, and t_2b bonding molecular orbitals. Two d electrons and eight σ electrons fill the t_2a orbitals and half fill the e orbitals. Two of the ligand electrons fill the a_1 orbital (not shown). The a_1 ligand orbital does not mix with the metal d orbital, but would mix strongly with the metal $4s$ orbital.

fill the bonding orbital, as shown, and half-fill the nonbonding e orbital. The ground state is $|e^2\,^3A_2\rangle$, as before, and Δ is the energy difference between e and the antibonding orbital t_2a. The excited states are ^1A and ^1E from e^2; $^3T_1$, $^3T_2$, $^1T_1$, and $^1T_2$ from $(e)(t_2a)$; and $^3T_1$, ^1E, $^1T_2$, and $^1A_1$ from $(t_2a)^2$. Now, however, the energies of these states are no longer expressible in terms of only A, B, C, and Δ. The parameters A, B, and C coming from matrix elements containing t_2a will be multiplied by powers of N_a, and new parameters arise because of the admixture of ψ. A new type of transition is also possible in which an electron in the bonding orbital t_2b is excited to e or to t_2a. This is called a charge transfer transition because it corresponds to partial transfer of an electron from the ligands to the metal. These transitions generally occur at short wavelengths in the blue or ultraviolet.

In an actual calculation π bonding would not be neglected, and the empty s and p metal orbitals would be included. A complete calculation is quite complicated and must be done numerically. Usually some parameters must be set equal to experimental values. Data have usually been analyzed according to the single ion model, when possible, because of its relative simplicity.

OPTICAL SPECTRA

In contrast to the rare earths the energy levels of the d transition metal ions depend strongly on the host crystal. The optical spectra consist of sharp lines and broad bands. In general, transitions between terms from different strong field configurations $(e)^n(t_2)^m$ produce broad bands because of the sensitivity of the transition frequency to Δ. Broadening is caused by thermal or site-to-site variations of Δ. Transitions between terms of the same strong field configuration are seen as sharper lines. Electric dipole transitions may be forced by odd terms in the static crystal field when the site lacks inversion symmetry or by vibrational modes of odd symmetry; transitions that violate the rule $\Delta S = 0$ are usually very weak. Magnetic dipole transitions are also observed.

Another result of the stronger coupling to the lattice is the lifting of orbital degeneracy by the Jahn–Teller effect. This leads to considerable modification of the optical spectra in some cases. The Jahn–Teller effect and electron-vibrational coupling for orbitally non-degenerate states are discussed in Chapter 3.

The charge transfer transitions are electric dipole allowed and are orders of magnitude more intense than transitions within a d^n configuration. They can be divided into two types: those in which the central metal ion accepts an electron from the nearby ligands and those in which the metal ion acts as a donor and the electron is transferred from metal to ligand. In ionic crystals the first type is more common.

The spectra of the $4d$ and $5d$ ions are more complex than those of the $3d$ ions and have not been studied as extensively. The $4d$ and $5d$ electrons form covalent bonds with ligands more readily than the $3d$. The Δ is larger, the spin-orbit interaction is larger, and the ratio of spin-orbit energy to Δ is also larger. Analysis of spectra is thus more difficult. Nevertheless, data have been interpreted using both these models. Charge transfer transitions are more important in the spectra than for iron-group ions, probably because of the tendency of these metals to form highly positive ions such as Mo^{6+} and Pt^{4+}.

THE ℱ CENTER IN ALKALI HALIDES

The ℱ center is a native defect in an alkali halide crystal. In consists of an electron trapped at an anion vacancy. The structure has been verified by ENDOR experiments, which map the density of the trapped electron in the region of the vacancy through its hyperfine interaction with the nuclear moments of the lattice ions.[23] The main feature of the optical spectrum is a strong absorption band in the visible and an associated emission band at lower energy. There are also weaker absorption bands at higher energy, but assignment of these bands to particular excited states is less certain. The discussion in this section is confined to the ground state and the first excited state. The ℱ center electron is, of course, very strongly coupled to the lattice, all its potential energy being contained in $\mathcal{H}_c$. The electronic structure of the center has been studied extensively. Theoretical models vary from that of an hydrogen atom in a dielectric continuum to molecular-orbital approaches.[24] After excitation of the electron to its first excited state a lattice relaxation occurs, reducing the energy of the excited state and leading to a more diffuse electronic state. The relaxed excited state has successfully been treated by a semicontinuum model, but the ground state and the unrelaxed excited state (the relevant states in an absorption experiment) are too localized for such an approach. Here we discuss the theory of Gourary and Adrian,[25] which is based in part on a point-ion approximation of the lattice. This theory is simpler than the molecular-orbital theories because of the large number of ions that must be included in the latter, and it adequately explains many experimental results.

The alkali halides are ionic crystals. Each cation or anion has a tightly bound core of electrons in filled spherical shells and an effective charge of $+e$ or $-e$. The Hamiltonian operator $\mathcal{H}$ for the $\mathcal{F}$ electron consists of the kinetic energy term and a term $\mathcal{H}_c$ representing the interaction with the lattice. On the point-ion model $\mathcal{H}_c$ is taken to be the sum of the electrostatic interactions between the $\mathcal{F}$ electron at position $\mathbf{r}$ and each ion considered a point charge q_i at $\mathbf{R}_i$.

$$\mathcal{H}_c = \sum_i \frac{-|e|q_i}{|\mathbf{r}-\mathbf{R}_i|}$$

$$= e^2 \sum_{p,s,t} (-1)^{p+s+t}\left[(x-pa)^2+(y-sa)^2+(z-ta)^2\right]^{-\frac{1}{2}} \quad (1.80)$$

a is the nearest-neighbor distance. The sum is over all positive and negative integers and zero except for the point $p=s=t=0$. Only lattices with the NaCl structure are considered. Each term can be expanded in spherical harmonics.

$$\frac{1}{|\mathbf{r}-\mathbf{R}_i|} = 4\pi \sum_{l=0}^{\infty} \sum_{m=-l}^{l} (2l+1)^{-1}\frac{r_<^l}{r_>^{l+1}} Y_l^{m*}(\theta_i,\varphi_i)Y_l^m(\theta,\varphi) \quad (1.81)$$

$r_<$ and $r_>$ are the lesser and greater of r and R_i. Because $\mathcal{H}_c$ must transform as A_{1g} of O_h, the symmetry group of the site, many terms in the expansion cancel to give a final result in the form

$$\mathcal{H}_c = V_0(r)F_0(\theta,\varphi) + V_4(r)F_4(\theta,\varphi) + V_6(r)F_6(\theta,\varphi)+\cdots \quad (1.82)$$

The $F_l(\theta,\varphi)$ are the combinations of the $Y_l^m(\theta,\varphi)$ (see Table 1.4 and

Equation 1.35) that transform as A_{1g}. The first three are

$$F_0 = Y_0^0$$

$$F_4 = Y_4^0 + \sqrt{\frac{5}{14}} \left(Y_4^4 + Y_4^{-4} \right) \tag{1.83}$$

$$F_6 = Y_6^0 - \sqrt{\frac{7}{2}} \left(Y_6^4 + Y_6^{-4} \right)$$

In the expansion of $\mathcal{H}_c$ in Equation 1.33 we assumed that $r/a \ll 1$. In this case there is no attractive central ion core. We do not make this restriction, because the electron wavefunction may extend past the radius $r = a$. The first term in the expansion of $\mathcal{H}_c$ is

$$V_0(r)F_0 = |e|^2 \sum a^{-1}(-1)^{p+s+t} \left[p^2 + s^2 + t^2 \right]^{-1/2}, \qquad r < a$$

$$= -\frac{|e|^2}{a}(1.74756) \tag{1.84}$$

$$V_0(r)F_0 = |e|^2 \left[-\frac{1.74756}{a} + \frac{6}{a} - \frac{6}{r} \right], \qquad a < r < \sqrt{2}\, a$$

$$V_0(r)F_0 = |e|^2 \left[-\frac{1.74756}{a} + \frac{6}{a} - \frac{12}{a\sqrt{2}} + \frac{12}{r} \right], \qquad \sqrt{2}\, a < r < \sqrt{3}\, a$$

and so on

For $r < a$, $V_0(r)F_0$ is a negative constant. For $r > a$, it initially increases with increasing r as $-r^{-1}$ and then varies erratically within increasingly narrow limits, the "average" behavior for $r > a$ resembling $-r^{-1}$.

The solutions $|\psi\rangle$ of the Schroedinger equation $\mathcal{H}|\psi\rangle = E|\psi\rangle$ are basis functions for the irreducible representations Γ of O_h and are written as $|\Gamma i\rangle$. The higher-order terms in (1.82) are small compared with $V_0(r)F_0$. They can be dropped and treated later as perturbations. The Schroedinger equation is now separable because $\mathcal{H}_c$ is independent of θ and φ. The solutions are of the form $R_l(r) Y_l^m(\theta, \varphi)$, and the radial equation is

$$\left\{ \frac{1}{r^2} \frac{d}{dr} \left(r^2 \frac{d}{dr} \right) + \left[\frac{2m}{\hbar^2} (E - V_0(r)F_0) - \frac{l(l+1)}{r^2} \right] \right\} R(r) = 0 \tag{1.85}$$

For $r < a$, $V_0(r)$ is constant and the solution is $R_l(r) = j_l(kr)$, the modified spherical Bessel function with $k^2 = (2m/\hbar^2)(E - V_0 F_0)$. Gourary and Adrian found the solutions $|\Gamma i\rangle$ by a variational method. An analytical form of $|\Gamma i\rangle$ is guessed which contains adjustable parameters. The parameters are varied to minimize

$$E(\Gamma) = \left\langle \Gamma i \left| \left[-\frac{\hbar^2}{2m} \nabla^2 + V_0(r) F_0 \right] \right| \Gamma i \right\rangle, \qquad \langle \Gamma i | \Gamma i \rangle = 1 \qquad (1.86)$$

Three types of wavefunction were tried. The lowest energy for the ground state was obtained with the function

$$|a_{1g}\rangle = R_0(r) Y_0^0(\theta, \varphi) \qquad (1.87)$$

$$R_0(r) = \left(\frac{N_0}{\eta} \right) j_0\left(\xi \frac{r}{a} \right) \exp(-\eta), \qquad r < a$$

$$= \left(\frac{N_0 a}{\eta r} \right) j_0(\xi) \exp\left(-\eta \frac{r}{a} \right), \qquad r > a$$

$$\eta = -\xi \cot \xi$$

ξ is the variational parameter and N_0 a normalization constant. The best wavefunction for the first excited state was found to be

$$|t_{1u} 0\rangle = R_1(r) Y_1^0(\theta, \varphi) \qquad (1.88)$$

$$R_1(r) = N_1 j_1\left(\xi' \frac{r}{a} \right) \exp(-\eta'), \qquad r < a$$

$$= \left(\frac{N_1 r}{a} \right) j_1(\xi') \exp\left(-\eta' \frac{r}{a} \right), \qquad r > a$$

$$\eta' = 3 - \left[\frac{\xi'^2}{1 - \xi' \cot \xi'} \right]$$

For NaCl, for example, $\eta = 2.38$ and $\eta' = 3.40$. The wavefunctions are rather well localized and are not so very different in shape from hydrogenic $1s$ and $2p$ functions. Because the potential is not a simple r^{-1} Coulomb potential, we cannot expect the $2p$like state to be degenerate in energy with a $2s$like state, however.

Finally, these wavefunctions are made orthogonal to the tightly bound core orbitals $|n\rangle$ of the lattice ions in order to satisfy the exclusion principle. The orthogonalized functions are

$$|\tilde{\Gamma}i\rangle = N\left(|\Gamma i\rangle - \sum_n \langle n|\Gamma i\rangle|n\rangle\right), \qquad N = \left[1 - \sum_n (\langle \Gamma i|n\rangle)^2\right]^{-\frac{1}{2}} \quad (1.89)$$

The functions $|n\rangle$ are much more localized then $|\Gamma i\rangle$. They superimpose a rapid spatial modulation on the envelope $|\Gamma i\rangle$ in the regions near the lattice sites.

The calculated energy difference $E(t_{1u}) - E(a_{1g})$ agrees with the $\mathscr{F}$ band absorption transition energy to within 15%. By means of ENDOR spectroscopy the hyperfine interactions of the $\mathscr{F}$ electron in the ground state with the nuclear magnetic moments of the lattice ions have been measured for the first eight shells of neighbors of the vacancy in many alkali halides.[26] The isotropic part of the interaction of the electron spin S with the nuclear spin I_i of nucleus i is $c_i \mathbf{I}_i \cdot \mathbf{S}$, where c_i is proportional to $|\langle a_{1g}|a_{1g}\rangle|^2$ evaluated at the nucleus i. Calculated values of c agree well with experimental values for the first four shells and fall below the experimental values for the more distant shells. Good agreement can be obtained if a small amount of g state ($l = 4$) is mixed into the $l = 0$ ground state by $V_4(r)F_4(\theta,\varphi)$. The admixture is too small to affect the energy appreciably.[27] The wavefunction of Equation 1.89 for the t_{1u} state also predicts the observed sign of the spin-orbit splitting, which is opposite to that of an electron in an atomic p state.[28]

In this model all interelectron electrostatic interactions (see Equation 1.60) between the $\mathscr{F}$ center electron and the core electrons are neglected. These extra terms in the energies of the a_{1g} and t_{1u} states are, apparently, very nearly equal so that they do not affect the transition energy. They must be included if the ionization energy of the $\mathscr{F}$ electron—the difference between $E(a_{1g})$ and the bottom of the conduction band—is to be calculated, however. They have been included in more detailed treatments of the problem, in which other approximations are also necessary in order to make the calculations tractable.[29,30]

Many features of the optical spectra of the $\mathscr{F}$ center are completely dominated by the strong coupling of the $\mathscr{F}$ electron with lattice vibrations. The vibrational coupling in the excited states is very complex and remains a subject of active research.

BIBLIOGRAPHY

Abragam, A. and Bleaney, B., *Electron Paramagnetic Resonance of Transition Ions* (Clarendon, Oxford, 1970).

Ballhausen, C. J. and Gray, H. B., *Molecular Orbital Theory* (Benjamin, New York, 1965).

Ballhausen, C. J., *Introduction to Ligand Field Theory* (McGraw-Hill, New York, 1962).

Dieke, G. H., *Spectra and Energy Levels of Rate Earth Ions in Crystals* (Wiley-Interscience, New York, 1968).

Ferguson, J., "Spectroscopy of 3d Complexes," in *Progress in Inorganic Chemistry*, Vol. 12, S. Lippard, Ed. (Wiley-Interscience, New York, 1970) p. 159.

Fowler, W. B., Ed., *Physics of Color Centers* (Academic, New York, 1968).

Griffith, J. S., *The Theory of Transition-Metal Ions* (Cambridge, Cambridge, 1961).

Heine, V., *Group Theory in Quantum Mechanics* (Pergamon, New York, 1960).

McClure, D. S., "Electronic Structure and Spectra of Impurities in the More Ionic Crystals," in *Treatise on Solid State Chemistry*, Vol. 2, N. B. Hannay, Ed. (Plenum, New York, 1975) Ch. 1.

Sugano, S., Tanabe, Y., and Kamimura, H., *Multiplets of Transition Metal Ions in Crystals* (Academic, New York, 1970).

Tinkham, M., *Group Theory and Quantum Mechanics* (McGraw-Hill, New York, 1964).

Watanabe, H., *Operator Methods in Ligand Field Theory* (Prentice-Hall, Englewood Cliffs, N.J., 1966).

REFERENCES

1. C. Froese, *J. Chem. Phys.* **45**, 1417 (1966).

2. E. Clementi, IBM *J. Res. Dev.* **9**, 2 (1965).

3. V. Heine, *Group Theory in Quantum Mechanics* (Pergamon, New York, 1960).

4. M. Tinkham, *Group Theory and Quantum Mechanics* (McGraw-Hill, New York, 1964).

5. G. F. Koster, J. O. Dimmock, R. G. Wheeler, and H. Statz, *Properties of the Thirty-two Point Groups* (MIT Press, Cambridge, Mass., 1963).

6. Y. Onodera and M. Okazaki, *J. Phys. Soc. Japan* **21**, 2400 (1966).

7. J. C. Eisenstein and M. H. L. Pryce, *Proc. Roy. Soc.* **A255**, 181 (1960).

8. Y. Ebina and N. Tsuya, *Rep. Res. Inst. Elect. Commun. Tohoku Univ.* **B12**, 1 (1960).

9. M. Rotenberg, R. Bivins, N. Metropolis, and J. K. Wooten, Jr., *The 3j and 6j Symbols* (Technology Press, Cambridge, Mass., 1959).

10. Z. J. Kiss, *Phys. Rev.* **127**, 718 (1962).

11. H. A. Weakliem and Z. J. Kiss, *J. Chem. Phys.* **41**, 1507 (1964).

12. A. J. Freeman and R. E. Watson, *Phys. Rev.* **127**, 2058 (1962).

13. B. Bleaney, *Proc. Roy. Soc.* **A277**, 289 (1964).

14. M. T. Hutchings in *Solid State Physics*, Vol. 16, F. Seitz and D. Turnbull, Eds. (Academic, New York, 1964) Ch. 3.

15. J. D. Axe, Jr., *Phys. Rev.* **136A**, 42 (1964).

16. Y. Ayant, E. Bélorizky, and J. Rosset, *J. Phys. Radium* **23**, 14 (1962).

17. G. H. Dieke and H. M. Crosswhite, *Appl. Opt.* **2**, 675 (1963).

18. G. H. Dieke, *Spectra and Energy Levels of Rare Earth Ions in Crystals* (Wiley-Interscience, New York, 1968).

19. S. S. Bishton, M. M. Ellis, D. J. Newman, and J. Smith, *J. Chem. Phys.* **47**, 4133 (1967).

20. J. S. Griffith, *The Theory of Transition-Metal Ions* (Cambridge, Cambridge, 1961).

21. H. Watanabe, *Operator Methods in Ligand Field Theory* (Prentice-Hall, Englewood Cliffs, N.J., 1966).

22. R. K. Watts, *Phys. Lett.* **27A**, 469 (1968).

23. W. C. Holton and H. Blum, *Phys. Rev.* **125**, 89 (1962).

24. W. B. Fowler in *Physics of Color Centers*, W. B. Fowler, Ed. (Academic, New York, 1968) Ch. 2.

25. B. S. Gourary and F. J. Adrian, *Phys. Rev.* **105**, 1180 (1957).

26. H. Seidel and H. C. Wolf in *Physics of Color Centers*, W. B. Fowler, Ed. (Academic, New York, 1968) Ch. 8.

27. B. S. Gourary and A. E. Fein, *J. Appl. Phys.* **33**, (Suppl.), 331 (1962).

28. D. Y. Smith, *Phys. Rev.* **137A**, 574 (1965).

29. T. Kojima, *J. Phys. Soc. Japan* **12**, 908 (1957).

30. J. K. Kübler and R. J. Friauf, *Phys. Rev.* **140A**, 1742 (1965).

Defects with Loosely Bound Electrons

In Chapter 2 the states of electrons bound tightly by more than an electron volt to point defects were discussed and found to be similar in many respects to the states of a free atom. The discussion of spectra was largely confined to transitions between localized states. It is usually not possible to calculate accurately or to measure the position of an energy level of a highly localized state with respect to a band edge. We now see in this chapter that in the opposite limit of very weak binding (about a tenth of an electron volt or less) a unified theoretical approach is also possible. For intermediate cases, on the other hand, the situation is more complex, and usually a molecular orbital approach must be taken.[1] The simplification in the case of weak binding arises because $\langle r \rangle$, the expected value of the distance of the loosely bound electron or hole from the center to which it is bound, may be two orders of magnitude larger than for tightly bound electrons, allowing a description in terms of modified conduction band or valence band waves. Because in discussions of shallow states it is customary to use electron volts or milli-electron volts as the unit of energy rather than inverse centimeters, this convention is followed here. The conversion factor is 1 meV $= 8.066$ cm^{-1}.

In a perfect crystal the one-electron Hamiltonian is

$$\mathcal{H} = -\frac{\hbar^2}{2m}\nabla^2 + V(\mathbf{r}) \tag{2.1}$$

$V(\mathbf{r})$ is the periodic crystalline potential. The solutions of $(\mathcal{H} - E)\phi = 0$ are the Bloch waves

$$\phi_{q\mathbf{k}}(\mathbf{r}) = u_{q\mathbf{k}}(\mathbf{r})\exp(i\mathbf{k}\cdot\mathbf{r}) \tag{2.2}$$

and the corresponding eigenvalues are $E_q(\mathbf{k})$, the energy bands of the crystal. The term $u_{q\mathbf{k}}(\mathbf{r})$ has the periodicity of the lattice. We are interested only in semiconductors and insulators, not in metals. In these two cases the Fermi level lies in an energy region between two bands, an upper, lowest-conduction band, numbered $q=0$, and a lower, highest-valence band. In cases of interest the conduction band contains at most a small number of electrons, and these have energies near the bottom of the band, about which $E_0(\mathbf{k})$ can be expanded in even powers of k_x, k_y, and k_z. Let us first consider the case in which the minimum of $E_0(\mathbf{k})$ occurs at $k=0$, where the $q=0$ band is nondegenerate, except for the twofold spin degeneracy, and $E_0(\mathbf{k})=E_0(k)$ at this point. Then for $\mathbf{k}$ near [000] the first term in the expansion is

$$E_0(k) = \frac{\hbar^2 k^2}{2m_e^*} \tag{2.3}$$

m_e^* is the effective mass, which in the general case is a tensor with elements

$$m_{ij}^* = \hbar^2 \left[\frac{\partial^2 E_0}{\partial k_i \, \partial k_j} \right]^{-1} \tag{2.4}$$

The effective mass for holes will be written m_h^*. When other subscripts, such as i,j in Equation 2.4 are necessary and it is obvious which type of particle is meant, the subscripts e,h are dropped. The mass of a free electron is designated by m.

DONORS AND ACCEPTORS

If now an atom of the host crystal is replaced by a donor, in the simplest case an atom from the next column of the periodic table to the right, this atom has an extra electron not needed for bonding, which may be bound to the excess positive charge of the donor nucleus and core electrons. Far from the donor, the extra electron at position $\mathbf{r}$ will be affected by a potential $|e|/\varepsilon r$, where ε is a dielectric constant and represents shielding of the donor charge at $r=0$ by the induced polarization. For small r the potential will not be shielded in this way, but variation in the small central region will be neglected at first. If the orbits are large, the electron density in this central region may be small. It has been shown that the wavefunction of the donor electron is then given approximately by an expansion in Bloch waves of the lowest band.[2]

$$\psi(\mathbf{r}) = \sum_{\mathbf{k}} A_{\mathbf{k}} \phi_{0\mathbf{k}}(\mathbf{r}) \approx F(\mathbf{r}) u_{00}(\mathbf{r}) \qquad (2.5)$$

$A_{\mathbf{k}}$ is a constant coefficient in the expansion. The function $F(\mathbf{r})$ is a solution of the equation[2]

$$\left[-\left(\frac{\hbar^2}{2m_e^*} \right) \nabla^2 - \frac{e^2}{\varepsilon r} \right] F(\mathbf{r}) = E F(\mathbf{r}) \qquad (2.6)$$

$u_{00}(\mathbf{r})$ is the function in Equation 2.2 for $q = 0$, $k = 0$.

Equation 2.6 is the Schroedinger equation of a one electron atom. The solutions are

$$F_n(\mathbf{r}) = R_{nl}(r) Y_l^m(\theta, \varphi) \qquad (2.7)$$

$$E_n = -\frac{m_e^* e^4}{2\varepsilon^2 \hbar^2 n^2}$$

The zero of energy is set at the bottom of the conduction band. The binding energy E_D is $-E_1$; E_D is also called the ionization energy, because it is the energy that must be expended to remove the electron from the impurity ground state and place it in the conduction band minimum far from the donor. The ionization energy of a hole on an acceptor is called E_A. The "ladder" of impurity energy levels is conventionally grafted onto the band diagram of the perfect crystal as shown, for example, in Figure 7.7. The levels converge to an ionization limit at the band extremum with which they are associated by Equation 2.5 or its analogue.

The orbit of an electron bound to a donor may be very large because of the large value of the dielectric constant and small value of effective mass in many materials. The expected value of r for the electron in the state with principal quantum number n is $\langle r \rangle_n = n^2 a_D$, where a_D is given by $a_D = (m\varepsilon/m_e^*)a_0$. The a_0 is the Bohr radius $\hbar^2/me^2 = 0.53\text{Å}$. For GaAs the effective Bohr radius of the donor electron is $a_D = 200 a_0$. For the ground state $\langle r \rangle_1$ is very large, and for the excited states with $n \geqslant 2$, $\langle r \rangle_n$ is even larger.

The effective mass theory of impurity states exemplified by Equations 2.5 and 2.6 is based on several assumptions. The worst of these is the assumption that the impurity potential $V(\mathbf{r})$, approximated by a Coulomb term in Equation 2.6, varies slowly over a distance of the order of the lattice spacing. If $V(\mathbf{r})$ varies slowly and smoothly, Bloch functions from other bands are not coupled into the impurity wave-function, and only the lowest band need be considered in Equation 2.5. The function $u_{00}(\mathbf{r})$ varies rapidly near the atoms of the crystal and is modulated by the slowly varying envelope $F(\mathbf{r})$. Because $F(\mathbf{r})$ varies slowly in r, its Fourier expansion would have appreciable terms only for small k. This is why the approximation (2.3) can be used in Equation 2.6, where the kinetic energy of the electron is written $E_0(-i\nabla)$.

The approximation of representing the potential energy by $e^2/\varepsilon r$ is generally good for the p states ($l=1$) and other excited states with $l \neq 0$. These have zero amplitude at the origin and so are little affected by departures of the potential from the Coulomb form for small r. Energies of these states are generally rather accurately given by Equation 2.7. The s states have nonzero amplitude at $r=0$; their energies are often not well predicted by Equation 2.7. The largest errors occur for the $1s$ ground state. It is most compact and most sensitive to details of the potential at small r. In Si and Ge there are large discrepancies for the $1s$ states, with the binding energy varying considerably from donor to donor. All donors in the same crystal should have the same ionization energy according to Equation 2.7. But in GaAs and CdS the binding energy is nearly the same for all donors. These data are shown in Figures 7.6 and 7.17 and in Tables 8.3 and 9.3. There have been many attempts to improve the theory for the s states. Some workers have scaled the wavefunctions with the aid of experimental data.[2,3] Others replace $V(\mathbf{r})$ by $e^2/\varepsilon r + U(\mathbf{r})$ where $U(\mathbf{r})$ is a correction which is nonzero for r less than some small value of the order of the nearest-neighbor separation.[4]

In the case of isoelectronic impurities, atoms from the same column of the periodic table as the host atoms for which they substitute, the effective charge of the impurity is zero. There is no Coulomb potential, only the short-range potential $U(\mathbf{r})$. Some isoelectronic impurities are able to bind charge carriers nevertheless. For example, Bi substitutional for P in GaP can bind a hole.[5] Whether an isoelectronic impurity will bind a hole or an electron can be predicted from the sign of the difference between its electronegativity and that of the atom it replaces. The electronegativity of Bi is less than that of P. Phillips[6] has tried to explain the variation of $U(\mathbf{r})$, the "central cell correction," in terms of a dielectric theory of bonding in crystals.

In an optical absorption experiment transitions from the $1s$ state to the excited states can be observed. The transition probability is proportional to the square of the matrix element of the electric dipole operator between the ground state ψ_{1s} and some excited state ψ_i. This matrix element reduces[2] to a matrix element between the envelope functions F_{1s} and F_i, $\langle F_{1s}|\mathbf{r}|F_i\rangle$. Only transitions to other states i of opposite parity (p states or f states) are allowed. However, in the semiconductors of interest the site group of the impurity does not contain the inversion operation, and the true potential $V(\mathbf{r})$ will not be spherically symmetric. The $V(\mathbf{r})$ will contain terms of odd parity, such as xyz in the case of a substitutional site in the diamond or zincblende structure. Parity will not be a good quantum number, and transitions to states i of even "parity" may be also observed in absorption. The other terms in $V(\mathbf{r})$ are small compared with the spherical term, however. The strongest lines correspond to transitions to states of opposite parity. Because in absorption spectra the line or band corresponding to the transition from the $1s$ ground state to the band edge is generally much broader than the lines corresponding to transitions to the localized excited states, E_D is often evaluated by adding to the transition energy of $1s \rightarrow 2p$ the calculated binding energy of the $2p$ state.

We have been dealing with a conduction band with a nondegenerate (except for the twofold spin degeneracy, which we have suppressed) minimum at $k = 0$ isotropic in $\mathbf{k}$ space near the minimum. Let us consider now the slightly more complicated case of CdS, in which the conduction band minimum is at $k = 0$ but the surfaces of constant energy in $\mathbf{k}$ space about this minimum are ellipsoids rather than spheres. CdS has the hexagonal wurtzite structure, each atom having four nearest neighbors. There are four atoms per unit cell. The Brillouin zone is also hexagonal and is shown in Figure 2.1 with some points of high symmetry indicated and labeled in the conventional way. These points are of interest because

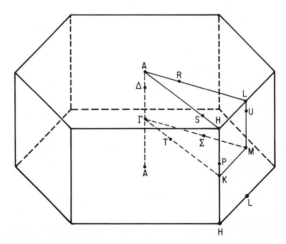

Figure 2.1 Brillouin zone of a crystal with the wurtzite structure. Some of the points of high symmetry are indicated. A point not at the intersection of two or more lines is meant to be a general point anywhere on the line.

for **k** extending from the origin Γ to one of them the secular matrix of the Hamiltonian (2.1) can be factored. This Hamiltonian is invariant under the operations of the space group of the lattice—the group of all translations and rotations that leave the crystal invariant. The rotations alone represent the point group of the crystal, the group C_{6v}. This should not be confused with the point group of a substitutional impurity, which is C_{3v}. The character tables of these groups are given in Tables 2.1 and 2.2. Table 2.3 shows for which irreducible representations of one group basis functions for the other group form bases.

Table 2.1 Character Table of C_{6v}; Because This Is the Group of Γ, the Representations Are Conventionally Labeled Γ_i

	E	$\mathcal{R}$	C_2 $C_2\mathcal{R}$	$2C_3$	$2C_3\mathcal{R}$	$2C_6$	$2C_6\mathcal{R}$	$3\sigma_d$ $3\sigma_d\mathcal{R}$	$3\sigma_v$ $3\sigma_v\mathcal{R}$	Some bases
Γ_1	1	1	1	1	1	1	1	1	1	z
Γ_2	1	1	1	1	1	1	1	-1	-1	l_z
Γ_3	1	1	-1	1	1	-1	-1	1	-1	$x(x^2-3y^2)$
Γ_4	1	1	-1	1	1	-1	-1	-1	1	$y(y^2-3x^2)$
Γ_5	2	2	-2	-1	-1	1	1	0	0	$x,y;l_x,l_y$
Γ_6	2	2	2	-1	-1	-1	-1	0	0	$\Gamma_3\times\Gamma_5$
Γ_7	2	-2	0	1	-1	$\sqrt{3}$	$-\sqrt{3}$	0	0	$|\tfrac{1}{2},\tfrac{1}{2}\rangle,|\tfrac{1}{2},-\tfrac{1}{2}\rangle$
Γ_8	2	-2	0	1	-1	$-\sqrt{3}$	$\sqrt{3}$	0	0	$\Gamma_7\times\Gamma_3$
Γ_9	2	-2	0	-2	2	0	0	0	0	$|3/2,3/2\rangle,|3/2,-3/2\rangle$

Table 2.2 Character Table of C_{3v}

		E	$\mathcal{R}$	$2C_3$	$2C_3\mathcal{R}$	$3\sigma_v$	$3\sigma_v\mathcal{R}$	Some bases
A_1	Γ_1	1	1	1	1	1	1	z
A_2	Γ_1	1	1	1	1	-1	-1	l_z
E	Γ_3	2	2	-1	-1	0	0	$x,y;l_x,l_y$
E'	Γ_4	2	-2	1	-1	0	0	$|\tfrac{1}{2},\tfrac{1}{2}\rangle,|\tfrac{1}{2},-\tfrac{1}{2}\rangle$
A_1'	Γ_5	1	-1	-1	1	i	$-i$	$|3/2,-3/2\rangle - i|3/2,3/2\rangle$
A_2'	Γ_6	1	-1	-1	1	$-i$	i	$-|3/2,3/2\rangle - i|3/2,-3/2\rangle$

Table 2.3 Compatibility Table for C_{6v} and C_{3v}; Basis Functions of Γ_i of C_{6v} Form Bases for the Corresponding Representation of C_{3v}

C_{6v}	Γ_1	Γ_2	Γ_3	Γ_4	Γ_5	Γ_6	Γ_7	Γ_8	Γ_9
C_{3v}	A_1	A_2	A_1	A_2	E	E	E'	E'	$A_1'+A_2'$

Each element of the point group C_{6v} acting on the vector **k** will, in general, rotate it to a new position in the zone. These rotated **k** vectors form a pattern called the "star of **k**," but if **k** ends upon a point of high symmetry, many of the elements will leave it invariant. These form a subgroup called the "group of **k**." The **k** vectors differing by a reciprocal lattice vector are considered identical. The operations of the group of **k** leave $\phi_k(\mathbf{r}) = u_k(\mathbf{r}) \exp(i\mathbf{k} \cdot \mathbf{r})$ unchanged or transform it into a new function $u'_k(\mathbf{r}) \exp(i\mathbf{k} \cdot \mathbf{r})$. These several $u_k(\mathbf{r})$ belonging to the same **k** form a basis for an irreducible representation of the group of **k**. The symmetry points shown in Figure 2.1 all correspond to groups of **k** containing several elements in addition to the identity operation. For a more detailed discussion of the transformation properties of Bloch functions the reader is referred to reference 7.

Figure 2.2a shows schematically the bands near the energy gap in CdS. The gap is direct; that is, the conduction band minimum and the valence band maximum occur at the same point Γ ($k = 0$). The group of Γ is the complete point group C_{6v}, because all rotations leave **k** invariant if $k = 0$. In the tight binding approximation the valence and conduction band states are linear combinations of Cd $5s$ and $5p$ and S $3s$ and $3p$ orbitals. The valence band states consist largely of S orbitals (bonding) and the conduction band, largely of Cd orbitals (antibonding). The Γ_1 and Γ_5 valence bands shown in Figure 2.2a are degenerate in the cubic zincblende structure. Their splitting is due to the extra field along the c axis of the wurtzite structure. When the electron spins are included the representations are labled as shown in Figure 2.2b. The further splitting is due to the spin-orbit interaction.

In this hexagonal crystal the dielectric tensor and the effective mass tensor have axial symmetry with principal values $\varepsilon_\parallel$, $\varepsilon_\perp$, $\varepsilon_\perp$, and $m_\parallel$, $m_\perp$, $m_\perp$. The Hamiltonian operator of the effective mass Schroedinger equation corresponding to Equation 2.6 is

$$\mathcal{H} = -\frac{1}{2}\hbar^2 \left[\frac{1}{m_\perp^*}\left(\frac{\partial^2}{\partial x^2} + \frac{\partial^2}{\partial y^2} \right) + \frac{1}{m_\parallel^*} \frac{\partial^2}{\partial z^2} \right]$$

$$- e^2 \left(\varepsilon_\parallel \varepsilon_\perp x^2 + \varepsilon_\parallel \varepsilon_\perp y^2 + \varepsilon_\perp^2 z^2 \right)^{-\frac{1}{2}} \tag{2.8}$$

For CdS the anisotropies in the effective mass and dielectric tensors are small. Equation 2.8 is the same as the Hamiltonian for the envelope function of a free exciton in a uniaxial crystal if the hole mass is allowed to

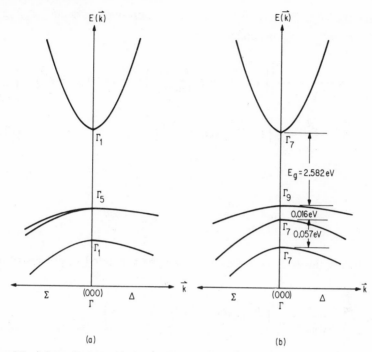

Figure 2.2 Schematic representation of the variation of the CdS bands near the gap about the Γ point along the Σ and Δ directions in **k** space. In (*a*) spin has been disregarded, and the orbital representations Γ_1 and Γ_5 are written. Spin is included in (*b*). The spin-orbit interaction leads to further splitting of the valence band.

become infinite. Therefore, the perturbation treatment of Wheeler and Dimmock[8] for the exciton case can be adapted to find the eigenvalues of Equation 2.8. $\mathcal{H}$ is separated into parts $\mathcal{H} = \mathcal{H}_0 + \mathcal{H}_\alpha$, where

$$\mathcal{H}_0 = -\frac{1}{2}\hbar^2\left[\frac{1}{m^*_\perp}\left(\frac{\partial^2}{\partial x^2} + \frac{\partial^2}{\partial y^2}\right) + \frac{1}{m^*_\parallel}\frac{\partial^2}{\partial z^2}\right] - \frac{e^2}{\sqrt{\varepsilon_\parallel \varepsilon_\perp}}\left(x^2 + y^2 + \frac{z^2 m^*_\parallel}{m^*_\perp}\right)^{-\frac{1}{2}}$$

$$(2.9)$$

$$\mathcal{H}_\alpha = \frac{-e^2}{\sqrt{\varepsilon_\parallel \varepsilon_\perp}}\left(x^2 + y^2 + \frac{z^2 \varepsilon_\perp}{\varepsilon_\parallel}\right)^{-\frac{1}{2}} + \frac{e^2}{\sqrt{\varepsilon_\parallel \varepsilon_\perp}}\left(x^2 + y^2 + \frac{z^2 m^*_\parallel}{m^*_\perp}\right)^{-\frac{1}{2}}$$

The last term has been added to $\mathcal{H}$ in $\mathcal{H}_\alpha$ and subtracted from $\mathcal{H}$ in $\mathcal{H}_0$.

Under the transformation to a new coordinate system $x', y', z' = x, y, z(m_{\parallel}^*/m_{\perp}^*)^{\frac{1}{2}}$, $\mathcal{H}_\alpha$ becomes

$$\mathcal{H}_\alpha = -\frac{e^2}{\sqrt{\varepsilon_\parallel \varepsilon_\perp}} \left\{ \left(x'^2 + y'^2 + z'^2 \frac{m_\perp^* \varepsilon_\perp}{m_\parallel^* \varepsilon_\parallel} \right)^{-\frac{1}{2}} - (x'^2 + y'^2 + z'^2)^{-\frac{1}{2}} \right\}$$

$$= \frac{e^2}{\sqrt{\varepsilon_\parallel \varepsilon_\perp}} \frac{1}{r'} \left\{ 1 - (1 - \alpha\cos^2\theta')^{-\frac{1}{2}} \right\} \tag{2.10}$$

where

$$\alpha = 1 - \frac{m_\perp^* \varepsilon_\perp}{m_\parallel^* \varepsilon_\parallel}, \qquad z'/r' = \cos\theta'$$

For small α the term $(1 - \alpha\cos^2\theta')^{-\frac{1}{2}}$ can be expanded in powers of α to give

$$\mathcal{H}_\alpha = -\frac{e^2}{\sqrt{\varepsilon_\parallel \varepsilon_\perp}} \frac{1}{r'} \left(\frac{1}{2}\alpha\cos^2\theta' + \frac{3}{8}\alpha^2\cos^4\theta' + \cdots \right)$$

$$= -\frac{e^2}{\sqrt{\varepsilon_\parallel \varepsilon_\perp}} \frac{1}{r'} \left\{ \frac{1}{6}\alpha + \frac{3}{40}\alpha^2 + \left(\frac{1}{3}\alpha + \frac{3}{14}\alpha^2 \right) \sqrt{\frac{4\pi}{5}} \, Y_2^0(\theta', \varphi') \right.$$

$$\left. + \frac{3}{35}\alpha^2 \sqrt{\frac{4\pi}{35}} \, Y_4^0(\theta', \varphi') + \cdots \right\} \tag{2.11}$$

Under the same transformation $\mathcal{H}_0$ becomes

$$\mathcal{H}_0 = -\frac{\hbar^2}{2m_\perp^*} \left[\frac{\partial^2}{\partial x'^2} + \frac{\partial^2}{\partial y'^2} + \frac{\partial^2}{\partial z'^2} \right] - \frac{e^2}{\sqrt{\varepsilon_\parallel \varepsilon_\perp}} (x'^2 + y'^2 + z'^2)^{-\frac{1}{2}} \tag{2.12}$$

Equation 2.12 is the Hamiltonian of an hydrogen atom with electron mass $m_\perp^*$. The zero order eigenvectors are

$$|nlm\tfrac{1}{2}m_s\rangle = R_{nl}(r') Y_l^m(\theta', \varphi') |\tfrac{1}{2}, m_s\rangle \tag{2.13}$$

$|\tfrac{1}{2}, m_s\rangle$ stands for the s like conduction band edge function u_{00}, which transforms as $|\frac{1}{2}, \pm\frac{1}{2}\rangle$. The eigenvalues of $\mathcal{H}_0$ are

$$E_n^0 = -\frac{m_\perp^* e^4}{2\hbar^2 \varepsilon_{\parallel} \varepsilon_\perp n^2} = -\frac{\hbar^2 e^2}{2m_\perp^* a_D^2 n^2} \tag{2.14}$$

The fundamental unit of length $a_D = a_0 m \sqrt{\varepsilon_{\parallel} \varepsilon_\perp} / m_\perp^*$. The s functions have axial, not spherical, symmetry. The dependence of $|1s0\frac{1}{2}m_s\rangle$ on the coordinates is $\exp(-r'/a_D) = \exp[-(x^2 + y^2 + z^2 m_{\parallel}^*/m_\perp^*)^{\frac{1}{2}}/a_D]$. The first-order corrections to the energies $\langle nlm\frac{1}{2}m_s|\mathcal{H}_\alpha|nlm\frac{1}{2}m_s\rangle$ are easily found from Equation 1.37 and a table of $3j$ symbols, together with the expression for the expected value of $1/r'$,

$$\left\langle \frac{1}{r'} \right\rangle = \int_0^\infty R_{nl}^2(r')r'\,dr' = \frac{1}{n^2 a_D}, \qquad l = n - 1 \tag{2.15}$$

The first few corrected energies to second order in α are

$$E(1s) = E_1^0 - \frac{e^2}{a_D \sqrt{\varepsilon_{\parallel} \varepsilon_\perp}} \left(\frac{1}{6}\alpha + \frac{3}{40}\alpha^2 \right) = E_1^0 \left(1 + \frac{1}{3}\alpha + \frac{3}{20}\alpha^2 \right)$$

$$E(2s) = E_2^0 \left(1 + \frac{1}{3}\alpha + \frac{3}{20}\alpha^2 \right)$$

$$E(2p0) = E_2^0 \left(1 + \frac{3}{5}\alpha + \frac{9}{28}\alpha^2 \right)$$

$$E(2p\pm1) = E_2^0 \left(1 + \frac{1}{5}\alpha + \frac{9}{140}\alpha^2 \right) \tag{2.16}$$

The p state is split by the axial perturbation.

The selection rules are easily found by the methods of Chapter 1. From Table 2.2 it can be seen that different selection rules will obtain for different polarizations, because x, y, and z do not form bases for the same irreducible representation. For light polarized along the z or c axis the electric dipole transition from the ground state to $2p0$ is allowed. For polarization perpendicular to c, transitions from the ground state to $2p\pm1$ are allowed.

Magneto-optical studies have been very useful in identifying shallow centers. Thus it is of interest to calculate the Zeeman splitting of these energy levels. The spin-orbit coupling is small and can be neglected. Then the interaction of the orbital angular momentum of the electron with a magnetic field $\mathbf{H}$ is incorporated in the Hamiltonian by replacing

$-\hbar^2(\partial^2/\partial x^2)$ in $\mathcal{H}_0$ by $-\hbar^2(\partial^2/\partial x^2) - i2(e\hbar/c)\mathfrak{A}_x(\partial/\partial x) + (e^2/c^2)\mathfrak{A}_x^2$ and making similar substitutions for the y and z components of momentum; $(e = |e|)$. The $\mathfrak{A}$ is the vector potential $\frac{1}{2}(\mathbf{H} \times \mathbf{r})$. The small diamagnetic terms in $\mathfrak{A}_i^2$ are neglected. For $\mathbf{H}$ perpendicular to the c axis the Zeeman term is

$$\mathcal{H}_z = -i\left(\frac{e\hbar}{2c}\right)H\left[\frac{1}{m_\perp^*}\left(-z\frac{\partial}{\partial y}\right) + \frac{1}{m_\parallel^*}y\frac{\partial}{\partial z}\right] + g_\perp \beta H s_x$$

$$= -i\frac{e\hbar H}{2c\sqrt{m_\parallel^* m_\perp^*}}\left(y'\frac{\partial}{\partial z'} - z'\frac{\partial}{\partial y'}\right) + g_\perp \beta H s_{x'}$$

$$= \frac{e\hbar}{2c\sqrt{m_\parallel^* m_\perp^*}}Hl_{x'} + g_\perp \beta H s_{x'} \qquad (2.17)$$

If $\mathbf{H}$ is parallel to the c axis $\mathcal{H}_z$ is given by

$$\mathcal{H}_z = -i\left(\frac{e\hbar}{2cm_\perp^*}\right)H\left(-y'\frac{\partial}{\partial x'} + x'\frac{\partial}{\partial y'}\right) + g_\parallel \beta H s_{z'}$$

$$= \left(\frac{e\hbar}{2m_\perp^* c}\right)Hl_{z'} + g_\parallel \beta H s_{z'} \qquad (2.18)$$

$\mathcal{H}_z + \mathcal{H}_\alpha$ can now be considered the perturbation of $\mathcal{H}_0$ and the splitting in a magnetic field calculated. With $\mathbf{H}$ along the c axis the $1s$ and $2p \pm 1$ states split according to

$$E\left(1s, \pm\tfrac{1}{2}\right) = E(1s) \pm \tfrac{1}{2}g_\parallel \beta H$$

$$E\left(2p \pm 1, \pm\tfrac{1}{2}\right) = E(2p \pm 1) \pm \frac{m}{m_\perp^*}\beta H \pm \tfrac{1}{2}g_\parallel \beta H$$

$$E\left(2p \pm 1, \mp\tfrac{1}{2}\right) = E(2p \pm 1) \pm \frac{m}{m_\perp^*}\beta H \mp \tfrac{1}{2}g_\parallel \beta H \qquad (2.19)$$

β is the Bohr magneton. Without more exact knowledge of the band edge function u_{00} the effective g factors $g_\parallel$ and $g_\perp$ cannot be calculated. The splittings for $\mathbf{H}$ perpendicular to the axis are found in the same way.

From the Zeeman effects in the luminescence and electronic Raman spectra Henry and Nassau[9] find for the Cl donor in CdS the values $E_D = -32.7$ meV, $m_\perp^* = 0.190m$, $m_\parallel^* = 0.180m$, $g_\parallel = g_\perp = 1.75$, $\alpha = 0.054$. The transitions observed by Raman scattering were $1s \rightarrow 2s$ and $1s \rightarrow 2p0$. Luminescence transitions from excited states due to an exciton bound to the donor to $2p \pm 1$, $1p0$, $2s$, and $1s$ were seen.

In some crystals—Si, Ge, and GaP, for example—the conduction band has several equivalent minima away from $k = 0$. In this case the wavefunction $F_{nj}(\mathbf{r})$ at each minimum $j = 1, 2, \ldots, N$ satisfies an equation of the form of Equation 2.6 with an anisotropic effective mass or, like equation 2.9, with an isotropic dielectric tensor, because Si, Ge, and GaP have higher symmetry than CdS. This sort of band represents the next level of complexity. There are several equivalent minima, each with ellipsoidal surfaces of constant energy. This case is discussed in Chapter 7, but we describe the essential features here.

The total wavefunction is a linear combination of the functions $F_{nj}(\mathbf{r})u_{0j}(\mathbf{r})\exp(i\mathbf{k}_j \cdot \mathbf{r})$. These N functions form bases for a representation of the point group of the donor site, which is reducible into a few irreducible representations Γ_i. These Γ_i states have the same energy in the effective mass approximation, but the degeneracy is lifted when corrections to the approximate theory are introduced. In Si there are six equivalent minima along the $\langle 100 \rangle$ directions in $\mathbf{k}$ space. The sixfold degeneracy of the $1s$ state is reducible into A_1, E, and T_2 components. The A_1 state has the totally symmetric form

$$\psi(A_1) = \frac{1}{\sqrt{6}} \sum_{j=1}^{6} F_{0j}(\mathbf{r})u_{0j}(\mathbf{r})\exp(i\mathbf{k}_j \cdot \mathbf{r}) \qquad (2.20)$$

and lies lowest for the substitutional donors P, As, and Sb. This state is characterized by an isotropic EPR spectrum. The electron density has been mapped by ENDOR spectroscopy through the interaction of the donor electron with the ^{29}Si nuclear moments.[10] For the interstitial donor Li the T_2 state is lowest.[11]

Other kinds of donor have also been studied. A center consisting of two electrons bound to the singly charged donor core can exist.[12] This is the analogue of the H^- ion. The binding energy of the second electron is quite small. It can be roughly estimated to be $(0.75/13.6)E_D$. The ionization energy of the H atom is 13.6 eV, and the energy required to remove an electron from H^- is 0.75 eV. Double donors such as sulfur in Si also exist, and they may bind one or two electrons. More complex defects such as clusters or associated simple defects may also behave as donors or acceptors.

Although the discussion has concentrated on electrons, donors, and conduction bands, the same formalism applies also to holes, acceptors, and valence bands. Often the highest valence band is degenerate at the maximum. The theory can be modified to apply to this case also[2] and leads to a set of coupled differential equations in place of Equation 2.6.

In many crystals with the diamond or zincblende structure the upper valence band maximum is at $k=0$. When electron spin is considered, the band is split by spin-orbit coupling into an upper fourfold degenerate band Γ_8 and a lower doublet Γ_7. These bands are shown schematically in Figure 7.7. At least four bands (Γ_8) contribute to the acceptor wavefunction, and at worst six ($\Gamma_8 + \Gamma_7$) contribute, depending on whether the $\Gamma_8 - \Gamma_7$ splitting is much larger than the acceptor ionization energy, comparable with it, or smaller. In general, then, a set of six coupled differential equations must be solved for the acceptor states.

Let us consider the simpler case in which the spin-orbit splitting is large enough that the lower Γ_7 band may be neglected. Near the maximum at $k=0$ the expansion corresponding to Equation 2.3 is

$$E_{\pm}(\mathbf{k}) = -E_g - a\left\{ k^2 \pm \left[\left(b - \frac{6}{5}c \right)k^4 + \frac{12}{5}c(5b-c) \right. \right.$$

$$\left. \left. \times \left(k_x^2 k_y^2 + k_y^2 k_z^2 + k_z^2 k_x^2 \right) \right]^{\frac{1}{2}} \right\} \tag{2.21}$$

The constants, a, b, and c depend on the particular crystal considered. The Γ_8 state splits into two doublets for $k \neq 0$ and the splitting is anisotropic because of the term in $k_i k_j$. This cubic anisotropy term is similar to that in Equation 1.54.

As the splitting of a Γ_8 state in a magnetic field, Equation 1.54, may be represented by the eigenvalues of an effective Hamiltonian involving a fictitious spin $S=3/2$ (Equation 9.9), so also the effective mass Hamiltonian for the hole bound to an acceptor may be similarly written as

$$\mathcal{H} = -a\nabla^2 - \frac{e^2}{\varepsilon r} - \left(\frac{ab}{9\hbar^2} \right)(\mathbf{P}^{(2)} \cdot \mathbf{J}^{(2)}) + \left(\frac{ac}{9\hbar^2} \right)$$

$$\times \left\{ \frac{1}{5}\sqrt{70} \left[\mathbf{P}^{(2)} \times \mathbf{J}^{(2)} \right]_0^{(4)} \right.$$

$$\left. + \left[\mathbf{P}^{(2)} \times \mathbf{J}^{(2)} \right]_4^{(4)} + \left[\mathbf{P}^{(2)} \times \mathbf{J}^{(2)} \right]_{-4}^{(4)} \right\} \tag{2.22}$$

The products of the second-rank tensor operators are defined by

$$(\mathbf{P}^{(2)} \cdot \mathbf{J}^{(2)}) = \sum_q (-1)^q P_q^{(2)} J_{-q}^{(2)}$$

$$\left[\mathbf{P}^{(2)} \times \mathbf{J}^{(2)} \right]_q^{(4)} = 3(-1)^q \sum_{rs} \begin{pmatrix} 2 & 2 & 4 \\ r & s-q \end{pmatrix} P_r^{(2)} J_s^{(2)}$$

The quantity $(\begin{smallmatrix} 2 & 2 & 4 \\ r & s & -q \end{smallmatrix})$ is the $3j$ symbol that is used, for example, in coupling two angular momenta. The tensor components $P_q^{(2)}$ and $J_q^{(2)}$ are defined in terms of the components of the linear momentum $p_x = -i\hbar(\partial/\partial x)$ and the components of an angular momentum operator j_x for spin $j = 3/2$.

$$P_0^{(2)} = \sqrt{\frac{3}{2}}\ P_{zz}$$

$$P_{qr} = 3p_q p_r - p^2 \delta_{qr}$$

$$P_{\pm 1}^{(2)} = \mp(P_{xz} + iP_{yz})$$

$$q, r = x, y, z$$

$$P_{\pm 2}^{(2)} = \tfrac{1}{2}(P_{xx} - P_{yy} \pm 2iP_{xy})$$

Similar equations connect $J_q^{(2)}$ with J_{mn} where J_{mn} is given by

$$J_{mn} = \frac{3}{2}(j_m j_n + j_n j_m) - j^2 \delta_{mn} \qquad m, n = x, y, z$$

The usefulness of this way of writing the effective mass Hamiltonian is due to the small size of the terms in Equation 2.22, which have cubic symmetry, compared with the other terms which have spherical symmetry. Balderschi and Lipari[13] have solved the approximate Schroedinger equation for acceptors in several crystals retaining only the spherical first three terms of Equation 2.2 and treating the cubic terms by perturbation theory.

The first two terms of the spherical Hamiltonian are like those in the Hamiltonian of an hydrogen atom. The third term represents a peculiar sort of spin-orbit coupling, but does not destroy the spherical symmetry. The total angular momentum $\mathbf{l} + \mathbf{j}$ is a constant of the motion. The states can be labeled in the Russel–Saunders notation. The ground state is $2S_{3/2}$ and the $2p$ states are $2P_{1/2}$, $2P_{3/2}$, and $2P_{5/2}$. When the cubic term is taken into account, the states should be classified by the O_h symmetry of Equation 2.22. The $1S_{3/2}$ becomes $1S_{3/2}\Gamma_{8g}$ and the p states become $2P_{1/2}\Gamma_{6u}$, $2P_{3/2}\Gamma_{8u}$, and $2P_{5/2}\Gamma_{8u}$ and $2P_{5/2}\Gamma_{7u}$. The $2P_{5/2}$ state splits in the lower symmetry. Some of these states for acceptors in Ge are shown in Figure 7.18.

In the opposite limit in which the $\Gamma_8 - \Gamma_7$ splitting of the valence bands is negligibly small a similar treatment is possible, but in the general case in which the six bands cannot be considered degenerate and Γ_7 cannot be neglected, no satisfactory solution of the acceptor problem has yet been given. In all calculations the acceptor potential has been approximated by the Coulomb term. The same question of the proper treatment of the potential at small r exists for acceptors as for donors. But because of the more complex nature of the acceptor problem, even when the Coulomb approximation is made, less progress has been made toward quantitative explanation of observed ionization energies and spectra. The experimentally observed ionization energies do, in general, vary from acceptor to acceptor in the same material, just as for donors. This is evident from Figure 7.18.

In polar crystals, like CdS and GaAs, for example, it is not so obvious what value to use for the dielectric constant in an expression like Equation 2.7. In such materials the dielectric function $\varepsilon(\omega)$ has resonances at high frequencies $E_g/\hbar$ and higher, corresponding to electronic excitations, and at much lower frequencies ω_{to}, corresponding to excitation of transverse optical vibrations of the charged ions. The E_g is the band gap energy. At frequencies lower than $E_g/\hbar$, but larger than ω_{to}, ε is nearly independent of frequency and has the value ε_o, the optical dielectric constant. Below ω_{to}, ε approaches the larger value ε_s, the static dielectric constant. (For a homopolar crystal like Si, $\varepsilon_o = \varepsilon_s$.) When $E_D/\hbar \gg \omega_{to}$, it seems reasonable to use ε_o, because the electron moves so rapidly that only electronic polarization is effective. In the other extreme $E_D/\hbar \ll \omega_{to}$ the electron moves so slowly that the ionic polarization contributes to the shielding also. Here ε_s would seem to be the logical choice.

The problem is actually quite complex. A slowly moving electron will drag along with it a deformation of the lattice. The deformation is usually represented as virtual phonons that are created and destroyed as the electron moves. The excitation consisting of the electron and the phonons is called a polaron. The stronger the electron-phonon coupling, the more phonons there will be in the "cloud" surrounding the electron. The electron interacts most strongly with longitudinal optical phonons because of the large electric fields produced by them. Examples of such coupling are very often seen in the form of sidebands due to these phonons in exciton spectra and in donor-acceptor pair spectra. The two parameters that characterize the interaction are the ratio $E_D/(\hbar\omega_{lo})$ and the dimensionless electron-phonon coupling constant α,

$$\alpha = \frac{e^2}{\hbar}\left(\varepsilon_o^{-1} - \varepsilon_s^{-1}\right)\left(\frac{m_e^*}{2\hbar\omega_{lo}}\right)^{\frac{1}{2}} \tag{2.23}$$

The effective dielectric constant $\bar{\varepsilon}$, where $\bar{\varepsilon}^{-1} = \varepsilon_o^{-1} - \varepsilon_s^{-1}$, characterizes the lattice polarizability. Dispersion of the longitudinal optical phonons has been neglected, and they are all assumed to have the frequency ω_{lo}. $\alpha/2$ is the average number of phonons in the cloud. For many semiconductors α is not large. It is ~ 0.3 for GaAs and CdTe but in more ionic crystals it is larger.

Larsen[14] has calculated by a variational method the ground state energy of a polaron bound in a Coulomb field. For $\alpha < 6$ and $E_D/(\hbar\omega_{lo}) \lesssim 1$ the ground state energy is given by

$$E = -\alpha\hbar\omega_{lo} - \left[\frac{1+\alpha/12}{1-\alpha/12}\right]E_D$$

$$E_D = \frac{m_e^* e^4}{2\varepsilon_s^2\hbar^2}$$

The ionization energy would then be

$$E_D' = \left[\frac{1+\alpha/12}{1-\alpha/12}\right]E_D$$

because the free polaron has an energy that is depressed below the energy of an electron in the conduction band by the amount $\alpha\hbar\omega_{lo}$. The quantity in square brackets may be viewed as an enhancement factor of the effective mass which represents the extra inertia of the lattice polarization. The enhancement amounts to only a few percent in many III-V semiconductors, but is larger in II-VI compounds. The polaron effective mass will be written as m_e^p. In the alkali halides α is about 3 and $E_D/(\hbar\omega_{lo}) \gg 1$. The polaron radius is not large compared with the lattice spacing, and the treatment just described, based on effective mass theory and a continuum rather than a discrete lattice, is inappropriate.

BOUND EXCITONS

The excited states of defects discussed in the previous sections have corresponded to excitations of particles bound to the defect and present with lower energy in the ground state. Other excited states, which are the subject of this section, are also often possible, corresponding to the excitations of electron-hole pairs. The electron-hole pair is bound to the defect only in the excited state and is not present in the ground state of the defect. If in the ground state n particles are bound to the defect, then in these excited states there are $n + 2m$ particles, m being the number of bound electron-hole pairs. These electron-hole pairs are often called bound excitons by analogy with free excitons. The theory of bound excitons is in an early state of development. It has not been elaborated to the extent of the crystal field theory, for example. In semiconducting crystals the luminescence and absorption spectra of bound excitons, occurring at slightly longer wavelengths than the free exciton lines, are rich in detail.

Much valuable information has been obtained, largely through magneto-optical experiments.

Electron-hole pairs have been observed bound to ionized donors or acceptors, neutral donors or acceptors, isoelectronic impurities, and more complex defects. Let us consider first the simplest case: an electron-hole pair bound to an ionized shallow donor or acceptor. In the case of an ionized donor, written D^+, in a crystal with isotropic valence and conduction bands at the extrema at $k=0$, it is natural in view of the discussion of the previous section to try to describe the system $D^+e^-h^+$ by an effective mass Hamiltonian

$$\mathcal{H} = -\frac{\hbar^2}{2m_e^*}\nabla_e^2 - \frac{\hbar^2}{2m_h^*}\nabla_h^2 - \frac{e^2}{\varepsilon r_e} + \frac{e^2}{\varepsilon r_h} - \frac{e^2}{\varepsilon r_{eh}} \qquad (2.24)$$

m_e^* and m_h^* are the effective masses at the conduction and valence band extrema. The $\mathbf{r}_e$ and $\mathbf{r}_h$ are the positions of the electron and hole measured from the singly charged donor at the origin; $r_{eh} = |\mathbf{r}_e - \mathbf{r}_h|$. Let now m_h^* be very much larger than m_e^*. The second term in Equation 2.24 can be neglected. If the transformation $\mathbf{r}' = \mathbf{r}/a_D$ is made, where $a_D = \varepsilon \hbar/e^2 m_e^*$, then Equation 2.24 becomes

$$\frac{\mathcal{H}}{2E_D} = -\frac{1}{2}\nabla_e'^2 - \frac{1}{r_e'} - \frac{1}{r_{eh}'} + \frac{1}{r_h'} \qquad (2.25)$$

$E_D = \hbar^2/2m_e^* a_D^2$ is the donor binding energy. Equation 2.25 is very similar to the Hamiltonian for the H_2^+ ion,

$$\frac{\mathcal{H}}{2E_H} = -\frac{1}{2}\nabla'^2 - \frac{1}{r_a'} - \frac{1}{r_b'} + \frac{1}{r_{ab}'} \qquad (2.26)$$

where $r' = r/a_0$

$$E_H = \hbar^2/2ma_0^2$$

The r_a and r_b are the distances separating the electron from the two protons. The Schroedinger equation with this Hamiltonian can be solved exactly in a system of ellipitical coordinates[15], and the ground state energy $E_G(H_2^+)$ agrees with the experimental value. If the H_2^+ ground state energy is $E_G(H_2^+) = -\eta E_H$, the ground state eigenvalue associated with Equation 2.25 would be $E_G = -\eta E_D$. From the H_2^+ problem $\eta = 1.21$. Thus when the hole effective mass is much larger than the electron effective mass, binding occurs with binding energy $1.21\,E_D$.

Approximate ground state solutions of the Schroedinger equation of the Hamiltonian (2.24) have been found by variational methods.[16] The solution $E_G(D^+e^-h^+)$ has been obtained as a function of the mass ratio m_e^*/m_h^*. Binding is predicted, that is $-E_G(D^+e^-h^+) > E_D$ if $m_e^*/m_h^* < 0.4$. This result can be understood in the following way.[17] When m_e^*/m_h^* is large, the light hole has a large orbit and the heavy electron, a small one. The positive charge of the donor is nearly canceled by the negative charge of the electron, and the hole is not bound. It should always be more likely that the complex $D^+e^-h^+$ dissociate into a neutral donor and a free hole than into an ionized donor and a free exciton, because the donor binding energy is greater than that of the free exciton, E_x. This can be seen by treating the free exciton on the effective mass theory as a single particle of reduced mass μ orbiting about the center of mass of the electron and hole.

$$E_x = \left(\frac{\mu}{m_e^*}\right) E_D$$

$$\frac{1}{\mu} = \frac{1}{m_e^*} + \frac{1}{m_h^*}$$

Because $\mu < m_e^*$, we have $E_x < E_D$. When the electron-hole pair is annihilated, if the decay is radiative, a photon of energy $h\nu = E_g - |E_G(D^+e^-h^+)|$ is emitted. The E_g is the band gap energy.

The Hamiltonian describing the system $A^-e^-h^+$, an ionized acceptor interacting with an electron-hole pair, is

$$\mathcal{H} = \frac{\hbar^2}{2m_h^*}\nabla_h^2 - \frac{\hbar^2}{2m_e^*}\nabla_e^2 + \frac{e^2}{\varepsilon r_e} - \frac{e^2}{\varepsilon r_h} - \frac{e^2}{\varepsilon r_{eh}}$$

This Hamiltonian differs from (2.24) only through the interchange of the subscripts e and h. Therefore binding occurs in this case if $m_h^*/m_e^* < 0.4$, or if $m_e^*/m_h^* > 2.5$. In the intermediate region $0.4 < m_e^*/m_h^* < 2.5$ excitons can be bound neither to ionized donors nor to ionized acceptors. This simple model applies to no real crystal. In cubic crystals the valence band is fourfold degenerate at the maximum. Perhaps it might apply best to a crystal of lower symmetry, like CdS, where this degeneracy is lifted. Of course, the lower symmetry leads to anisotropy of the effective mass tensor —another complexity not incorporated in the theory. Nevertheless, it seems the best example. For CdS the electron mass is very nearly isotropic with $m_e^* \approx 0.19$. The hole mass tensor is very anisotropic: $m_{h\parallel}^* = 5$ and $m_{h\perp}^* = 0.7$. The average value $\overline{m}_h^*$ is given by

$$\frac{1}{\overline{m}_h^*} = \frac{1}{3}\frac{1}{m_{h\parallel}^*} + \frac{2}{3}\frac{1}{m_{h\perp}^*}$$

Then $m_e^*/\overline{m}_h^* = 0.19$. If the theory can be applied in this way, it would predict binding of excitons to ionized donors, but not to ionized acceptors. This prediction agrees with experiment. Ground-state energies for excitons bound to ionized donors in several III-V and II-VI compounds have been calculated by Mahler and Schröder[18] and by Elkomoss[19] taking into account the interaction with polar lattice vibrations. The results are in good agreement with experiment.

A neutral donor-bound exciton complex consists of the charged donor and three mobile particles $D^+e^-e^-h^+$. When $m_h^* \gg m_e^*$ the system is similar to the H_2 molecule and should be stable. It has already been mentioned that $D^+e^-e^-$ has a bound state. When $m_e^* \gg m_h^*$, the electrons are relatively near the donor, and the hole can be bound in the resulting attractive potential. The $D^+e^-e^-h^+$ complex is expected to be stable for all values of m_e^*/m_h^*.[17] Similarly, the neutral acceptor-bound exciton complex should be stable for all m_e^*/m_h^*. The most probable dissociation mode of $D^+e^-e^-h^+$ is apparently into a free exciton and a neutral donor, $D^+e^-e^-h^+ \rightarrow D^+e^- + e^-h^+$. When the bound exciton decays radiatively, a photon of energy $h\nu = E_g - E_x - Q$ is emitted. The E_x is the binding energy of the free exciton, and Q is the dissociation energy of $D^+e^-e^-h^+$ (or $A^-e^-h^+h^+$), which is about $0.1E_D$ (or $0.1E_A$) in Si and $0.15E_D$ in ZnSe.

The Q is sometimes also called the localization energy of the exciton, because the form $E_g - E_x - Q$ suggests that the energy of a free exciton, $E_g - E_x$, is depressed by the amount Q for an exciton localized on an impurity. The approximate relation Q/E_D or $Q/E_A \approx 0.1$ is called Haynes' rule for J. R. Haynes,[20] who discovered this relation for donors and acceptors in silicon. The emission wavelength is nearer that corresponding to decay of the free exciton than in the case of decay of the exciton bound to an ionized donor or acceptor. If the neutral donor is left in an excited state, this excitation energy must also be subtracted, leading to a somewhat longer emission wavelength. Such a transition is sometimes called a two-electron transition. One of the two electrons recombines with the hole, and the second is excited to some higher donor level. Two-electron transitions are shown in Figure 9.3 (dashed lines). The second electron might also be removed from the donor entirely and given the large amount of energy $h\nu$ in the form of kinetic energy. This nonradiative Auger process is often much more efficient than the radiative recombination process that competes with it.

BOUND EXCITONS IN CdS

A simple phenomenological theory has been successfully used to explain the magneto-optical spectra of bound excitons in many semiconductors.

We illustrate its use with two examples, bound excitons in a crystal of wurtzite structure (CdS, CdSe, ZnO) and in GaP, which has the cubic zincblende structure. An exciton bound to an ionized donor in CdS is considered first. The exciton is composed of a hole in the p-like Γ_9 valence band and an electron in the s-like Γ_7 conduction band. From Tables 2.1, 2.2, and 2.3 the hole wavefunction transforms as $|j = 3/2, m = \pm 3/2\rangle$ and the electron wavefunction as $|j' = 1/2, m' = \pm 1/2\rangle$. The states of the electron-hole pair are the four possible combinations of these.

Now an analogy is made with the states of a single atom. From atomic theory, when the interelectron electrostatic interaction is much smaller than the spin-orbit interaction, the jj coupling scheme is appropriate because the spin-orbit interaction is diagonal in this representation. In the p^5s configuration of the noble gases $j_1 = 3/2$ or $1/2$ from the hole in the p shell is coupled with $j_2 = 1/2$ of the s electron.[21] The $j_1 = 3/2$, $j_2 = 1/2$ combination yields a lower state with $J = 2$ and an upper state with $J = 1$. The splitting comes from the exchange integral $G^1(sp)$.

The spin-orbit interaction has already been included in calculating the band structure, and will be shown to be much larger than the electron-hole interaction. With the aid of a table of Wigner coefficients or $3j$ symbols we can write the coupled states in terms of the states $|JM\rangle$ with $J = 1$ or 2. They are

$$|j' = 1/2, m' = 1/2\rangle |j = 3/2, m = 3/2\rangle = |J = 2, M = 2\rangle$$

$$|1/2, -1/2\rangle |3/2, -3/2\rangle = |2, -2\rangle$$

$$|1/2, -1/2\rangle |3/2, 3/2\rangle = 1/2|2, 1\rangle + \sqrt{3}\,/2|1, 1\rangle$$

$$|1/2, 1/2\rangle |3/2, -3/2\rangle = 1/2|2, -1\rangle - \sqrt{3}\,/2|1, -1\rangle \qquad (2.27)$$

If the splitting due to the electron-hole electrostatic interaction $\mathcal{V}$ between the states of $J = 2$ and $J = 1$ is called 4Δ. then

$$\langle 2, M |\mathcal{V}| 2, M'\rangle = 0$$

$$\langle 1, M |\mathcal{V}| 1, M'\rangle = 4\Delta \delta_{MM'}$$

The matrix of $\mathcal{V}$ is also diagonal among the states (2.27), the energies of the first two states being zero and the second two, Δ.

In a magnetic field **H** the splitting of the states can be calculated by diagonalizing along with $\mathcal{V}$ the field dependent terms

$$\beta g_\parallel^e j_z' H_z + \beta g_\perp^e \left(j_x' H_x + j_y' H_y \right) + \beta g_\parallel^h j_z H_z \qquad (2.28)$$

It is not necessary to include terms in j_x and j_y, because their matrix elements are zero. For **H** parallel to the z or c axis the energies are

$$E_{1,2} = \pm \tfrac{1}{2}\left(g_{\parallel}^e + 3g_{\parallel}^h \right) \beta H$$

$$E_{3,4} = \Delta \pm \tfrac{1}{2}\left(g_{\parallel}^e - 3g_{\parallel}^h \right) \beta H$$

For **H** perpendicular to the axis the four states split into two doublets of energies

$$E_{\pm} = \tfrac{1}{2}\Delta \pm \tfrac{1}{2}\left[\Delta^2 + \left(g_{\perp}^e \, \beta H \right)^2 \right]^{\frac{1}{2}} \tag{2.29}$$

The splitting of these four states in a magnetic field can be described by the effective Hamiltonian

$$\mathcal{H}_{\text{eff}} = -\frac{2}{3}\Delta j_z j_z' + \beta g_{\parallel}^e j_z' H_z + \beta g_{\perp}^e \left(j_x' H_x + j_y' H_y \right) + \beta g_{\parallel}^h j_z H_z \tag{2.30}$$

The Zeeman effect of the hole is completely anisotropic, whereas that of the electron is found to be isotropic. For an unidentified donor in CdS Thomas and Hopfield[22] find $g_{\parallel}^e = g_{\perp}^e = 3g_{\parallel}^h = 1.74$, $\Delta = 0.31\text{meV}.\Delta$ is much smaller than the spin-orbit splitting of the valence band. This justifies the use of jj coupling. The ground state of the complex consists of the ionized donor D^+ alone, transforms as A_1, and, of course, does not split in a magnetic field.

A closer connection can be made between the phenomenological theory just described and the effective mass formalism by analogy with the effective mass theory for donors and acceptors and for free excitons. When the effective mass Hamiltonian for an exciton is derived from the Hamiltonian for the many electron system, the small exchange term $\mathcal{V}$ appears.[23,24] For a bound exciton the Hamiltonian for the envelope function is $\mathcal{H}_0 + \mathcal{V}$, where $\mathcal{H}_0$ is given by equation 2.24 in the case of isotropic masses. The effect of $\mathcal{V}$ can be calculated by perturbation theory. The zero-order ground state function is $F(\mathbf{r}_e, \mathbf{r}_h)|j', m'\rangle|j, m\rangle$. The F is the envelope function, the solution of $(\mathcal{H}_0 - E)F = 0$. Only the ground-state function, which goes over to the $1s$ solution of the H_2^+ problem as m_e^*/m_h^* approaches zero, has been calculated.[16] Isotropic masses are assumed. The matrix elements of $\mathcal{V}$ are nonzero only between functions

$$\psi_a = F(\mathbf{r}_e, \mathbf{r}_h)\phi_c(\mathbf{r}_e)|m_s'\rangle\phi_v(\mathbf{r}_h)|m_s\rangle \tag{2.31}$$

and ψ_b, which have $m_s^a + m_s'^a = m_s^b + m_s'^b = 0$. (Equation 2.31 has been written in a different form to separate the orbital and spin functions.) These correspond to the states $|1/2, -1/2\rangle|3/2, 3/2\rangle$ and

$|1/2, 1/2\rangle|3/2, -3/2\rangle$ of Equation 2.27. Skettrup, Suffczynski, and Gorzkowski[17] express this matrix element as

$$\Delta = J_0 \sum_{\mathbf{R}} |F(\mathbf{R}, \mathbf{R})|^2 \qquad (2.32)$$

The sum is over the positions of all unit cells and represents the probability that the electron and hole occupy the same cell. The J_0 is the exchange integral of the conduction and valence band functions of (2.31),

$$J_0 = \frac{e^2}{\varepsilon} \int\int \phi_c^*(\mathbf{r}_1)\phi_v^*(\mathbf{r}_2)|\mathbf{r}_1 - \mathbf{r}_2|^{-1}\phi_c(\mathbf{r}_2)\phi_v(\mathbf{r}_1)d\tau_1 d\tau_2 \qquad (2.33)$$

Using their calculated envelope function F and the experimental value of J_0 for free excitons, they obtain $\Delta = 0.49$ meV for CdS and $\Delta = 0.83$ meV for ZnO. The experimental value for ZnO is 0.9 meV.[25]

The oscillator strength f of the transition corresponding to decay of the bound exciton at frequency ν is related to the oscillator strength per unit cell f_x for decay of the free exciton at frequency ν_x by[16]

$$f = f_x\left(\frac{\nu_x}{\nu}\right)\left(\frac{\pi a_x^3}{V}\right)|\sum_{\mathbf{R}} F(\mathbf{R}, \mathbf{R})|^2 \qquad (2.34)$$

a_x is the radius of the free exciton state, and V is the volume of a unit cell. For m_e^*/m_h^* in the range 0.1 to 0.4, $f \approx 10^4 f_x$. Furthermore, because bound excitons are not mobile, their spectral lines are not broadened by thermal fluctuations to the extent that free exciton lines are. These two facts mean that bound exciton spectra are comparable in intensity with free exciton spectra at low impurity concentrations.

The spectrum of an exciton trapped at a neutral donor $D^+e^-e^-h^+$ can be explained by assuming that the two electrons are paired in the same orbital state with opposite spin. The Zeeman splitting is then due only to the hole and is anisotropic. The ground state is the neutral donor D^+e^- with the isotropic Zeeman splitting of the electron. For the exciton trapped at a neutral acceptor, $A^-h^+h^+e^-$, the situation is opposite. The excited state splits isotropically and the ground state, anisotropically. By observing the population distribution between the two ground Zeeman levels as a function of temperature in a low-temperature absorption experiment, one can tell whether the ground or excited state splits anisotropically. This determines whether the spectrum is due to an exciton bound to a neutral donor or to a neutral acceptor.

BOUND EXCITONS IN GaP

GaP has the cubic zincblende structure. The band gap is indirect with the valence band maximum at $k = 0$ and conduction band minima along the $\langle 100 \rangle$ axes. The point group of a substitutional impurity is T_d. At $k = 0$ the upper p-like valence bands are split by spin-orbit coupling into two components. The hole functions of the higher energy component transform as the four functions $|j = 3/2, \pm 3/2\rangle$ and $|j = 3/2, \pm 1/2\rangle$ and form bases for Γ_8. The wavefunction for an electron in the conduction band minima is complicated because of the several minima or valleys. However, measurements can be satisfactorily explained by assuming that for a donor on a phosphorus site the electron is in the lowest singlet orbital A_1 state.[26] (See Equation 8.2.)

A very weakly bound exciton with the electron in the indirect minima could decay only with phonon cooperation for conservation of momentum. Hence the absorption or emission spectrum would contain no "zero phonon" line. As the binding to the defect increases, the exciton wavefunction, becoming more localized, spreads in $\mathbf{k}$ space. Some component of the $k = 0$ part of the band is mixed into the wavefunction, and a zero phonon line appears. This mixing occurs only for the A_1 orbital state. If one or both of the particles are tightly bound, there may be a continuous acoustical phonon wing in the spectrum, as well as discrete optical phonon replicas.

For the exciton bound to a neutral donor $D^+e^-e^-h^+$ the electrons are paired with opposite spin, and only the hole contributes to magnetic effects. The Zeeman splitting of the hole states $|j = 3/2, m\rangle$ is given by an expression of the form of Equation 1.54. In the ground state D^+e^- only the electron is present. The Zeeman splitting is that of a Kramers doublet. An example is the exciton bound to the neutral sulfur donor.[26] The case of the exciton bound to a neutral acceptor is complicated because of the larger degeneracies of the states. The particles are less tightly bound, and the zero phonon lines are weak;[27] the spectra are not understood in detail.

The spectrum of the exciton trapped at an ionized donor or acceptor is due to the transitions between the eight states of $D^+e^-h^+$ or $A^-e^-h^+$ formed by coupling the electron and hole functions $|1/2, m'\rangle$ and $|3/2, m\rangle$ and the Γ_1 ground state D^+ or A^-. In the jj coupling scheme states of total $J = 2$ and $J = 1$ result from coupling $j = 3/2$ and $j = 1/2$, and these are split by the exchange interaction, $J = 2$ lying lower. The $j = 3/2$ states transform as Γ_8 and the $j = 1/2$ states, as Γ_6. From Table 1.6, $\Gamma_6 \times \Gamma_8 = \Gamma_3 + \Gamma_4 + \Gamma_5$. The $J = 1$ states transform as Γ_5. The $J = 2$ states transform as $\Gamma_3 + \Gamma_4$. The T_d crystal field can, therefore, split $J = 2$ into a doublet and a triplet. This splitting is usually very small. In a magnetic field the $J = 1$ and $J = 2$ states split into three and five Zeeman components, respectively.

Electric dipole transitions are allowed between the $J = 0$ ground state and the $J = 1$ levels. The $J = 0 \leftrightarrow J = 2$ transitions are generally also observed, but with much less intensity. These can become weakly allowed through admixture of $J = 1$ states by strain and the magnetic field. The $J = 0 \leftrightarrow J = 1, M = 0$ transition radiation is polarized parallel to the magnetic field and the $J = 0 \leftrightarrow J = 1, M = \pm 1$ transitions are polarized perpendicular to the field. The same sort of spectrum results from decay of an exciton bound to an isoelectronic impurity.

ISOELECTRONIC IMPURITIES

An isoelectronic impurity has the same number of valence electrons as the atom it replaces. Isoelectronic impurities are of two types. Those of the first type give rise to no localized states. Selenium in ZnTe is an example. As the selenium content x in the mixed crystal $ZnTe_{1-x}Se_x$ is varied, the band gap of the material shifts continuously, but no localized state results. Most isoelectronic impurities are of this type. Those of the second type produce localized states. Oxygen in ZnTe is an example. Those isoelectronic impurities that form localized states may be divided into isoelectronic donors and isoelectronic acceptors on the basis of electronegativity differences. An isoelectronic donor is less attractive to electrons than the atom for which it substitutes. Bismuth in GaP is an isoelectronic donor. From Table 2.4, bismuth is less electronegative than phosphorus. It can lose an electron (bind a hole), and subsequently an electron can be bound to the hole by Coulomb attraction. Similarly, oxygen in ZnTe is an isoelectronic acceptor. It first binds an electron, then a hole.

The electronegativity, a measure of $U(\mathbf{r})$, does not alone determine the binding of the first particle. Although the N-P electronegativity difference is much larger than the P-Bi difference, a hole is much more strongly bound to Bi in GaP than an electron to N in GaP. Lattice relaxation is also apparently important. All isoelectronic impurities that have been found to produce localized states have been anions, with the exception of the $\mathcal{V}_{kA}$ center in the alkali halides, in which a hole is trapped near a

Table 2.4 Isoelectronic impurities. Phillips electronegativities are listed for some elements. Isoelectronic traps producing localized states are ZnTe:O, GaP:Bi, GaP:N, CdS:Te, ZnS:Te, and several alkali halides as, for example, KCl:Br.

N	3.00	O	3.50	F	4.00
P	1.64	S	1.87	Cl	2.10
As	1.57	Se	1.79	Br	2.01
Sb	1.31	Te	1.47	I	1.63
Bi	1.24				

foreign alkali ion. No localized states have been observed for isoelectronic impurities in silicon or germanium.

The extreme sensitivity of the existence of a bound state to the electronegativity difference can be roughly understood by imagining that $U(\mathbf{r})$ is a square well potential energy of depth $-U_0$ and range r_0. Then there is no bound state unless $U_0 r_0^2 \geqslant \pi^2 \hbar^2 / (8m_e^*)$. This is in contrast with the long-range Coulomb field $-c/r$, which leads to bound states for c arbitrarily small. Because those isoelectronic impurities which produce bound states differ considerably from the host atom not only in electronegativity but also in size, the maximum solubility of the impurity is typically low. However, the associated luminescence can be quite efficient. The same model, which leads to $J=1$ and $J=2$ states of the exciton, describes well three isoelectronic impurities in compounds of zincblende structure— GaP:N, GaP:Bi, InP:Bi, and ZnTe:O. Because of stronger coupling to phonons in the case of ZnS:Te, the two zero phonon lines predicted are not seen.

When the exciton is destroyed, radiation of frequency ν is emitted. The binding energy of the system is $E_g - h\nu$. Often the quantity $E_g - E_x - h\nu$ is called the binding energy instead. However, we call this quantity the exciton localization energy, as in the case of the excitons bound to donors or acceptors.

Figure 2.3 shows schematically the observed energy levels when excitons are bound to the isoelectronic impurity nitrogen substitutional for phosphorus in GaP.[28] Because an exciton consists of two particles, the exciton energy levels are not easily displayed on a (one particle) band diagram in the fashion of the donor or acceptor levels. The first excited states are due to a bound exciton Ne^-h^+, and the second group of excited states, to two bound excitons $Ne^-h^+e^-h^+$. The states of the bound exciton molecule are those that result if the electrons are paired in the valley orbit singlet state and the holes are coupled to form states $J=2$ and $J=0$. These are the only values of J allowed by the Pauli principle for identical particles. The $\Gamma_3 - \Gamma_5$ crystal field splitting is observed in this case. (The triplet coming from $J=2$ is labeled Γ_4 in one case and Γ_5 in the other because the parity of the wavefunctions is opposite in the two cases. See Table 1.4.) Transitions were observed between the states of Ne^-h^+ and N and between $Ne^-h^+e^-h^+$ and Ne^-h^+.[28]

Excitons bound to pairs of isoelectronic impurities have also been studied.[29] For NN pairs in GaP the absorption spectrum consists of groups of lines that converge at higher energy to the lines due to the exciton bound to a single isolated N. Spectra from nearest-neighbor pairs to tenth-nearest-neighbor pairs are seen, the nearest-neighbor lines lying at the lowest transition energy. An exciton bound to two nitrogen atoms is

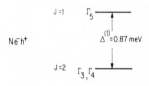

$Ne^-h^+e^-h^+$ $J=0$ Γ_1 0.11 meV $\Delta^{(2)}$ = 0.17 meV

$J=2$ Γ_5 0.16 meV

Γ_3

Ne^-h^+ $J=1$ Γ_5

$\Delta^{(1)}$=0.87 meV

$J=2$ Γ_3,Γ_4

N $J=0$ Γ_1 ———

Figure 2.3 Exciton and exciton molecule trapped at the isoelectronic impurity nitrogen in GaP. The Δ is the electron-hole exchange splitting. The crystal field splitting of the $J=2$ states of the exciton molecule is 0.16 meV (from Merz, Faulkner, and Dean, reference 28).

more tightly bound than an exciton bound to a single nitrogen atom. As the N-N spacing increases, the binding energy approaches that of the single nitrogen. The $J=1$ and $J=2$ states are split in the lower symmetry of the pair.[29] The nature of the splitting and the symmetry depend upon the position of the pair axis with respect to the crystallographic axes.

DONOR-ACCEPTOR PAIRS

In a crystal containing both donors and acceptors some donors are compensated by acceptors so that both are ionized, D^+ and A^-. Let us consider a donor-acceptor pair with separation R. When R is small, of the order of the nearest-neighbor separation, the pair is a complex defect, and, in general, its properties and energy levels cannot be predicted by the methods of this chapter. When R is large, the donor and acceptor can be considered nearly independent; the ionized donor can trap an electron, and the acceptor can trap a hole. For R large compared to the radii of the donor and acceptor effective mass states, the energy released as a photon when the hole on the acceptor recombines with the electron on the donor is approximately[30]

$$hv = E_g - E_D - E_A + \frac{e^2}{\varepsilon R} \tag{2.35}$$

This formula can be understood with reference to Figure 2.4, which shows the electron on the donor and the hole on the acceptor. After recombination the electron fills the hole on the acceptor, and the energy released is $E_g - E_D - E_A$. However, after recombination the attractive Coulomb interaction between the positively charged donor core D^+ and the negatively charged acceptor core A^- represents a final state energy $-e^2/\varepsilon R$. Subtraction gives Equation 2.35. In the case of a compound semiconductor the static dielectric constant is used for ε.

Equation 2.35 must be modified for the smaller R by the addition of a correction term $f(R)$. The term $f(R)$ vanishes as R becomes large.

$$hv = E_g - E_D - E_A + \frac{e^2}{\varepsilon R} + f(R) \tag{2.36}$$

The term $f(R)$ is composed of several contributions from initial state interactions.[31] We consider separations R that are not so large that overlap of the charge clouds of the bound electron and hole can be neglected. There are repulsive contributions, an interaction $E(D^+, h^+)$ between the donor core and the hole charge cloud and an interaction $E(A^-, e^-)$ between the acceptor core charge and the electron charge cloud. There are also two attractive interactions $E(D^+, A^-)$ between the core charges of the (neutral) donor and acceptor and $E(h^+, e^-)$ between the electron and the hole. At large separations R these four contributions add to zero, because the centers have no net charge. But because of the different dependences on separation of the interactions, the sum is positive at large R and becomes negative at small R.

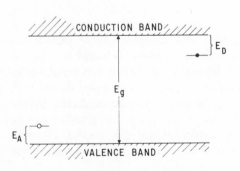

Figure 2.4 Initial state of a well-separated donor-acceptor pair before recombination of the electron on the donor with the hole on the acceptor.

In addition, there is a contribution from the mutually induced polarization of the neutral donor and neutral acceptor; that is, the attraction between the electron and hole distributions pulls the two charge clouds toward each other slightly so that they are no longer centered on the core charges. The shift of the charge distributions is largest when the volume of overlap of the two charge distributions is about equal to the volume of the nonoverlapping parts of the charge distributions.[32] For R large enough that overlap can be neglected, this van der Waals interaction can be expanded in inverse powers of R.[33,34] The leading term is $(6.5)(e^2/\varepsilon^2 R)(b/R)^5$; b is a characteristic length. The most detailed analyses of the various contributions to $f(R)$ have been made for GaP and are discussed in Chapter 8. Because the important length is R/a_D or R/a_A, for a given R the term $f(R)$ is generally more important the larger a_D or a_A, or the smaller the ionization energies E_D and E_A.

In some materials the emission spectra described by Equation 2.36 are quite striking, consisting of more than a hundred sharp lines of erratically varying intensity converging to a broad band at lower energy.[35] (In the case of the exciton bound to a pair of isoelectronic impurities the sharp lines converge to a limit at higher energy instead.) Each line corresponds to the zero phonon transition of a donor-acceptor pair with a certain separation R. As R becomes very large, the spacing between adjacent lines decreases and they merge into a band. Pairs with separations near the average value are not isolated; that is, the separation R is not small compared with the distance from the pair to another donor or acceptor. Emission from these pairs lies in the band to which the line spectra converge. Generally, phonon replicas of this band are also seen. The coupling to phonons increases with increasing separation R.[36]

Because the donors and acceptors occupy particular sites in the lattice, the separations can take on only certain discrete values R_n. To decide which line corresponds to which separation R_n, if there are more donors than acceptors, one calculates the average number of donors $N(R_n)$ at distance R_n from an acceptor. A random distribution is assumed usually. A model for the incorporation is chosen—donor and acceptor substitutional on cation sites, donor interstitial and acceptor substitutional on anion sites, and so on. $N(R_n)$ is converted to $N(\nu_n)$ by Equation 2.35 or 2.36; $N(\nu_n)$ would represent the emission intensity if transition strength and capture cross section were independent of R. These are slowly varying functions of R compared with the rapid fluctuations of $N(R)$, however, so that their R dependence can be neglected and $N(\nu_n)$ can be compared with experimental data directly.

In the case in which both donor and acceptor occupy the same type of substitutional site in the zincblende lattice, whether cation or anion, the separation R_n is given by [37]

$$R_n = \left(\tfrac{1}{2}n\right)^{\frac{1}{2}}a \qquad (a = \text{lattice constant})$$

except for a few shells $n = 14, 30, 46, \ldots$ where no lattice points occur. These missing values of n produce gaps in the spectra, labeled G in Figure 8.4. The gaps make it easy to distinguish this type-I spectrum from a type-II spectrum in which donors and acceptors occupy different types of substitutional site. No such gaps appear in type-II spectra for which R_n is given by

$$R_n = \left(\tfrac{1}{2}n - \tfrac{5}{16}\right)^{\frac{1}{2}}a$$

for the same zincblende lattice.

Generally no pair spectra are observed for n less than some minimum value n_0. When the problem is viewed as that of an exciton bound to two oppositely charged centers, it is possible to predict theoretically this minimum value of n_0 in simple cases.[17,38] The smaller the effective Bohr radii, the smaller n_0 will be.

In some crystals $e^2/\varepsilon R$ is larger than $E_D + E_A$ for many values of R. No emission can then be observed from these pairs. Sometimes the lines are broadened by strong coupling to lattice vibrations, and a broad band is seen, rather than discrete lines.[31,39]

The recombination transition probability decreases with increasing separation R as the overlap of donor and acceptor wavefunctions decreases.[31] For large R this decrease depends exponentially on R because the tails of the wavefunctions are exponential. Thus emission from closer pairs may have a characteristic time constant smaller by several orders of magnitude than that of pairs with much larger R. This effect can be seen in the time resolved spectrum after flash excitation.[40] In the case of a broad band, the peak of the emission intensity shifts to lower energies with increasing time after the flash.[41] Under continuous excitation the shape of the band depends on excitation intensity because of the different saturation properties of close and distant pairs. The papers by Williams[31] and Dean[39] contain good reviews of donor-acceptor pair properties.

BIBLIOGRAPHY

Dean, P. J., "Interimpurity Recombinations in Semiconductors," in *Progress in Solid State Chemistry*, Vol. 8, J. O. McCaldin and G. Somorjai, Eds. (Pergamon, Oxford, 1973) Ch. 1.

Greenaway, D. L. and Harbeke, G., *Optical Properties and Band Structure of Semiconductors* (Pergamon, Oxford, 1968).

Kohn, W., "Shallow Impurity States in Silicon and Germanium," in *Solid State Physics*, Vol. 5, F. Seitz and D. Turnbull, Eds. (Academic, New York, 1957) ch. 4.

Long, D. L., *Energy Bands in Semiconductors* (Wiley-Interscience, New York, 1968).

Tinkham, M., *Group Theory and Quantum Mechanics* (Mcgraw-Hill, New York, 1964) ch. 8.

Williams, F. E., "Donor-Acceptor Pairs in Semiconductors," *Phys. Stat. Sol.* **25**, 493 (1968).

REFERENCES

1. R. P. Messmer and G. D. Watkins, *Phys. Rev.* **B7**, 2568 (1973).
2. W. Kohn in *Solid State Physics*, Vol. 5, F. Seitz and D. Turnbull, Eds. (Academic, New York, 1957) Ch. 4.
3. H. B. Bebb, *Phys. Rev.* **185**, 1116 (1969).
4. J. S. Y. Wang and C. Kittel, *Phys. Rev.* **7B** (1973) and references therein.
5. P. J. Dean, J. D. Cuthbert, and R. T. Lynch, *Phys. Rev.* **179**, 754 (1969).
6. J. C. Phillips, *Phys. Rev.* **B1**, 1540 (1970).
7. M. Tinkham, *Group Theory and Quantum Mechanics* (McGraw-Hill, New York, 1964) Ch. 8.
8. R. G. Wheeler and J. O. Dimmock, *Phys. Rev.* **125**, 1805 (1962).
9. C. H. Henry and K. Nassau, *Phys. Rev.* **B2**, 997 (1970).
10. E. B. Hale and R. L. Mieher, *Phys. Rev.* **184**, 739 (1969) and references therein.
11. G. D. Watkins and F. S. Ham, *Phys. Rev.* **B1**, 4071 (1970).
12. D. D. Thornton and A. Honig, *Phys. Rev. Lett.* **30**, 909 (1973).
13. A. Baldereschi and N. O. Lipari, *Phys. Rev.* **B8**, 2697 (1973); *ibid.*, **B9**, 1525 (1974).
14. D. M. Larsen, *Phys. Rev.* **187**, 1147 (1969).
15. E. Teller, *Z. Physik* **61**, 458 (1930); D. R. Bates, K. Ledsham, and A. L. Stewart, *Phil. Trans. Roy. Soc.* **A246**, 215 (1953).
16. T. Skettrup, M Suffczynski, and W. Gorzkowski, *Phys. Rev.* **B4**, 512 (1971) and references therein.
17. J. J. Hopfield in *Proceedings of the Seventh International Conference on the Physics of Semiconductors, Paris, 1964* (Dunod, Paris, 1964) p. 725.
18. G. Mahler and U. Schröder, *Phys. Rev. Lett.* **27**, 1358 (1971).
19. S. G. Elkomoss, *Phys. Rev.* **B4**, 3411 (1971).
20. J. R. Haynes, *Phys. Rev. Lett.* **4**, 361 (1961).
21. E. U. Condon and G. H. Shortley, *The Theory of Atomic Spectra* (Cambridge, Cambridge, 1957) Ch. 13.
22. D. G. Thomas and J. J. Hopfield, *Phys. Rev.* **128**, 2135 (1962).
23. R. S. Knox in *Solid State Physics*, Suppl. 5, F. Seitz and D. Turnbull, Eds. (Academic, New York, 1963).
24. K. Dos, A. Haug, and P. Rohner, *Phys. Stat. Sol.* **30**, 619 (1968).
25. D. C. Reynolds, C. W. Litton, and T. C. Collins, *Phys. Rev.* **140A**, 1726 (1965).
26. D. G. Thomas, M. Gershenzon and J. J. Hopfield, *Phys. Rev.* **131A**, 2397 (1963).

27. P. J. Dean, R. A. Faulkner, S. Kimura, and M. Ilegems, *Phys. Rev.* **B4**, 1926 (1971).

28. J. L. Merz, R. A. Faulkner, and P. J. Dean, *Phys. Rev.* **188**, 1228 (1969).

29. D. G. Thomas and J. J. Hopfield, *Phys. Rev.* **150**, 680 (1966).

30. P. J. Dean in *Applied Solid State Science*, Vol. 1, R. Wolfe and C. J. Kriessman, Eds. (Academic, New York, 1969) p. 1.

31. F. E. Williams, *Phys. Stat. Sol.* **25**, 493 (1968).

32. R. Bindemann and K. Strebe, *Phys. Stat. Sol.* **B56**, 563 (1973).

33. W. Kauzmann, *Quantum Chemistry* (Academic, New York, 1957) p. 509.

34. P. J. Dean, R. A. Faulkner, and S. Kimura, *Phys. Rev.* **B2**, 4062 (1970).

35. D. G. Thomas, M. Gershenzon, and F. A. Trumbore, *Phys. Rev.* **133A**, 269 (1964).

36. H. L. Malm and R. R. Haering, *Can. J. Phys.* **49**, 2970 (1971).

37. J. D. Wiley and J. A. Seman, *Bell Syst. Tech. J.* **50**, 355 (1970).

38. G. Munschy and B. Strebe, *Phys, Stat. Sol.* **B59**, 525 (1973); B. Strebe and G. Munschy, *Phys. Stat. Sol.* **B60**, 133 (1973).

39. P. J. Dean in *Progress in Solid State Chemistry*, Vol. 8, J. O. McCaldin and G. Somorjai, Eds. (Pergamon, Oxford, 1973) Ch. 1.

40. D. G. Thomas, J. J. Hopfield, and K. Colbow, in *Radiative Recombination in Semiconductors* (Seventh International Conference on the Physics of Semiconductors, Vol. 4), (Dunod, Paris, 1965) p. 67.

41. S. Shionoya, K. Era, and Y. Washizawa, *J. Phys. Soc. Japan* **21**, 1624 (1966).

Chapter 3

Vibrational Properties of Defects

Lattice vibrations are of importance in the study of point defects for two reasons. They may modify, in some cases radically, the electronic transitions of the defect. This aspect is treated in the second and third sections of this chapter. Transitions between vibrational states of a defect, in the absence of any electronic excitation, may also be observed and information deduced about the symmetry of the defect and its position in the lattice. The first section deals with these localized vibrational states. Other effects of electronic-vibrational coupling are thermal broadening and shifting of lines in electronic spectra and radiationless transitions. If the radiative lifetime of an electronic state is rather long, it may be more probable for the state to decay by phonon emission, even though energy conservation may require the emission of many phonons. These effects are not as relevant to study of the atomic nature of point defects, however, and are not discussed further.

LOCALIZED MODES

When an impurity atom or a more complex defect is introduced into a crystal lattice, the lattice vibrations are altered. In analogy with the electronic states some new energy levels, or vibration frequencies, may appear outside the allowed bands of the pure crystal. To discuss these special defect vibrational states it is helpful first to recall a few properties of the vibrations of a perfect lattice.[1]

Let $u_\alpha(lb)$ be the α component of displacement from the equilibrium position $\mathbf{R}(lb)$ of the atom at site b in unit cell l. The α takes the three values x, y, z. $\mathbf{R}(lb) = \mathbf{R}(l) + \mathbf{R}(b)$. $\mathbf{R}(l)$ is the location of the unit cell l, and

$R(b)$ is the position in a cell of the bth atom of mass M_b, referred to an origin at some site in the cell. In the harmonic approximation the vibrational Hamiltonian is

$$\mathcal{H} = \frac{1}{2} \sum_{l,b,\alpha} M_b \dot{u}_\alpha^2(lb) + \frac{1}{2} \sum_{\substack{l,b,\alpha \\ l',b',\beta}} A_{\alpha\beta}(lb,l'b')u_\alpha(lb)u_\beta(l'b') \qquad (3.1)$$

The As are the force constants. They depend only on the difference $R(lb) - R(l'b')$ rather than on the separate values. When the time dependence of u is taken to be $\exp(i\omega t)$, the equations of motion are

$$-\omega^2 M_b u_\alpha(lb) + \sum_{\beta,l',b'} A_{\alpha\beta}(lb,l'b')u_\beta(l'b') = 0 \qquad (3.2)$$

If a plane wave solution,

$$u_\alpha(lb) = w_\alpha^b M_b^{-\frac{1}{2}} \exp\left[i\mathbf{k}\cdot\mathbf{R}(lb)\right] \qquad (3.3)$$

is assumed and substituted into Equation 3.2, the frequency ω must satisfy

$$\omega^2 w_\alpha^b = \sum_{\beta,b',l'} (M_b M_{b'})^{-\frac{1}{2}} A_{\alpha\beta}(lb,l'b')w_\beta^{b'}$$

$$\times \exp\left[i\mathbf{k}\cdot\mathbf{R}(l'b') - i\mathbf{k}\cdot\mathbf{R}(lb)\right] \qquad (3.4)$$

If there are p atoms per unit cell, there will be $3p$ solutions $\omega_j^2(\mathbf{k})$ for each allowed value of $\mathbf{k}$; $j = 1,2,\dots,3p$. These are the $3p$ branches of the phonon dispersion relation, usually displayed as a plot of $\omega_j(\mathbf{k})$ versus some component of $\mathbf{k}$, which takes on all allowed values in the Brillouin zone. Three are acoustic branches with $\omega_j(0) = 0$, and the others are optical branches.

Under a transformation to normal coordinates Q,

$$u_\alpha(lb) = (NM_b)^{-\frac{1}{2}} \sum_{j\mathbf{k}} \chi_\alpha^b(j\mathbf{k})Q(j\mathbf{k})\exp\left[i\mathbf{k}\cdot\mathbf{R}(l)\right] \qquad (3.5)$$

the Hamiltonian becomes

$$\mathcal{H} = \frac{1}{2} \sum_{j\mathbf{k}} \dot{Q}^*(j\mathbf{k})\dot{Q}(j\mathbf{k}) + \frac{1}{2} \sum_{j\mathbf{k}} \omega_j^2(\mathbf{k})Q^*(j\mathbf{k})Q(j\mathbf{k}) \qquad (3.6)$$

N is the number of unit cells in the crystal. The eigenvalues of (3.6) are

$$E = \sum_{j\mathbf{k}} \hbar\omega_j(\mathbf{k})\left[n_j(\mathbf{k}) + \frac{1}{2}\right] \qquad (3.7)$$

The positive integer $n_j(\mathbf{k})$ is the number of quanta of excitation in the mode $j\mathbf{k}$ of energy $\hbar\omega_j(\mathbf{k})$.

Suppose now that the crystal contains a small number of substitutional impurities. In general, the masses and force constants of these are different from the values for the pure crystal. The equations of motion are now

$$-\omega^2 M_{lb} u_\alpha(lb) + \sum_{\beta,l',b'} A'_{\alpha\beta}(lb,l'b')u_\beta(l'b') = 0 \tag{3.8}$$

This can be rewritten in another form to display explicitly the differences between the masses and force constants of the perfect and imperfect lattices:

$$\omega^2 M_b u_\alpha(lb) - \sum_{\beta,l',b'} A_{\alpha\beta}(lb,l'b')u_\beta(l'b')$$

$$= \sum_{\beta,l',b'} \Delta_{\alpha\beta}(lb,l'b')u_\beta(l'b') \tag{3.9}$$

$$\Delta_{\alpha\beta}(lb,l'b') = (M_b - M_{lb})\omega^2\delta_{ll'}\delta_{bb'}\delta_{\alpha\beta}$$

$$- A_{\alpha\beta}(lb,l'b') + A'_{\alpha\beta}(lb,l'b')$$

If there is only one impurity atom in the crystal, Δ has very few nonzero elements. Equation 3.9 can be written, using Equation 3.2, in the form

$$u_\alpha(lb) = \sum_{\substack{\beta,\gamma,l' \\ l'',b',b''}} G_{\alpha\beta}(lb,l'b')\Delta_{\beta\gamma}(l'b',l''b'')u_\gamma(l''b'') \tag{3.10}$$

or $\mathbf{u} = \mathbf{G}\Delta\mathbf{u}$ in matrix notation. The matrix $\mathbf{G}$ has elements[2,3]

$$G_{\alpha\beta}(lb,l'b') = \frac{1}{N\sqrt{M_b M_{b'}}} \sum_{j\mathbf{k}} \left[\frac{w_\alpha^b(j\mathbf{k})w_\beta^{b'*}(j\mathbf{k})}{\omega^2 - \omega_j^2(\mathbf{k})} \right]$$

$$\times \exp\left[i\mathbf{k}\cdot\mathbf{R}(lb) - i\mathbf{k}\cdot\mathbf{R}(l'b') \right] \tag{3.11}$$

If in Equation 3.10 we take α, l, and b on the left-hand side successively equal to the γ, l'', and b'' of the expression on the right-hand side, we have a small number of equations for only those $u_\alpha(lb)$ that are directly affected by the impurity. The vibrations of an atom lb are "directly affected" if the mass of the atom is changed (the impurity) or at least one of the force constants characterizing its interaction with another atom is changed from the value for the perfect lattice. Only the force constants of the impurity and its near neighbors are altered.

A simple example is an impurity at site $\mathbf{R}(lb) = \mathbf{R}(01)$ in the diamond lattice, for which $p = 2$. The impurity site (01) is taken as the origin. If the impurity has mass $M' \neq M_1 = M_2$, but the same force constants as the atom it replaces, then

$$\Delta_{\alpha\beta}(lb, l'b') = (M_1 - M')\omega^2 \delta_{l0}\delta_{l'0}\delta_{b'1}\delta_{b1}\delta_{\alpha\beta} \tag{3.12}$$

There are only three nonzero elements. The three equations obtained from Equation 3.10 are

$$u_\alpha(01) = (M_1 - M')\omega^2 \sum_\beta G_{\alpha\beta}(01,01)u_\beta(01) \tag{3.13}$$

$$G_{\alpha\beta}(01,01) = \frac{1}{NM_1} \sum_{j\mathbf{k}} \frac{w_\alpha^1(j\mathbf{k})w_\beta^{1*}(j\mathbf{k})}{\omega^2 - \omega_j^2(\mathbf{k})}$$

$G_{\alpha\beta}(01,01)$ can be simplified.[4] It is invariant under the symmetry operations of the point group of the impurity. The twofold rotations C_2^x, C_2^y, and C_2^z of T_d can be applied to show that $G_{\alpha\beta}(01,01)$ is zero if $\alpha \neq \beta$. Application of the C_3 rotations yields $G_{xx} = G_{yy} = G_{zz}$. If the labels of the two sublattices are interchanged, $G_{\alpha\alpha}(01,01) = G_{\alpha\alpha}(02,02)$, because $M_1 = M_2$. Thus, using these properties,

$$G_{\alpha\alpha}(01,01) = \frac{1}{6} \sum_\alpha \left[G_{\alpha\alpha}(01,01) + G_{\alpha\alpha}(02,02) \right]$$

$$= \frac{1}{6NM_1} \sum_{j\mathbf{k}\alpha} \left[\omega^2 - \omega_j^2(\mathbf{k}) \right]^{-1} \left[w_\alpha^1(j\mathbf{k})w_\alpha^{1*}(j\mathbf{k}) + w_\alpha^2(j\mathbf{k})w_\alpha^{2*}(j\mathbf{k}) \right]$$

$$= \frac{1}{6NM_1} \sum_{j\mathbf{k}} \frac{1}{\omega^2 - \omega_j^2(\mathbf{k})} \tag{3.14}$$

The last line follows from the orthonormality relation

$$\sum_{\alpha,b} w_\alpha^b(j\mathbf{k})w_\alpha^{b*}(j'\mathbf{k}) = \delta_{jj'} \tag{3.15}$$

With this value for G, Equations 3.13 reduce to

$$u_x(01) = \frac{\epsilon\omega^2}{6N} \sum_{j\mathbf{k}} \frac{u_x(01)}{\omega^2 - \omega_j^2(\mathbf{k})} \tag{3.16}$$

and two other equations with x replaced by y or z. The ϵ is the mass difference parameter $\epsilon = 1 - M'/M_1$. The term $u_x(01)$ can, of course, be eliminated from Equation 3.16. The displacements $u_x(01)$, $u_y(01)$, and $u_z(01)$ form a basis for the T_2 irreducible representation of the T_d point group of the impurity. The three modes are degenerate; that is, they have the same frequency ω, given by Equation 3.16.

It is possible to transform the Hamiltonian for the imperfect lattice to a form like that of Equation 3.6 by a normal coordinate transformation. The impurity destroys the translational invariance of the lattice, so that the normal coordinates Q_f cannot be specified by $\mathbf{k}$. The index f takes as many values as the two indices j and $\mathbf{k}$. The transformation is

$$u_\alpha(lb) = \frac{1}{\sqrt{NM_{lb}}} \sum_f \chi_\alpha^b(f,l) Q_f \tag{3.17}$$

The allowed values of $\mathbf{k}$ in the perfect lattice are so closely spaced that the sum in Equation 3.16 can be replaced by an integral and easily evaluated if ω lies outside the range of frequencies of the perfect lattice. If $g(x)\,dx$ is the fraction of modes of the perfect lattice with frequencies x in the range $(x, x+dx)$, then Equation 3.16 can be written[4,5]

$$1 = \epsilon \omega^2 \int_0^{\omega_M} \frac{g(x)\,dx}{\omega^2 - x^2} \tag{3.18}$$

ω_M is the maximum phonon frequency of the perfect lattice. If $g(x)$ is known, the frequency ω of the perturbed modes can be found from Equation 3.18 as a function of ϵ. For $M' < M_1$ the frequency of the triply degenerate mode, which is designated by $f = F$, is larger than ω_M if ϵ is greater than a small threshold value (0.075 for the silicon lattice[6]) and increases with increasing ϵ. This is a localized mode (actually three modes, but because they are degenerate, they are often denoted collectively in this way.) The amplitude $|\chi_\alpha^b(Fl)|^2$ is strongly dependent on b and l, decreasing rapidly with increasing distance from the impurity site l, $b = 0, 1$. The amplitude $|\chi_\alpha^1(F0)|^2$ also increases with increasing ϵ. If $M' > M_1$, the frequency of this mode is less than ω_M. In this case a broad peak at the mode frequency may be observed in an absorption experiment. The larger M'/M_1, the narrower the peak and the lower the frequency ω_F. The absorption line is always narrower when $\omega_F > \omega_M$, however.

The other modes of the imperfect crystal lie below ω_M in frequency. Their frequencies differ by very small amounts, of the order of the frequency separation ω_M/N of the modes, from the frequencies of the modes of the perfect crystal. In the diamond lattice (diamond, Si, Ge), because the atoms are not charged, there is no optical absorption by single phonons in the perfect crystal. When an impurity is present, absorption becomes possible for $\omega < \omega_M$ and also at the local mode frequency ω_F. The absorption strength of the local mode is roughly the same as that distributed over all the other modes. Only those modes of the perfect crystal are perturbed and allowed in the defect-activated absorption spectrum whose space group representation contains in its reduction the point group representation of the impurity vibration, T_2 of T_d in this case. Decompositions for phonons with **k** ending at high symmetry points in the zone are given for a few cubic groups by Loudon.[7] This is quite analogous to reduction of the group of **k** for electrons, described in Chapter 2. These high symmetry points are important because they often are associated with peaks in $g(\omega)$ which will appear in the spectrum. For the diamond lattice, however, all phonons at high symmetry points in the zone are allowed in the defect-activated spectrum.[7] For Si $\omega_M = 523$ cm^{-1}. Local modes are observed,[8] for example, from boron at 618 cm^{-1} (^{11}B with $\epsilon = 0.61$) and 642 cm^{-1} (^{10}B with $\epsilon = 0.65$). The intensities are in the ratio of the natural abundances of the isotopes.

Isotope effects are a valuable tool for deciding which atoms take part in the oscillation and are therefore present in the defect center. From the intensities it is sometimes possible to tell how many atoms of a particular kind are present. Local mode spectra yield more information about defects than spectra of the perturbed band modes because the lines are often very narrow in the local mode spectra. For some crystals there is a forbidden gap between the highest allowed acoustic frequency and the lowest allowed frequency of an optical branch; GaP, InAs, and ZnTe with the zinc blende lattice structure are examples. In these cases impurities may in principle produce modes whose frequencies lie in the gap.

Because the amplitude of oscillation in the localized mode is large only very near the impurity, it is often a good approximation when dealing with such a mode to consider vibrations of the impurity and its nearest neighbors only. For very light impurities a good description can sometimes be obtained by considering only the vibrations of the impurity. The lighter the impurity compared with the atom it replaces, the larger its vibration amplitude in the local mode compared with the amplitudes of the neighbors in this mode. This approximation has been applied by Hayes and co-workers[9-13] to very light impurities at tetrahedral sites: H substitutional for F in CaF$_2$, Be substitutional for Cd in CdTe, and interstitial Li in GaAs. Let us consider this model in more detail. The equilibrium position of the impurity is taken to be the origin. The displacements u_x, u_y, and u_z of the impurity are then X, Y, and Z. The vibrational Hamiltonian is approximated by [12]

$$\mathcal{H} = \frac{1}{2M}P^2 + \frac{1}{2}AR^2 + BXYZ + C_1(X^4 + Y^4 + Z^4)$$

$$+ C_2(X^2Y^2 + Y^2Z^2 + Z^2X^2) \qquad (3.19)$$

The last three anharmonic terms are necessary to describe the observation of harmonics and splittings of the vibrational levels. Each term transforms as A_1 of T_d. The anharmonic terms are small, and their effect can be calculated by perturbation theory. The zero-order Hamiltonian consisting of the first two terms describes an isotropic three-dimensional harmonic oscillator. The eigenstates are

$$|mnp\rangle = \psi_m(X)\psi_n(Y)\psi_p(Z) \tag{3.20}$$

$$\psi_m(X) = \left(\frac{a}{2^m m! \sqrt{\pi}}\right)^{\frac{1}{2}} H_m(aX) \exp\left(\frac{-a^2 X^2}{2}\right)$$

$$a^4 = \frac{AM}{\hbar^2} = \frac{M^2 \omega^2}{\hbar^2}$$

$$H_m(y) = (-1)^m e^{y^2} \frac{d^m}{dy^m} e^{-y^2}$$

H_m is an Hermite polynomial. The eigenvalues are

$$E(m+n+p) = \hbar\omega(m+n+p+3/2) \tag{3.21}$$

The ground state is

$$|000\rangle = \left(\frac{a^{1/2}}{\pi^{1/4}}\right)^3 \exp\left(\frac{-a^2 R^2}{2}\right) \tag{3.22}$$

It transforms as A_1. There are three degenerate states $|100\rangle$, $|010\rangle$, $|001\rangle$, corresponding to excitation of each of the three oscillators by one vibrational quantum. They have the form

$$|100\rangle = \sqrt{2}\left(\frac{a^{1/2}}{\pi^{1/4}}\right)^3 aX \exp\left(\frac{-a^2 R^2}{2}\right) \tag{3.23}$$

These three states form bases for T_2. There are six degenerate states with energy $E(2)$. They transform as A_1, E, and T_2 and are

$$A_1 \qquad \frac{1}{\sqrt{3}}(|200\rangle + |020\rangle + |002\rangle) \tag{3.24}$$

$$E \qquad \frac{1}{\sqrt{2}}(|200\rangle - |020\rangle), \frac{1}{\sqrt{6}}(2|002\rangle - |200\rangle - |020\rangle)$$

$$T_2 \qquad |011\rangle, |101\rangle, |110\rangle$$

Similarly, there are ten states with energy $E(3)$ which form bases for A_1, T_1, and two T_2 representations. Inclusion of the anharmonic perturbation splits apart the states belonging to different representations, as shown in Figure 3.1.

The electric dipole operator $q\mathbf{R}$ transforms as T_2 so that transitions from the ground state to the T_2 states can be observed in absorption. The same selection rule holds for Raman scattering, because the polarizability tensor components XY, YZ, ZX also transform as T_2. In the absence of the term $BXYZ$ the second harmonic would not be observable because $\langle 000|X|011\rangle = 0$. This term mixes some of the state $|111\rangle$ into $|000\rangle$ and some of $|100\rangle$ into $|011\rangle$, making the transition allowed:

$$|0A_1\rangle = |000\rangle + \frac{\langle 111|BXYZ|000\rangle}{E(0) - E(3)}|111\rangle \qquad (3.25)$$

$$|2T_2x\rangle = |011\rangle + \frac{\langle 100|BXYZ|011\rangle}{E(2) - E(1)}|100\rangle$$

The levels shown in Figure 3.1 which are degenerate in T_d symmetry may split into several levels when the symmetry is lowered by association with another defect[12] or by the application of external uniaxial stress or

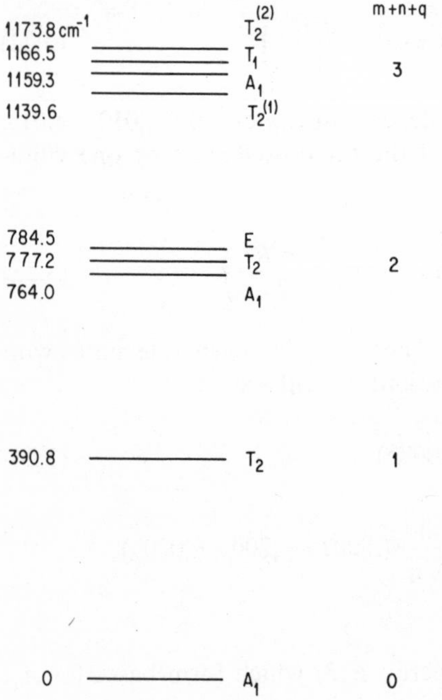

1173.8 cm⁻¹	$T_2^{(2)}$	m+n+q
1166.5	T_1	
1159.3	A_1	3
1139.6	$T_2^{(1)}$	

(This is Figure 3.1 — reproduced as a diagram of energy levels)

| 390.8 | T_2 | 1 |

Figure 3.1 Positions and symmetries of the vibrational energy levels corresponding to the localized mode of Be substitutional for Cd in CdTe. Transitions from the ground level to the four T_2 levels were observed and the positions of the other levels calculated by Hayes and Spray[9] (from Hayes and Spray, reference 9).

electric fields.[11] For example, suppose that some perturbation lowers the symmetry of the defect from tetrahedral T_d to trigonal C_{3v}. C_{3v} has two onefold irreducible representations A_1 and A_2 and one twofold representation E. From Table 3.1 Z is a basis function for A_1 and X and Y form a basis for E. Therefore, a T_2 state in T_d symmetry would split into two states A_1 and E in the lower trigonal symmetry. The transition from the ground state to A_1 is polarized parallel to the trigonal axis, and that to the E state is polarized perpendicular to the axis. Vibrations of impurity pairs are discussed in Chapter 7. Transitions forbidden in the higher symmetry may become allowed in the lower symmetry.[11] All these effects have been observed experimentally and confirm the usefulness of the model.

Table 3.1 Some Basis Functions for the Group C_{3v}; the z Axis Is the Trigonal Axis

A_1	z	
A_2	l_z	
E	x, y;	l_x, l_y

Often local modes due to the vibrations of a molecule or molecular ion trapped in a crystal are observed. The spectrum usually is similar to that of the free molecule. OH^- is such a species and is commonly found in many types of crystal. Another example is $(BO_3)^{3-}$, which can substitute for a halogen in an alkali halide.[14]

VIBRONIC TRANSITIONS BETWEEN ORBITALLY NONDEGENERATE ELECTRONIC STATES

The electronic and nuclear states have to this point been treated separately. Usually, the electronic and nuclear motions can be separated because of the great difference between the nuclear and electronic masses. In two cases this separation is not justified. One case is that of a weakly bound electron whose orbital motion is not rapid compared with the motion of the vibrating nuclei.[15] The electron drags lattice polarization along with it, and the excitation is called a bound polaron. Polaron effects should be taken into account in calculating some detailed properties[16] of the weakly bound electrons and holes treated in Chapter 2. However, many properties can be adequately described without this complication, and no further discussion of bound polarons is given. The other case is that of an orbitally degenerate electronic state. The effects of electron-phonon coupling are dramatically different and are quite important for the study of defect structure. This case is discussed in a following section.

Let us turn now to consider the system composed of the lattice and a point defect with an orbitally nondegenerate ground state and no orbital degeneracy in the excited electronic states of interest. The Schroedinger equation for the total system of nuclei and electrons is $\mathcal{H}\Psi = E\Psi$, where $\mathcal{H}$ is given by

$$\mathcal{H} = T_e + T_n + V_n + V_e + V_{en} \tag{3.26}$$

T stands for kinetic energy and V for potential energy. The subscripts indicate whether the term is a function of electronic coordinates $\mathbf{r}$, nuclear coordinates $\mathbf{R}$, or both. According to the zeroth order Born–Oppenheimer approximation[17] the approximate wavefunction of the total system, called the vibronic wavefunction, is

$$\Psi(\mathbf{r}, \mathbf{R}) = \Phi(\mathbf{r}, \mathbf{R})\psi(\mathbf{R}) \tag{3.27}$$

where the electronic wavefunction Φ is a solution of

$$(T_e + V_e + V_{en} + V_n)\Phi(\mathbf{r}, \mathbf{R}) = \epsilon(\mathbf{R})\Phi(\mathbf{r}, \mathbf{R}) \tag{3.28}$$

The nuclear kinetic energy term has been omitted. Thus Φ corresponds to the limit of infinitely heavy nuclei. Usually, in practice one must use electronic states $\phi(\mathbf{r})$, which are functions only of the electronic coordinates. All the electronic states used as zero-order approximations in the first two chapters are of this type. The V_n is just a constant that changes the zero of electronic energy. The nuclear vibrational state ψ is a solution of

$$(T_n + \epsilon)\psi(\mathbf{R}) = E\psi(\mathbf{R}) \tag{3.29}$$

The slow nuclei move in the effective potential $\epsilon(\mathbf{R})$ of the rapidly moving electrons. Their motion thus depends on the electronic state.

Let the defect have at least two electronic states, a ground state ϕ_g and an excited state ϕ_s. These states may have spin degeneracy but no orbital degeneracy. The ϕ_g and ϕ_s are eigenstates of $T_e + V_e$ or of $T_e + V_e + V^{(0)}$, where $V^{(0)}$ is defined in Equation 3.31. The vibrational state ψ is a product of harmonic oscillator functions $\psi(n_f, Q_f)$, one for each normal coordinate Q_f. The term n_f is the number of excitation quanta in the mode f. In general, the equilibrium positions of the atoms are different, especially near the defect, depending whether the defect is in the state ϕ_g or the state

ϕ_s. The two corresponding vibrational states are

$$\psi_g = \prod_f \psi(n_f, Q_f) \tag{3.30}$$

$$\psi_s = \prod_f \psi(n_f', Q_f - Q_f^0)$$

The vibrational force constants could also differ in the two states, but are assumed to be equal as a simplification. This amounts to retaining only the first two terms in the expansion of V_{en} in the Qs,

$$V_{en} = V^{(0)}(\mathbf{r}) + \sum_f V_f^{(1)}(\mathbf{r})Q_f + \cdots \tag{3.31}$$

As before, $\mathbf{r}$ is a short way of writing the coordinates of the defect electrons of interest. In Chapter 1 $V^{(0)}$ is called V_c.

If the defect is at a site with inversion symmetry and ϕ_g and ϕ_s have the same parity, the transition $\phi_g\psi_g \leftrightarrow \phi_s\psi_s$ is electric dipole forbidden if ψ_g and ψ_s have the same parity and magnetic dipole forbidden if ψ_g and ψ_s have opposite parity. The normal modes must be expressed as standing waves, rather than running waves, to display their parity with respect to inversion in the impurity site. If f is an odd mode, then $V_f^{(1)}$ is an odd function of $\mathbf{r}$ so that the sign of V_{en} is not changed on inversion. As mentioned in Chapter 1, even though ϕ_g and ϕ_s have the same parity, interaction with an odd mode can force an electric dipole transition. The absorption intensity is strongly temperature dependent in this case. This can occur by the mixing into ϕ_s by V_{en} of another state ϕ_t of opposite parity:

$$\phi_s' = \phi_s + \phi_t Q_f \langle \phi_s | V_f^{(1)} | \phi_t \rangle (\epsilon_s - \epsilon_t)^{-1} \tag{3.32}$$

Then the transition matrix element for $\phi_g\psi_g \leftrightarrow \phi_s'\psi_s$ is proportional to $\langle \phi_g | \mathbf{r} | \phi_t \rangle \langle \psi_g | Q_f | \psi_s \rangle$. It may then happen that the zero phonon line, the line corresponding to the purely electronic transition, arises from a different mechanism from that of a phonon assisted line of the same spectrum. Even though the zero phonon transition is electric dipole forbidden, it may occur by the magnetic dipole mechanism if this is allowed. In the limit of weak coupling the matrix element of Q_f is zero unless ψ_g and ψ_s differ only by one quantum in mode f, if the force constant is the same in states g and s. Thus in this limit only one odd mode can participate; it may be any odd mode allowed by symmetry. If ϕ_g and ϕ_s have opposite parity, an electric dipole transition $\phi_g\psi_g \leftrightarrow \phi_s\psi_s$ is forbidden if ψ_g and ψ_s have opposite parity. In general a phonon may be coupled to an electronic transition if $\Gamma_g \times \Gamma \times \Gamma_s$ contains an irreducible representation of the space group representation of the phonon reduced to the point group of the impurity site. Γ is the irreducible representation according to which the electronic transition operator transforms.[7,18]

CONFIGURATION COORDINATE MODEL

The configuration coordinate model[19] is a very useful way of treating the electron-vibrational coupling, especially in the limit of strong coupling, where the situation is too complex otherwise to form a picture of the physical process. In this model the defect and a small number of near neighbors are considered rather than the whole crystal. These atoms are a small cluster, and the normal modes of the cluster are the only ones considered. Certainly the electrons of the defect are most affected by changes in position of the near neighbors. If Q_f represents a local mode of the defect the frequency ω_f is well defined, because f is a normal mode of the crystal. If f is a normal mode of the cluster only, Q_f can be expanded in the normal modes of the crystal and will have a distribution of frequencies. Nevertheless, a frequency ω_f is always assigned to each normal mode f of the cluster. But it may be necessary finally to consider this frequency poorly defined or "smeared." As an example we consider an impurity tetrahedrally coordinated by four nearest neighbors, as in Figure 1.1, and fix a right-handed coordinate system X_i, Y_i, Z_i on each of the four neighbors i, oriented such that each positive Z_i direction is radially outward along the $\langle 111 \rangle$ bond direction and each positive X_i direction is upward in the plane of Z_i and the Z axis of the figure. There are nine normal modes which transform as A_1, E, $T_2^{(1)}$, or $T_2^{(2)}$. These can be found in terms of the displacements from equilibrium X_i, Y_i, Z_i in a fashion similar to the way the ligand orbitals were found in Chapter 1. They are written out in Table 3.2, and some are shown in Figure 3.2.

V_{en} is expanded in a Taylor series in these Q_f. The first-order correction to the electronic energy of a state ϕ due to each linear term is $\langle \phi | V_f^{(1)} | \phi \rangle Q_f$. If ϕ transforms as the one-dimensional irreducible representation Γ (electron spin is not included) of the point group of the site of the impurity, then for this matrix element to be nonzero, to transform as A_1, $V_f^{(1)}$ must transform as a representation Γ_f that is contained in the product $\Gamma \times \Gamma$, that is, as A_1. Therefore, only Q_1 need be included in $\Sigma V_f^{(1)} Q_f$. Every term in the Hamiltonian must be invariant to a symmetry transformation of the coordinates $\mathbf{r}$ and $\mathbf{R}$, or $\mathbf{r}$ and Q. This implies that $V_f^{(1)}$ transforms according to the same irreducible representation as Q_f. The electronic energy ϵ_j corresponding to the electronic state ϕ_j ($j = g$ or s) is, from Equation 3.28

$$\epsilon_j = \epsilon_{0j} + \langle \phi_j | V_1^{(1)} | \phi_j \rangle Q_1 + \tfrac{1}{2} A_j Q_1^2 \tag{3.33}$$

Table 3.2 Normal Modes of a Tetrahedral Complex; the Atoms Are Numbered as in Figure 1.1

A_1		$Q_1 = \frac{1}{2}(\ Z_1 + Z_2 + Z_3 + Z_4)$
E	u	$Q_2 = \frac{1}{2}(\ X_1 + X_2 - X_3 - X_4)$
	v	$Q_3 = \frac{1}{2}(\ Y_1 + Y_2 - Y_3 - Y_4)$
$T_2^{(1)}$	x	$Q_4 = \frac{1}{2}(-Z_1 + Z_2 + Z_3 - Z_4)$
	y	$Q_5 = \frac{1}{2}(-Z_1 + Z_2 - Z_3 + Z_4)$
	z	$Q_6 = \frac{1}{2}(\ Z_1 + Z_2 - Z_3 - Z_4)$
$T_2^{(2)}$	x	$Q_7 = \frac{1}{4}(\ X_1 - X_2 - X_3 + X_4) + \frac{\sqrt{3}}{4}(\ Y_1 - Y_2 - Y_3 + Y_4)$
	y	$Q_8 = \frac{1}{4}(\ X_1 - X_2 + X_3 - X_4) + \frac{\sqrt{3}}{4}(-Y_1 + Y_2 - Y_3 + Y_4)$
	z	$Q_9 = \frac{1}{2}(\ X_1 + X_2 + X_3 + X_4)$

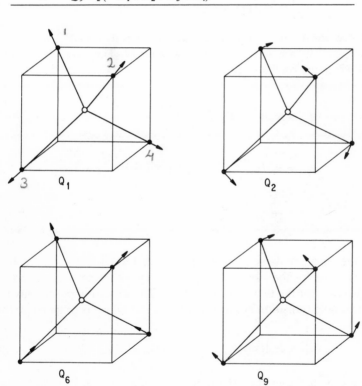

Figure 3.2 Representation of four of the normal modes of a tetrahedral cluster. The expressions for these normal modes in terms of displacements of the atoms are given in Table 3.2.

ϵ_{0j} contains contributions from T_e, V_e, and $V^{(0)}$. Only the first two terms in the expansion (3.31) are retained. The $\frac{1}{2}A_jQ_1^2$ is the contribution of Q_1 to the vibrational energy V_n in the harmonic approximation. In the equilibrium configuration, $Q_1 = Q_1^0$,

$$\left.\frac{\partial\epsilon_j}{\partial Q_1}\right)_{Q_1^0} = \langle\phi_j|V_1^{(1)}|\phi_j\rangle + A_jQ_1^0 = 0 \tag{3.34}$$

Equation 3.34 defines Q_1^0 and shows how Q_1^0 depends on the state ϕ_j; Q_1^0 can be set equal to zero for one of the states (ϕ_g), but will, in general, be different from zero for the other state (ϕ_s). With also a shift of the energy zero the two energies are

$$\epsilon_g(Q_1) = \frac{1}{2}A_gQ_1^2 \tag{3.35}$$

$$\epsilon_s(Q_1) = \epsilon_0 + \frac{1}{2}A_s(Q_1 - Q_1^0)^2$$

The force constants A_g and A_s are not necessarily equal. The vibrational states ψ_g and ψ_s are solutions of Equation 3.29. They are the one-dimensional harmonic oscillator functions

$$\psi_{gn} = N_nH_n(a_gQ_1)\exp\frac{-a_g^2Q_1^2}{2} \tag{3.36}$$

$$\psi_{sm} = N_mH_m(a_sQ_1 - a_sQ_1^0)\exp\frac{-a_s^2(Q_1 - Q_1^0)^2}{2}$$

$$a_j^2 = \frac{A_j}{\hbar\omega_j}$$

$\hbar\omega_j$ is the vibrational energy. It is proportional to $(A_j/M)^{\frac{1}{2}}$, where M is an effective mass for the vibration. Figure 3.3 shows the configuration coordinate diagram for the two electronic states. The transition probability for an electric dipole transition between the states gn and sm is proportional to

$$|\langle\phi_g\psi_{gn}|er|\phi_s\psi_{sm}\rangle|^2 = |\langle\phi_g|er|\phi_s\rangle|^2|\langle\psi_{gn}|\psi_{sm}\rangle|^2 \tag{3.37}$$

The overlap between the vibrational functions $\langle\psi_{gn}|\psi_{sm}\rangle$ is not zero because of the nonzero displacement Q_1^0 and the different frequencies $\omega_g \neq \omega_s$. At nonzero temperatures a number of the initial vibrational states n of the transition are populated. The observed shape of the band $\hbar(\nu)$ corre-

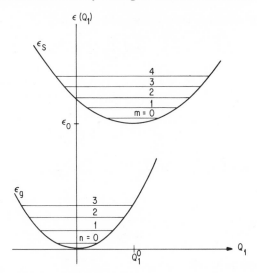

Figure 3.3 Configuration coordiante diagram for two electronic levels ε_s and ε_g as a function of Q_1. Several vibrational levels are indicated in each well.

sponding to the $g \rightarrow s$ transition is given by

$$\left| \langle \phi_g | e\mathbf{r} | \phi_s \rangle \right|^2 l_i(\nu)$$

$$= \left| \langle \phi_g | e\mathbf{r} | \phi_s \rangle \right|^2 \sum_n w_n \sum_m \left| \langle \psi_{gn} | \psi_{sm} \rangle \right|^2$$

$$\times \delta \left(\epsilon_0 + (m + \tfrac{1}{2})\hbar\omega_s - (n + \tfrac{1}{2})\hbar\omega_g - h\nu \right) \qquad (3.38)$$

w_n is the probability that vibrational level n is occupied. The absorption cross section $\sigma(\nu)$ is proportional to $\nu l_i(\nu)$ (Equation 5.10). The band is composed of lines at the transition energies $h\nu$ between the levels of the lower well, $E_n = (n + \tfrac{1}{2})\hbar\omega_g$, and the levels of the upper well, $E_m = \epsilon_0 + (m + \tfrac{1}{2})\hbar\omega_s$.

The electron-vibrational coupling strength has been characterized by Q_1^0 or by $\langle V_1^{(1)} \rangle$. It is convenient to introduce another measure of the coupling strength, the dimensionless Huang–Rhys factor S_j,

$$S_j = \left(Q_1^0 \right)^2 \frac{A_j}{2\hbar\omega_j} \qquad (3.39)$$

For simplicity, let us assume for a moment that the frequencies or curvatures of the two wells are identical: $\hbar\omega_g = \hbar\omega_s = \hbar\omega$. Then $S_g = S_s = S$

also. The square of the overlap integral is given by[20]

$$|\langle \psi_{gn}|\psi_{sm}\rangle|^2 = \exp(-S)\left(\frac{n!}{m!}\right)S^{(m-n)}L_n^{m-n}(S) \tag{3.40}$$

where $L_p^q(x)$ is the associated Laguerre polynomial. At temperatures low enough that only the $n=0$ initial state is significantly populated ($w_n = \delta_{n0}$),

$$|\langle \psi_{g0}|\psi_{sm}\rangle|^2 = \exp(-S)\left(\frac{S^m}{m!}\right) \tag{3.41}$$

This result is easily found by use of the generating function of (3.20).

$$\langle \psi_{g0}|\psi_{sm}\rangle = a^{-1}N_0N_m(-1)^m\int_{-\infty}^{\infty} e^{-(x-x_0)^2/2}e^{x^2/2}\frac{d^m}{dx^m}e^{-x^2}dx$$

$$x = aQ_1$$

Integration by parts m times gives, with $x_0^2/2 = S$,

$$\langle \psi_{g0}|\psi_{sm}\rangle = (m!)^{-1/2}S^{m/2}\exp\frac{-S}{2}$$

The absorption spectrum represented by Equation 3.41 consists of a number of lines at frequencies ν_{0m} given by $h\nu_{0m} = \epsilon_0 + m\hbar\omega$. This is illustrated in Figure 3.4 for three different values of S. The peak intensity occurs at $m \approx S$. As S increases, the zero-phonon line with intensity proportional to $\exp(-S)$ becomes relatively weaker. These results apply to temperature $T = 0$ K and all transitions originate from the level $n=0$. The result can be generalized to arbitrary temperature with Equation 3.38. The occupation probabilities w_n are given by Boltzmann factors. It is easy to see from Figure 3.3 that population of upper states of the lower well will lead to lines in the spectrum at lower energy than the $n=0\rightarrow m=0$ line. For strong coupling (large S) the envelope of the spectrum approaches a Gaussian band shape. For an isolated free molecule, or if Q_1 represents a strongly localized mode of a defect, the emission could consist of discrete lines. But often the frequency is not well defined and the observed line shape is a band with contour given by the envelope of the line spectrum.

As Q_1^0 increases, the minima of the two curves of Figure 3.3 move farther apart. For Q_1^0 sufficiently large (strong coupling) the ψ_{g0} and ψ_{s0} functions do not overlap appreciably, and there is no zero phonon or $m=p\leftrightarrow n=p$ transition. For weak coupling, on the other hand, this transition is strong. At a sufficiently low temperature, only the ψ_{g0} state is populated. In absorption, transitions take place from this state to the states ψ_{sm} with which there is overlap. The transitions are vertical with no change

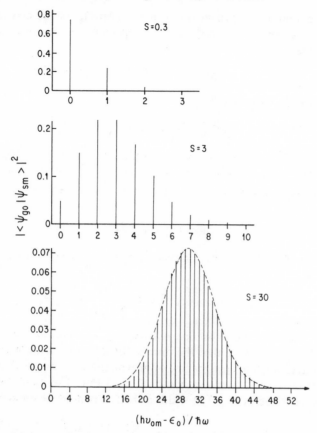

Figure 3.4 Transition probability for absorption from the state ψ_{g0} to the states ψ_{sm} of Figure 3.3 for three different values of electronic-vibrational coupling strength S. For $S=30$ a Gaussian envelope is included for comparison.

of Q_1 during the transition, which is assumed to occur rapidly with respect to lattice motion. The neighbors relax with emission of phonons until the new equilibrium configuration $Q_1 \approx Q_1^0$ is reached. Luminescence occurs at a longer wavelength from ψ_{s0} to the states ψ_{gm}, again without change of Q_1. The difference in transition energy between absorption and emission is called the Stokes shift. The lattice then relaxes with emission of phonons to the original position $Q_1 \approx 0$, and the vibronic state is again $\phi_g \psi_{g0}$.

Let us consider the emission process with strong coupling in a qualitative fashion. A similar treatment can apply to absorption. We remove the restriction $\omega_g = \omega_s$. If the coupling is strong, at low temperature the emission occurs from ψ_{s0} to ψ_{gn} where n is large. For large n the function ψ_{gn} has large amplitude only near the turning point Q_{1n} of the classical oscillator, where $\frac{1}{2}A_g Q_{1n}^2 = (n+\frac{1}{2})\hbar\omega_g$. If ψ_{gn} is approximated by a δ function $\delta(Q_1 - Q_{1n})$, then

$$\langle \psi_{s0} | \psi_{gn} \rangle \approx \text{constant} \times \psi_{s0}(Q_{1n} - Q_1^0) \qquad (3.42)$$

Because Q_{1n} is near Q_1^0,

$$\epsilon_g(Q_{1n}) - \epsilon_g(Q_1^0) \approx A_g(Q_{1n} - Q_1^0)Q_1^0$$

This amounts to approximating the lower parabola by a straight line near $Q_1 = Q_1^0$. If the separation between the $m=0$ level of the upper well and the point on the lower curve $\epsilon_g(Q_1^0)$ is called $h\nu_0'$ and the transition energy is $h\nu_n$, then

$$h\nu_0' - h\nu_n \approx A_g Q_1^0 (Q_{1n} - Q_1^0)$$

Substituting this into the Gaussian function ψ_{s0} of (3.42) yields

$$|\langle \psi_{s0} | \psi_{gn} \rangle|^2 \propto \exp \frac{-(h\nu_n - h\nu_0')^2}{2S_g(\hbar\omega_g)^3/\hbar\omega_s} \qquad (3.43)$$

Let us drop the subscript n from $h\nu_n$ and consider $h\nu$ a continuous variable. We assume that the discrete emission lines are broadened into a band and that the straight line approximation for the section of the lower parabola is good over the width of the band. The band has a Gaussian shape centered at $h\nu_0'$. The width at half maximum $W(0)$ is given by

$$[W(0)]^2 = \frac{8(ln2)S_g(\hbar\omega_g)^3}{\hbar\omega_s} \qquad (3.44)$$

At higher temperatures, other initial states ψ_{sm} are thermally populated and contribute to the spectrum. The width at higher temperatures is given by[19]

$$W(T) = W(0)\left(\coth\frac{\hbar\omega_s}{2kT}\right)^{\frac{1}{2}} \qquad (3.45)$$

The central frequency ν_0', from Figures 3.3 and 3.5 and Equations 3.35 and 3.39, satisfies the relation

$$h\nu_0' = \epsilon_0 + \frac{1}{2}\hbar\omega_s - \epsilon_g(Q_1^0)$$

$$= \epsilon_0 + \frac{1}{2}\hbar\omega_s - S_g\hbar\omega_g \qquad (3.46)$$

$S_g\hbar\omega_g$ is the vibrational energy given up as the system relaxes from the configuration $Q_1 \approx Q_1^0$ to the equilibrium configuration $Q_1 \approx 0$. The S_g is

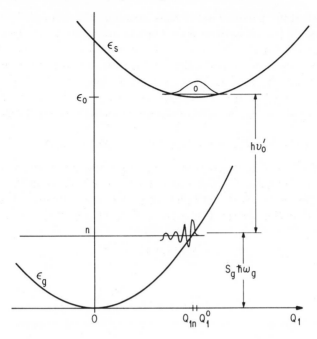

Figure 3.5 Configuration coordinate diagram illustrating emission with strong coupling at low temperature.

the average number of vibrational quanta produced in this process. (We neglect for large S_g the small zero point energy $\frac{1}{2}\hbar\omega_g$.) A similar treatment for the absorption spectrum shows that S_s is the average number of quanta involved in the relaxation of the upper state from $Q_1 \approx 0$ to $Q_1 \approx Q_1^0$. The width of the absorption band is also given by Equation 3.45 if the subscripts g and s are interchanged.

Another characteristic of the broad-band spectra describable by the configuration coordinate model is the shift of the peak position with changing temperature. If we call the peak position of the emission band $E_{em}(T)$, instead of $h\nu_0$, and the peak position of the absorption band $E_{ab}(T)$, these shifts are given by[21]

$$E_{em}(T) - E_{em}(0) = \left\{ \frac{\omega_g^2 - \omega_s^2}{\omega_s^2} + \frac{8\omega_g^4}{\omega_s^2(\omega_g^2 + \omega_s^2)} \left[-1 + \frac{E_{ab}(0)}{E_{em}(0)} \right] \right\} kT \quad (3.47)$$

$$E_{ab}(T) - E_{ab}(0) = \left[-1 + \frac{\omega_s^2}{\omega_g^2} \right] kT$$

The temperature dependences of peak position and width of the bands are commonly measured to determine whether an experimental spectrum can be described by a configuration coordinate model. A convenient method for studying these broad featureless bands is by analysis of the moments of the band shape. The second moment is proportional to $[W(T)]^2$. Effects of externally applied perturbations such as stress, electric, or magnetic fields are often studied by analysing the resulting changes in the moments.[22]

COUPLING TO MORE THAN ONE VIBRATIONAL MODE

The general problem of electronic coupling to the many modes implied by the form of Equations 3.30 is rather complex, especially in the regime of strong coupling in which multiphonon processes dominate.[23] The strength of coupling is characterized by a generalized Huang–Rhys factor S.

$$S = \sum_f (2n_f + 1)S^{(f)}$$

$$= \sum_f S^{(f)} \coth\left(\frac{\hbar\omega_f}{2kT}\right)$$

$S^{(f)}$ is the Huang–Rhys factor for mode f. The frequency ω_f and $S^{(f)}$ are assumed not to depend on the electronic state, and n_f and S increase with increasing temperature. The weak coupling limit is defined by $S \ll 1$. For an allowed transition in the weak coupling limit most of the intensity of the spectrum is in the zero-phonon line, corresponding to $n_f = n_f'$ for all f. In absorption this zero-phonon line is flanked on the high-frequency side by a band corresponding to creation of one phonon and extending in frequency from ν_0 to $\nu_0 + \omega_M/2\pi$. The term $\nu_0 = \omega_0/2\pi$ is the frequency of the zero-phonon line. The spectral shape of the absorption coefficient $\alpha_1(\omega_0 + \omega)$ due to this Stokes band is roughly[23]

$$\alpha_1(\omega_0 + \omega) \propto \frac{g'(\omega - \omega_0)}{\omega[1 - \exp(-\hbar\omega/kT)]} \tag{3.48}$$

$g'(\omega)$ is the spectral density of phonons in the imperfect lattice. It contains a strong peak at the frequency of any local mode with frequency less than ω_M. Its ordinate is shifted so that it is nonzero in the range $(\omega_0, \omega_0 + \omega_M)$ rather than the range $(0, \omega_M)$. Unless there are peaks due to local modes the spectral density of phonon steates of the perfect lattice is generally used in calculations rather than $g'(\omega)$. The shape of the anti-Stokes band, corresponding to destruction of one phonon, extends from ω_0 to $\omega_0 - \omega_M$

and is approximately given by[23]

$$\alpha_{-1}(\omega_0 - \omega) \propto \frac{g'(\omega + \omega_0)}{\omega[\exp(\hbar\omega/kT) - 1]} \qquad (3.49)$$

At low temperature the anti-Stokes band is much less intense than the Stokes band. In deriving Equations 3.48 and 3.49 a coupling $S^{(f)} \propto \omega_f^{-1}$ was assumed.[23] This may not always be the case, the actual coupling being much stronger to some modes and much weaker to others nearby in frequency (Figure 3.6). We have already seen that coupling to some modes is forbidden if the defect has inversion symmetry.

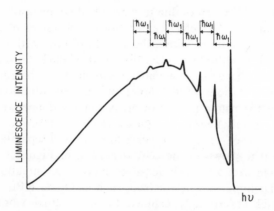

Figure 3.6 Hypothetical emission spectrum illustrating peaks due to coupling to a mode of energy $\hbar\omega_1$ superimposed on a smooth band due to coupling to other modes.

As S increases, multiphonon processes become important. In the region of intermediate coupling, $S \approx 1$, the zero phonon line is weaker compared with the bands. The bands are now composed of contributions from one, two, three, or more phonons and extend beyond the limits $\omega_0 + \omega_M$ and $\omega_0 - \omega_M$. The contribution from processes in which two phonons are created or destroyed lies in the intervals $(\omega_0, \omega_0 + 2\omega_M)$ and $(\omega_0, \omega_0 - 2\omega_M)$; whereas those from processes in which one phonon is created and one is destroyed lie in the same spectral range as the one phonon processes. The band maxima are nearer the zero phonon line relative to the band width. In the strong coupling limit, $S \gg 1$, the zero-phonon line has negligible intensity compared with the bands, which have a skewed Gaussian shape. The shape function for a p phonon process is obtained by convolving the shape function for the one phonon process with itself p times.[23,24] The peak occurs at $h\nu_p$ given by

$$h\nu_p = h\nu_0 \pm p\hbar\bar{\omega}$$

The upper sign applies to absorption and the lower to emission. The $\bar{\omega}$ is the frequency obtained by averaging ω over the shape function of the one phonon process. The spectrum consists of the sum of the bands for all p. The stronger the coupling, the larger is the number of multiphonon processes that make nonnegligible contributions. For details of the calculation of these spectral shapes the reader is referred to the paper by Pryce.[23]

OBSERVED BAND SHAPES

A wide variety of line shapes due to vibrational coupling occurs, depending on the nature of the defect center and the character of the coupling with the lattice. Transitions within the $4f$ shell of rare-earth impurities are weakly coupled to vibrations. Usually, a very thick crystal sample is necessary to see the weak one phonon sideband in absorption even at rare-earth concentrations of the order of 1%. No vibrational structure appears in the absorption spectra of many shallow donors and acceptors. Figure 7.8 shows only lines due to purely electronic transitions.

The "picket fence" spectrum consisting of well-separated lines, for example, like that shown in the central portion of Figure 3.4, can result from interaction with a strongly localized mode. This situation occurs for small molecular impurities in a crystal, such as O_2^- in KBr, in which the electronic states are strongly coupled to the internal vibration of the molecule.[25] At low temperature the "edge emission" in CdS, the origin of which is not yet certain, contains separated longitudinal optical phonon replicas and is characterized by $S \approx 1$.[26,27] The broad, roughly or very closely Gaussian absorption bands with accompanying Stokes shifted emission bands are observed in many different host crystals, including alkali halides,[19] II–VI compounds,[21] and III–V compounds.[28]

More complicated spectra also occur. Figure 3.6 illustrates a case of fairly strong coupling to many modes leading to a broad band. In addition there is a special coupling to a mode of energy $\hbar\omega_1$ which leads to sharp lines superimposed on the band. Such shapes occur for $f \to d$ transitions of rare earths in alkali halides[29] and in bound exciton spectra in III–V compounds.[30]

Figure 3.7 shows the Stokes portion of an hypothetical absorption spectrum characterized by weak coupling to the allowed lattice modes and stronger coupling to a localized mode of frequency $\omega_L > \omega_M$. Coupling to the lattice modes gives a phonon wing of shape $g'(\omega - \omega_0)/\omega$, and the

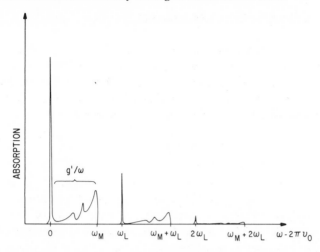

Figure 3.7 Hypothetical absorption spectrum (Stokes portion) of a defect characterized by moderate coupling to a local mode of frequency ω_L and weak coupling to other phonons of density of states $g(\omega)$. The zero-phonon line occurs at $2\pi\nu_0$.

coupling to the local mode is as shown in the upper portion of Figure 3.4. The zero-phonon line at ω_0 with the wing from ω_0 to $\omega_0 + \omega_M$ is repeated, removed from ω_0 by multiples of the local mode frequency ω_L and with intensity determined by the strength of the coupling to the local mode. Such shapes occur for small molecules in crystals. The similarity of the repeated patterns is called the similarity law by Rebane[25] and follows from the product form of the vibrational wavefunction, Equation 3.30.

Isotope effects are sometimes observable in the sharp lines of vibrational spectra as well as in local mode spectra. When they are present, more information about the structure of the defect center is available. Sometimes isotope shifts are seen even in the zero-phonon line.[31,32] From the configuration coordinate diagram we see that the frequency of the zero-phonon line is not really independent of vibrational effects but depends on the zero-point vibrational energy $\frac{1}{2}\hbar\omega$, which in turn depends on the mass. Let us suppose that the nucleus in question is surrounded by a number of neighbor atoms which we picture as point charges. On this point-charge model it is easy to see that the larger the zero-point vibrational amplitude of the central nucleus, the nearer it would come to these charges, and the larger would be the average electrostatic interaction between the electrons on the central atom and these point charges. If the central nucleus has a number of naturally abundant isotopes, each would have a slightly different vibrational amplitude and a slightly different interaction. If the line is narrow enough, these small effects can be observed.

ORBITALLY DEGENERATE ELECTRONIC STATES

The situation envisaged in the previous section is somewhat artificial. Rarely are both states between which an optical transition occurs orbitally nondegenerate if the site symmetry is high. When orbital degeneracy is present, the problem is more complicated because of the Jahn–Teller effect.[33] In this case Equation 3.27 must be modified to

$$\Psi = \sum_k \Phi_k \psi_k \qquad (3.50)$$

$\Phi_k(\mathbf{r}, Q)$ or $\phi_k(\mathbf{r})$ is the kth basis function of a particular irreducible representation Γ of the symmetry group of the site of the defect. With each electronic state a vibrational state $\psi_k(Q)$ is associated. In this case the matrix element $\langle \phi_k | V_f^{(1)} | \phi_l \rangle$ is nonzero for other distortions as well as Q_1. Because the wavefunctions ϕ_k can be chosen real, as in Chapter 1, the matrix element can be written

$$\langle \phi_k | V_f^{(1)} | \phi_l \rangle = \int \frac{1}{2}(\phi_k \phi_l + \phi_l \phi_k) V_f^{(1)} \, d\tau \qquad (3.51)$$

The function in parentheses is the symmetric product[34] of ϕ_k and ϕ_l. A product of two functions can be written as the sum of a symmetric product and an antisymmetric product. The matrix element is nonzero if Γ_f, the irreducible representation of $V_f^{(1)}$, is contained in the representation based on the symmetric products and written $[\Gamma \times \Gamma]_s$. The symmetric product representation is, in general, reducible and contains fewer irreducible representations than the product representation $\Gamma \times \Gamma$, because the products $\phi_k \phi_l$, and $\phi_l \phi_k$ cannot be considered different. The character of the symmetric product representation for the element R, called $[\chi(R)]_s$, is given by

$$2[\chi(R)]_s = (\chi^\Gamma(R))^2 + \chi^\Gamma(RR) \qquad (3.52)$$

Suppose, for example, that the defect is a substitutional impurity in T_d symmetry with a single electron or hole in the d shell, and that the states of interest are the doubly degenerate pair $|eu\rangle$ and $|ev\rangle$, whose angular dependences are given in Table 1.4. The following development is not dependent on this specialization, however. The two electronic functions may be any two states forming bases for E. The symmetric product $[E \times E]_s = A_1 + E$. These states can interact with the E distortion (Q_2 and Q_3) as well as the A_1 distortion. The interaction with the E mode is very different, however. From Figure 3.2 it can be seen that Q_1 does not change the symmetry of the complex. Hence $|eu\rangle$ and $|ev\rangle$ are shifted in energy by the interaction with Q_1, but remain degenerate. However, Q_2 reduces the symmetry to D_{2d}. In D_{2d} symmetry the basis functions of the E representation of T_d transform as basis functions for the one-dimensional A_1 and B_1 representations of D_{2d}. The degeneracy of $|eu\rangle$ and $|ev\rangle$ is removed by interaction with the E distortion. For Q_2 or Q_3 nonzero one state is raised in energy and the other depressed.

The states $|eu\rangle$ and $|ev\rangle$ are eigenstates of $T_e + V_e + V^{(0)}$ with eigenvalue E_0. The operator whose eigenvalues correspond to ϵ of Equation 3.28 is

$$\mathcal{H}' = E_0 + V_1 Q_1 + V_u Q_2 + V_v Q_3 + \tfrac{1}{2}A_1 Q_1^2 + \tfrac{1}{2}A\left(Q_2^2 + Q_3^2\right) \quad (3.53)$$

It is assumed that the splitting due to $V_e + V^{(0)}$ is very large so that only the states $|eu\rangle$ and $|ev\rangle$ need be considered. This splitting is largest in sixfold coordination. The following development is formally the same for an E state in cubic or tetrahedral symmetry. Because the terms $E_0 + V_1 Q_1 + \tfrac{1}{2}A_1 Q_1^2$ cause no splitting, they are dropped. The V_u and V_v transform in the same way as Q_2 and Q_3. Their dependence on the electronic coordinates can be taken to be, without loss of generality,

$$V_u = c'(3z^2 - r^2), \qquad V_v = c'\sqrt{3}\,(x^2 - y^2) \quad (3.54)$$

c' is a constant. The matrix elements of V_u and V_v can be found from Equation 1.37 and Table 1.4. They are

$$\langle eu|\,\mathcal{H}'\,|eu\rangle = cQ_2 + \tfrac{1}{2}A\left(Q_2^2 + Q_3^2\right) \quad (3.55)$$

$$\langle eu|\,\mathcal{H}'\,|ev\rangle = \langle ev|\,\mathcal{H}'\,|eu\rangle = -cQ_3$$

$$\langle ev|\,\mathcal{H}'\,|ev\rangle = -cQ_2 + \tfrac{1}{2}A\left(Q_2^2 + Q_3^2\right)$$

c is another constant, proportional to c'. The problem is simplified by a transformation to new coordinates ρ, θ given by

$$Q_2 = \rho\cos\theta \qquad Q_3 = \rho\sin\theta \quad (3.56)$$

Diagonalization of the matrix of $\mathcal{H}'$ yields the two eigenvalues and corresponding eigenfunctions

$$\epsilon_- = -c\rho + \tfrac{1}{2}A\rho^2 \quad (3.57)$$

$$\phi_- = \sin\left(\frac{\theta}{2}\right)|eu\rangle + \cos\left(\frac{\theta}{2}\right)|ev\rangle$$

$$\epsilon_+ = c\rho + \tfrac{1}{2}A\rho^2$$

$$\phi_+ = \cos\left(\frac{\theta}{2}\right)|eu\rangle - \sin\left(\frac{\theta}{2}\right)|ev\rangle$$

In Q_2, Q_3 space, or equivalently in ρ, θ space, $\epsilon(\rho)$ is a two-sheet surface with rotational symmetry about the axis $\rho = 0$. A section of this surface along the Q_2 axis is shown in Figure 3.8. The surface is often called the "Mexican hat" because of its shape. The amount by which the energy is depressed at the minimum is called the Jahn–Teller energy $E_{JT} = c^2/2A$ and may be as large as several thousand cm^{-1} in the case of octahedral coordination.[35] The position of the minimum $\rho_0 = |c|/A$ is another measure of the strength of the electronic-vibrational coupling in analogy with Q_1^0 for the nondegenerate electronic state.

The vibronic Hamiltonian of (3.26) is obtained by adding the kinetic energy of the nuclei to $\mathcal{H}'$. In the coordinates ρ, θ this is

$$\mathcal{H} = -\frac{\hbar^2}{2M}\left[\frac{1}{\rho}\frac{\partial}{\partial \rho}\left(\rho\frac{\partial}{\partial \rho}\right) + \frac{1}{\rho^2}\frac{\partial^2}{\partial \theta^2}\right] + \mathcal{H}' \qquad (3.58)$$

M is the mass appropriate to the vibration. It is the mass of one of the four neighbors if the rest of the crystal is neglected. In general, the solution Ψ of $\mathcal{H}\Psi = E\Psi$ is not separable into a product $\phi(\mathbf{r})\psi(\rho, \theta)$. We look instead for a solution of the form of (3.50),

$$\Psi = \phi_+ \psi_+ + \phi_- \psi_- \qquad (3.59)$$

The functions ϕ_+ and ϕ_-, as has been shown, diagonalize $\mathcal{H}'$. The term

$$\frac{1}{\rho}\frac{\partial}{\partial \rho}\left(\rho\frac{\partial}{\partial \rho}\right)$$

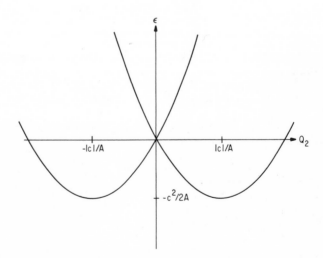

Figure 3.8 Section along the Q_2 axis of the "Mexican hat" potential.

acts only on the functions $\psi_{\pm}(\rho,\theta)$. But the term in $\partial^2/\partial\theta^2$ mixes ϕ_+ and ϕ_-, because, from (3.57),

$$\frac{\partial^2}{\partial\theta^2}\phi_{\pm}\psi_{\pm} = -\tfrac{1}{4}\phi_{\pm}\psi_{\pm} \mp \phi_{\mp}\frac{\partial}{\partial\theta}\psi_{\pm} + \phi_{\pm}\frac{\partial^2}{\partial\theta^2}\psi_{\pm} \qquad (3.60)$$

The Schroedinger equation, when multiplied by ϕ_+ or ϕ_- and integrated over the electronic coordinates, yields the two coupled equations for ψ_+ and ψ_-,

$$(T_\rho - E + \epsilon_+)\psi_+ - \frac{\hbar^2}{2M\rho^2}\left(-\tfrac{1}{4}\psi_+ + \frac{\partial^2}{\partial\theta^2}\psi_+ + \frac{\partial}{\partial\theta}\psi_-\right) = 0 \qquad (3.61)$$

$$(T_\rho - E + \epsilon_-)\psi_- - \frac{\hbar^2}{2M\rho^2}\left(-\tfrac{1}{4}\psi_- + \frac{\partial^2}{\partial\theta^2}\psi_- - \frac{\partial}{\partial\theta}\psi_+\right) = 0$$

T_ρ stands for the first term of T_n which depends only on ρ. The substitution $\psi_{\pm} = f^{\pm}(\rho)\exp(ij\theta)$ leads to the equations

$$-\frac{\hbar^2}{2M\rho^2}\left[\rho\frac{\partial}{\partial\rho}\left(\rho\frac{\partial}{\partial\rho}\right) - j^2 - \frac{1}{4}\right]f^+ + (c\rho + \tfrac{1}{2}A\rho^2 - E)f^+ = i\frac{\hbar^2 j}{2M\rho^2}f^-$$

$$(3.62)$$

$$-\frac{\hbar^2}{2M\rho^2}\left[\rho\frac{\partial}{\partial\rho}\left(\rho\frac{\partial}{\partial\rho}\right) - j^2 - \frac{1}{4}\right]f^- + (-c\rho + \tfrac{1}{2}A\rho^2 - E)f^- = -i\frac{\hbar^2 j}{2M\rho^2}f^+$$

The vibronic wavefunction then has the form

$$\Psi_j(\mathbf{r},\rho,\theta) = \left[f_j^+(\rho)\phi_+ + f_j^-(\rho)\phi_-\right]\exp(ij\theta) \qquad (3.63)$$

j must be half integral so that $\Psi_j(\mathbf{r},\rho,\theta+2\pi) = \Psi_j(\mathbf{r},\rho,\theta)$, because $\phi_{\pm}(\mathbf{r},\theta+2\pi) = -\phi_{\pm}(\mathbf{r},\theta)$. The solutions Ψ_j are degenerate in pairs.[36] The ground-state pair has $j = \pm\tfrac{1}{2}$, and they transform as the basis states of an E representation. Half the degenerate pairs transform as E, and the others are pairs of A_1 and A_2 states that are degenerate because of the cylindrical symmetry of $\epsilon(\rho)$.

When the electronic-vibrational coupling is strong so that $E_{JT}/\hbar\omega \gg 1$, where $\omega = (A/M)^{\frac{1}{2}}$, the sheets of Figure 3.8 are well separated near $\rho = \rho_0$, and the terms on the right-hand side of Equations 3.62 can be neglected. The lowest states are then approximately,[37] for positive c,

$$\Psi_j = f_j^- (\rho)\phi_- \exp(ij\theta) \tag{3.64}$$

The doubly degenerate ground state has $j = \pm\frac{1}{2}$. These states represent a circulation about the symmetry axis $\rho = 0$ and a vibration along the ρ direction about the minimum $\rho = \rho_0$. The circulation is accompanied by a change in the electronic state, which oscillates back and forth from $|eu\rangle$ to $|ev\rangle$. The electronic and vibrational motions are very closely connected in a way fundamentally different from that represented by a Born–Oppenheimer product for a nondegenerate state.

If the coupling is very large, it may be necessary to consider higher order terms in V_{en} than the linear ones.[38,39] When these are included, ϵ is no longer independent of θ. Three minima separated by saddle points occur at $\theta = 0$, $2\pi/3$, and $4\pi/3$ for a particular sign of the higher order terms, corresponding to distortions along the Z, Y, and X directions of Figure 1.1. If these wells are deep enough, the lowest state is a linear combination of the three states (for positive c),

$$\Psi_k = \phi_- (\mathbf{r}, \theta_k) f(\rho) h_k(\theta - \theta_k), \qquad k = x, y, z \tag{3.65}$$

θ_k is one of the three angles 0, $2\pi/3$, or $4\pi/3$; $h_k(\theta - \theta_k)$ is an oscillator function representing oscillation in the θ direction about the point $\theta = \theta_k$. Another consequence of the lowered symmetry of $\epsilon(\rho, \theta)$ is the splitting apart of the (formerly) degenerate A_1, A_2 pairs. The ground state remains a doublet E, but as the nonlinear coupling increases, the first excited state, a singlet, approaches it in energy. This can be viewed in the following way.[40]

The three states Ψ_x, Ψ_y, and Ψ_z are equivalent and would seem to have the same energy. But when overlap between the vibrational states $\psi_k = fh_k$ is taken into account, they split into a lower doublet E and an upper singlet A_1 or A_2, depending on the signs of the higher order terms in V_{en}, given by

$$\Psi(A) = (3 - 3\gamma)^{-\frac{1}{2}}(\Psi_x + \Psi_y + \Psi_z) \tag{3.66}$$

$$\Psi(Eu) = (2 + \gamma)^{-\frac{1}{2}}(\Psi_x - \Psi_y)$$

$$\Psi(Ev) = (6 + 3\gamma)^{-\frac{1}{2}}(2\Psi_z - \Psi_x - \Psi_y)$$

These are separated by the energy $3E_w\gamma$, where E_w is the depth of a well and γ is the vibrational overlap,

$$\gamma = \langle \psi_k | \psi_{k'} \rangle = \exp(-3E_w/\hbar\omega') \tag{3.67}$$

The second equality holds only for the lowest vibrational state corresponding to the zero-point vibration.[40] The ω' is the circular frequency of the two-dimensional oscillator in the well; a harmonic oscillator is assumed. As the well depth increases, the overlap decreases, and the singlet-doublet separation decreases, leading in the limit to a triply degenerate vibronic state.

When the coupling is strong, the sensitivity of the electronic orbitals to Q_2, Q_3 distortions is, by definition, great. Random strains in the crystal may cause the bottom of the $\theta = 0$ well to be lowest for one impurity, the $\theta = 2\pi/3$ well to be lowest for another, and so on, with the result that at low temperature one may find, from an EPR experiment, for example, an equal number of centers with axial symmetry about each of the $\langle 100 \rangle$ directions. At higher temperatures the vibrational states above the saddle points become thermally populated, and the resonance is isotropic. EPR spectra can be predicted theoretically[39] by including the electron spin in the vibronic states and taking into account spin-orbit and Zeeman interactions. It has been experimentally observed in several systems.[41]

Random strains can also lead to a mixing of the first excited singlet with the ground doublet and to a splitting of the ground doublet. The random nature of this splitting leads to a characteristic line shape[37] of the magnetic resonance of the doublet when the average strain splitting is larger than the Zeeman splitting. Such line shapes have been observed for several cases of weak electronic-vibrational coupling.[42, 43]

It is often possible to use this sensitivity to strain to prove that an observed low site symmetry is due to a strong Jahn–Teller effect rather than to association with another defect when no isotropic resonance can be observed at higher temperatures. An externally applied uniaxial stress along a $\langle 100 \rangle$ direction may raise or depress one well with respect to the others, causing a redistribution of population of centers oriented along the three $\langle 100 \rangle$ directions. If this redistribution occurs at a temperature low enough that an associated defect would not be mobile, the Jahn–Teller effect is shown to be the cause of the low symmetry. Because the g tensor of a center with the electron localized in one well is anisotropic, the resonance signals from centers with localization in different wells occur at different values of the magnetic field for a general orientation of the field, and the relative populations in the different wells are easily monitored by EPR.[44]

Optical transitions involving the Jahn–Teller split doublet can be treated by including the Mexican hat surface of Figure 3.8 in a configuration coordinate diagram, as in Figure 3.9. We neglect the three dimples in the hat. Transitions between the doublet and a singlet at low temperature are shown. The energy of the singlet depends on the coordinates Q_2, Q_3 only

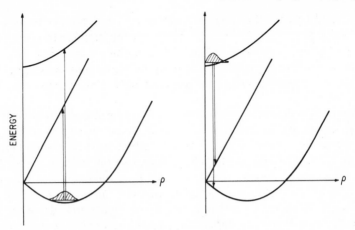

Figure 3.9   Transitions between a singlet and a Jahn–Teller split doublet at low temperature.

through the elastic energy term. Transitions are, in principle, possible from the lower sheet of the doublet to the upper sheet and to the singlet.[35] A two-peaked band should also result from transitions from singlet to doublet, as shown at right in the figure. The displacement of the arrows from $\rho=0$ comes about because of the two-dimensional configuration space. The peak of the probability distribution for the singlet oscillator state, even though the oscillator function is a Gaussian, occurs at a nonzero value of ρ. The probability density is $|\psi(\rho)|^2$, but the volume element is $\rho\,d\rho\,d\theta$, which is zero at $\rho=0$.

The preceding discussion applies equally well to sixfold or eightfold octahedral coordination, as can be seen from Tables 3.3 and 3.4. Only the forms of normal modes, Table 3.1, are different. For twelvefold octahedral coordination there are two E_g modes leading to a four-dimensional problem. A T_1 or T_2, electronic state can interact with both E and T_2 modes (or E_g and T_{2g} modes in O_h symmetry). Coupling only to a single E (or E_g) mode leads in the case of linear coupling to three equivalent minima along the $\langle 100 \rangle$ directions. The lowest vibronic state is triply degenerate. Coupling only to a single T_2 (or T_{2g}) mode leads to four equivalent minima corresponding to distortions along the $\langle 111 \rangle$ directions. The lowest vibronic state is triply degenerate in this case also.[35,45] The problem becomes quite complicated when interaction with more than one mode is considered. Unless the normal modes of these small clusters we have been discussing are local modes, the frequencies associated with them cannot really be discrete, but must be considered smeared out in some fashion. As in the configuration coordinate model, limiting the problem in this way usually leads to predictions in agreement with experiment and is, therefore, justified.

Table 3.3 Symmetries of the Normal Modes of the Cluster for Sixfold, Eightfold, and Twelvefold Octahedral Coordination (O_h Symmetry) and Fourfold Tetrahedral Coordination (T_d Symmetry)

O_h	6	$A_{1g}, E_g, T_{1g}, 2T_{1u}, T_{2g}, T_{2u}$
	8	$A_{1g}, A_{2u}, E_g, E_u, T_{1g}, 2T_{1u}, 2T_{2g}, T_{2u}$
	12	$A_{1g}, A_{2g}, A_{2u}, 2E_g, E_u, 2T_{1g}, 3T_{1u}, 2T_{2g}, 2T_{2u}$
T_d	4	$A_1, E, T_1, 2T_2$

Table 3.4 Symmetric Products of Some Representations of T_d and O_h

T_d	$[E \times E]_s = A_1 + E$
	$[T_1 \times T_1]_s = [T_2 \times T_2]_s = A_1 + E + T_2$
O_h	$[E_g \times E_g]_s = [E_u \times E_u]_s = A_{1g} + E_g$
	$[T_{1g} \times T_{1g}]_s = [T_{1u} \times T_{1u}]_s = [T_{2g} \times T_{2g}]_s$
	$= [T_{2u} \times T_{2u}]_s = A_{1g} + E_g + T_{2g}$

Another manifestation of the Jahn–Teller effect is seen in the reduction of certain matrix elements.[37,46] An operator O, which does not operate on the vibrational coordinates Q, when bracketed with two vibronic functions Ψ_k and $\Psi_{k'}$

$$\langle \Psi_k | O | \Psi_{k'} \rangle = \langle \phi_k | O | \phi_{k'} \rangle \langle \psi_k | \psi_{k'} \rangle \tag{3.68}$$

has its off-diagonal matrix elements reduced by the vibrational overlap $\langle \psi_k | \psi_{k'} \rangle$ compared with its value in the absence of the Jahn–Teller effect. This can lead to reductions in orbital contributions to g factors and splittings due to lower symmetry crystal fields and the spin-orbit interaction. The interaction represented by the operator O is assumed to be weaker than that of V_{en}. Figure 3.10 shows an example of such reduction effects, as observed by Kaufmann et al.[47] The levels are those coming from a $^3T_2$ excited state of Ni^{2+} in ZnO. The substitutional Ni^{2+} is tetrahedrally coordinated by four oxygen atoms, but there is a weak trigonal field in addition to the stronger field of T_d symmetry. The state is split by spin-orbit interaction and the trigonal field to give the levels shown in the figure. They are labeled by Γ_i to indicate the inclusion of spin. The identities of the states are established from the polarization properties of the absorption spectra. The measured energy level positions are compared with the positions calculated assuming no Jahn–Teller effect and the spin-orbit and trigonal field parameters measured for other states not

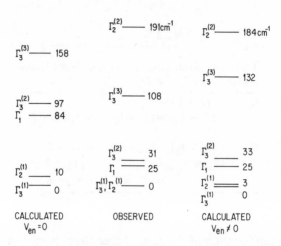

Figure 3.10 Some excited energy levels of Ni^{2+} in ZnO observed and calculated by Kaufman, Koidl, and Schirmer, reference 47. The zero of energy is arbitrarily set.

experiencing a Jahn–Teller effect, with the levels calculated using these same parameters but with a vibrational overlap $\langle \psi_k | \psi_{k'} \rangle = 0.29$ included in the appropriate matrix elements of the trigonal field and the spin-orbit terms.

In general, the Jahn–Teller effect can be expected to occur in an orbitally degenerate state whenever V_{en} is large enough. If the spin-orbit coupling is not small compared with V_{en}, orbital degeneracy remaining after application of the spin-orbit interaction may still be lifted by V_{en} as in the case of the p-like excited state of $\mathscr{F}$ centers in the cesium halides.[48] In the rare earths this residual degeneracy is not lifted because V_{en} is too small; this is expected since the static crystal field splittings are very small. Jahn–Teller effects have been most often observed in d^n transition metal ions. Strong effects have been observed for defects in covalent crystals in which the defect states are bonding or antibonding s and p orbitals.[44] Effects are also seen in the bound hole states of cubic semiconductors in which the hole has orbital degeneracy. They appear as reductions of orbital contributions to g factors and can lead to vibronic splittings of bound exciton J states.[49] These splittings, called crystal field splittings in Chapter 2 could also be due to the static crystal field, $V^{(0)}$. Apparently, V_{en} is derived largely from the short-range interaction $U(\mathbf{r})$ in this case.

BIBLIOGRAPHY

Born, M., and Huang, K., *Dynamical Theory of Crystal Lattices* (Oxford, Oxford, 1956).

Curie, D., *Luminescence in Crystals* (Wiley, New York, 1963).

Englman, R., *The Jahn–Teller Effect in Molecules and Crystals* (Wiley-Interscience, New York, 1972).

Maradudin, A. A., "Theoretical and Experimental Aspects of the Effects of Point Defects and Disorder on the Vibrations of Crystals," in *Solid State Physics*, Vol. 18, F. Seitz and D. Turnbull, Eds. (Academic, New York, 1966) Ch. 4.

Newman, R. C., *Infrared Studies of Crystal Defects* (Taylor and Francis, London, 1973).

Rebane, K. K., *Impurity Spectra of Solids* (Plenum, New York, 1970).

Stevenson, R. W. H., Ed., *Phonons in Perfect Lattices and in Lattices with Point Imperfections* (Plenum, New York, 1966).

REFERENCES

1. M. Born and K. Huang, *Dynamical Theory of Crystal Lattices*, (Oxford, Oxford, 1956).
2. A. A. Maradudin, E. W. Montroll, and G. H. Weiss, *Theory of Lattice Dynamics in the Harmonic Approximation*, (Academic, New York, 1963).
3. A. A. Maradudin in *Solid State Physics*, Vol. 18, F. Seitz and D. Turnbull, Eds., (Academic, New York, 1966) Ch. 4.
4. A. A. Maradudin in *Reports on Progress in Physics*, Vol. 28, A. C. Strickland, Ed. (Institute of Physics, London, 1965) Ch. 10.
5. P. G. Dawber and R. J. Elliott, *Proc. Roy. Soc.* **A273**, 222 (1963).
6. P. G. Dawber and R. J. Elliott, *Proc. Phys. Soc.* **81**, 453 (1963).
7. R. Loudon, *Proc. Phys. Soc.* **84**, 379 (1964).
8. J. F. Angress, A. R. Goodwin, and S. D. Smith, *Proc. Roy. Soc.* **A287**, 64 (1965).
9. W. Hayes and A. R. L. Spray, *J. Phys.* **C2**, 1129 (1969).
10. W. Hayes in *Localized Excitations in Solids*, R. F. Wallis, Ed. (Plenum, New York, 1968) p. 140.
11. W. Hayes and H. F. Macdonald, *Proc. Roy. Soc.* **A297**, 503 (1967).
12. R. J. Elliott, W. Hayes, G. D. Jones, H. F. Macdonald, and C. T. Sennett, *Proc. Roy. Soc.* **A289**, 1 (1965).
13. W. Hayes, *Phys. Rev.* **138A**, 1227 (1965).
14. S. C. Jain, A. V. R. Warrier, and H. K. Sehgal, *J. Phys.* **C5**, 1 (1972).
15. D. M. Larsen, *Phys. Rev.* **187**, 1147 (1969).
16. G. Mahler and U. Schröder, *Phys. Rev. Lett.* **27**, 1358 (1971).
17. M. Born and R. Oppenheimer, *Ann. Phys.* **84**, 457 (1927).

18. U. Kaufmann and O. F. Schirmer, *Optics Commun.* **4**, 234 (1971).

19. D. Curie, *Luminescence in Crystals* (Wiley, New York, 1963).

20. T. H. Keil, *Phys. Rev.* **140A**, 601 (1965).

21. S. Shionoya, T. Koda, K. Era, and H. Fujiwara, *J. Phys. Soc. Japan* **19**, 1157 (1964).

22. C. H. Henry and C. P. Slichter in *Physics of Color Centers*, W. B. Fowler, Ed. (Academic, New York, 1968) Ch. 6.

23. M. H. L. Pryce in *Phonons in Perfect Lattices and in Lattices with Point Imperfections*, R. W. H. Stevenson, Ed. (Plenum, New York, 1966) Ch. 15.

24. D. B. Fitchen in *Physics of Color Centers*, W. B. Fowler, Ed. (Academic, New York, 1968) Ch. 5.

25. K. K. Rebane and L. A. Rebane in *Optical Properties of Ions in Solids*, B. Di Bartolo, Ed. (Plenum, New York, 1975) Ch. 8.

26. J. J. Hopfield, *J. Chem. Phys. Solids* **10**, 110 (1959).

27. E. Gutsche and O. Goede, *J. Luminesc.* **1,2**, 200 (1970).

28. E. W. Williams, *Phys. Rev.* **168**, 922 (1968).

29. M. Wagner and W. E. Bron, *Phys. Rev.* **139A**, 223 (1965).

30. P. J. Dean in *Progress in Solid State Chemistry*, Vol. 8, J. O. McCaldin and G. Somorjai, Eds. (Pergamon, Oxford, 1973) Ch. 1.

31. J. L. Merz, K. Nassau, and J. W. Shiever, *Phys. Rev.* **B8**, 1444 (1973).

32. G. F. Imbusch, W. M. Yen, A. L. Schawlow, G. E. Devlin, and J. P. Remeika, *Phys. Rev.* **136A**, 481 (1964).

33. H. A. Jahn and E. Teller, *Proc. Roy. Soc.* **A161**, 220 (1937).

34. V. Heine, *Group Theory in Quantum Mechanics* (Pergammon, New York, 1960) p. 260.

35. M. D. Sturge in *Solid State Physics*, Vol. 20, F. Seitz, D. Turnbull, and H. Ehrenreich, Eds. (Academic, New York, 1967) Ch. 3.

36. H. C. Longuet-Higgins, U. Öpik, M. H. L. Pryce, and R. A. Sack, *Proc. Roy. Soc.* **A244**, 1 (1958).

37. F. S. Ham, *Phys. Rev.* **166**, 307 (1968).

38. M. S. Child and H. C. Longuet-Higgins, *Phil. Trans. Roy. Soc.* **254**, 259 (1961).

39. M. C. M. O'Brien, *Proc. Roy. Soc.* **A281**, 323 (1964).

40. I. B. Bersuker, *Sov. Phys. JETP* **16**, 933 (1963).

41. W. Low and J. T. Suss, *Phys. Lett.* **7**, 310 (1963) and references therein.

42. L. L. Chase, *Phys. Rev.* **B2**, 2308 (1970).

43. J. R. Herrington, T. L. Estle, and L. A. Boatner, *Phys. Rev.* **B3**, 2933 (1971).

44. G. D. Watkins and J. W. Corbett, *Phys. Rev.* **134A**, 1359 (1964).

45. R. Englman, *The Jahn–Teller Effect in Molecules and Crystals* (Wiley-Interscience, New York, 1972).

46. F. S. Ham, *Phys. Rev.* **138A**, 1727 (1965).

47. U. Kaufmann, P. Koidl, and O. F. Schirmer, *J. Phys.* **C6**, 310 (1973).

48. P. R. Moran, *Phys. Rev.* **137A**, 1016 (1965).

49. T. N. Morgan, *Phys. Rev. Lett.* **24**, 887 (1970).

Chapter 4

Defect Chemistry

One of the simplest defects is an impurity atom in an otherwise perfect host crystal. The simplest host crystal is composed of one kind of atom—Si or Ge, for example. The question of how much of an impurity B can be incorporated in such a crystal A is answered by the phase diagram of the $A - B$ system. Figure 4.1 is a phase diagram of temperature versus composition of an hypothetical $A - B$ system in which B is soluble to some extent $x_B^{(m)}$ or less in A, but A is not soluble in B. For readers not familiar with such diagrams a brief digression to discuss the meaning of Figure 4.1 may be helpful.

The diagram represents the thermodynamical equilibrium between the solid and liquid phases of a two-component system at various temperatures and compositions. The two components A and B are assumed to form no compound $A_p B_q$. Therefore, the diagram is really more typical of metallic systems than nonmetallic crystalline systems, which generally have more complicated phase diagrams.

The liquid is a single phase of randomly moving atoms regardless of composition. There are three solid phases indicated as $A(s)$, $A(s) + B(s)$, and $B(s)$ or $x_B = 1$. $A(s)$ and $B(s)$, in general, would have different crystal structures. $A(s)$ may contain at most the fraction $x_B^{(m)}$ of B atoms without a change of structure. For simplicity, we assume that the corresponding $x_A^{(m)}$ on the right side of the diagram is so small that the $B(s)$ phase can be represented by a single vertical line at $x_B = 1$. In a real system A would be slightly soluble in B as well, because a pure phase is thermodynamically unstable.

Let us suppose that a liquid of composition $x_B^{(2)}$ is cooled. When the temperature T_1 is reached, the solid of phase $A(s)$ of composition $x_B^{(1)}$ freezes and is in equilibrium with the liquid. Further cooling causes the two points indicated to move down the curves as the liquid loses A atoms

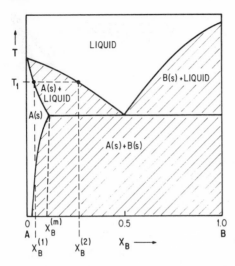

Figure 4.1 Phase diagram of A-B system; x_B is the atomic fractional concentration of B.

to the solid. The upper and lower curves are called liquidus and solidus curves, respectively. The composition of the solid is adjusted by diffusion as cooling proceeds. When the point where the upper curve intersects the horizontal line is reached, the liquid is in equilibrium with the two solid phases $A(s)$ of composition $x_B^{(m)}$ and $B(s)$. When the temperature decreases further, islands of phase $B(s)$ form in $A(s)$; the B content of $A(s)$ declines, and we move downward to the left along the right-hand boundary of the $A(s)$ region.

Often the phase diagram for a system of interest is not known at all or is not known to the accuracy desired. Even when it is known, one would like to know how the impurity is incorporated in the host lattice. Some general rules can be used to predict the mode of incorporation. Often interesting exceptions to these rules occur; they are only a crude guide. In ionic compounds the cations are small and the anions large. An impurity usually substitutes for the host ion of electronegativity nearest its own, even if the sizes of the two ions differ. In KCl, Ca and O would be expected to occupy the cation and anion sites, respectively. In more covalent compounds the electronegativities of the constituents of the host lattice may be nearly equal. In this case a foreign atom might be expected to substitute for the atom nearest it in size. Whether an impurity will occupy an interstitial site is even more difficult to predict. Most interstitial atoms are small, but even large atoms are sometimes found at interstitial sites.

The incorporation mechanism and the atomic structure of a defect can usually be definitely established only by analysis of spectra according to theories like those discussed in the previous chapters. In the unusual circumstance that enough measurements have been made on an imperfect crystal to characterize nearly completely the defect state the defect chemical theory sketched here can be applied to determine the equilibrium concentrations of the various defects as a function of temperature. The position of the Fermi level can also be found.

DEFECT EQUILIBRIA

Because the defect chemical theory is based on the mass-action law, it may be helpful to recall first some features of the chemical formalism. A general chemical reaction between m_1 moles of A_1 and m_2 moles of A_2 with reaction products m_3 moles of A_3 and m_4 moles of A_4 is written

$$m_1 A_1 + m_2 A_2 \rightarrow m_3 A_3 + m_4 A_4 \tag{4.1}$$

An example is the reaction

$$Al_2(SO_4)_3 + 3BaCl_2 \rightarrow 3BaSO_4 + 2AlCl_3 \tag{4.2}$$

in which $m_1 = 1$, $m_2 = m_3 = 3$, $m_4 = 2$. When A_3 and A_4 can also react with each other to give A_1 and A_2, the reaction is said to be reversible, and Equation 4.1 is written with forward and reversed arrows. Another way to write (4.1) is

$$\sum_i m_i A_i = 0 \tag{4.3}$$

In this case the numbers on the left, m_1 and m_2, are taken to be negative, and m_3 and m_4 remain positive. For a reversible reaction, equilibrium is reached when the rates of forward and reverse reactions become equal. At equilibrium

$$\sum_i m_i \mu_i = 0 \tag{4.4}$$

The m_i are defined as in (4.3) and μ_i is the electrochemical potential of species i, defined as the derivative of the Gibbs free energy of the system with respect to the number of particles n_i (atoms, ions, or molecules) of type i at constant pressure, temperature, and number of particles of other types,

$$\mu_i = \left(\frac{\partial G}{\partial n_i} \right)_{P,T,n_j} \quad \text{all} \quad j \neq i \tag{4.5}$$

For an ideal mixture μ_i is related to the molar fractional concentration of A_i, written $[A_i]$, by

$$\mu_i = \mu_i^0 + NkTln[A_i] \qquad (4.6)$$

μ_i^0 is the value of μ_i under standard conditions $P = 1$ atm, $T = 300$ K, and $[A_i] = 1$. If A_i is a gas, the partial pressure p_i replaces $[A_i]$ in Equation 4.6. N is Avogadro's number. Equations 4.6 and 4.4 can be combined to give the mass-action law,

$$\prod_i [A_i]^{m_i} = \exp \frac{-\sum_i m_i \mu_i^0}{NkT} \equiv K \qquad (4.7)$$

$\sum m_i \mu_i^0 / N$ is the change due to the reaction of Gibbs free energy $\Delta G = \Delta H - T\Delta S$ under standard conditions. The equilibrium constant K can be written

$$K = \exp \frac{-\Delta G}{kT} = \exp \frac{\Delta S}{k} \exp \frac{-\Delta H}{kT} \qquad (4.8)$$

ΔH and ΔS are the corresponding changes in enthalpy and entropy.

Kröger and Vink[1] have applied the mass-action formalism to the calculation of defect concentrations in crystals. As an example we consider the binary compound MX. In general, small deviations from perfect stoichiometry may occur for such a compound. The curve enclosing the region of compound existence on a phase diagram of temperature versus composition would not be just a line at $[M] = [X]$, but would enclose a small, nonzero area. Even with no impurities present, several types of native defect are possible. There may be interstitial atoms of either type, written M_i or X_i; vacancies on either type of site, V_M and V_X; and antistructure defects consisting of M atoms on X sites M_X or X atoms on M sites X_M. Antistructure defects should be rare in ionic crystals because of the large repulsive Coulomb interaction. Association of two or more defects is also, in general, possible. Since creation of a native defect requires only a finite amount of energy, some are present in any crystal at nonzero temperature. Native defects represent disorder, and the creation of defects causes the entropy to increase. The number of defects cannot increase indefinitely because of the enthalpy needed for defect formation. The equilibrium concentration is that which minimizes the Gibbs free energy change $\Delta H - T\Delta S$.

For the sake of simplicity we assume that the only disorder in the pure compound consists of intersititial M atoms and M atom vacancies, M_i and V_M (Frenkel disorder on the M lattice). That is, the change in Gibbs free energy necessary to form this type of disorder is assumed to be much less than for other types. Slight deviations from stoichiometry are brought about by interaction of the solid with another phase, in this case taken to be the gas phase consisting of single M atoms $M(g)$ and diatomic molecules $X_2(g)$. The temperature is assumed to be greater than room temperature. The solid-vapor equilibrium reaction is

$$M(g) \leftrightarrow M_i \qquad [M_i] p_M^{-1} = K_1 \qquad (4.9)$$

The mass action equation is written at the right. It is not necessary to include an equation for the interaction of $X_2(g)$ with the solid, because the partial pressures of the two components of the gas are not independent.[1] For the equilibrium between the native defects

$$0 \leftrightarrow V_M + M_2 \qquad [V_M][M_i] = K_2 \qquad (4.10)$$

The equations representing interactions among defects are treated in just the same fashion as equations of ordinary chemical reactions like (4.2). There is one difference. The change in electrochemical potential corresponding to creation of a vacancy or other defect is defined in (4.5) with all other defects unchanged in number. Because this definition in general, implies creation of an imbalance between the number of M and X sites, these potentials are called virtual potentials. In practice the distinction can be ignored because of the way μ is used. The book by Girifalco[2] contains a treatment of defect equilibria based on statistical mechanics rather than thermodynamics. This way of proceeding is more satisfying, but also more involved and requires a more complex notation than the simpler approach, which starts with the mass-action law.

The defects may become ionized. The M_i is assumed able to give up a single electron. The donor level is separated from the conduction band edge by an energy E_{D1}. V_M is considered an acceptor with a level E_A above the valence band edge. The ionization of these defects is represented by

$$M_i \leftrightarrow M_i^{(+)} + e^- \qquad [M_i^{(+)}][e^-][M_i]^{-1} = K_3 \qquad (4.11)$$

$$V_M \leftrightarrow V_M^{(-)} + h^+ \qquad [V_M^{(-)}][h^+][V_M]^{-1} = K_4 \qquad (4.12)$$

The charges in parentheses are effective charges relative to the host lattice. The symbol V_M, for example, represents the defect that results when an atom M is removed from the crystal and has zero effective charge. If the crystal were KCl, V_M would imply removal of a K^+ ion and an electron. The meaning of Equation 4.12 is that an atom near the vacancy releases a hole. An electron-hole pair can be created with an expenditure of energy E_g, the band gap energy.

$$0 \leftrightarrow e^- + h^+ \qquad [e^-][h^+] = K_5 \qquad (4.13)$$

If the crystal contains a total concentration $[I]_t$ of a nonvolatile donor impurity I that substitutes on M sites forming the defect $I_M^{(-)}$ and has a level E_{D2}, there are, in addition, the equations

$$I_M \rightarrow I_M^{(+)} + e^- \qquad [I_M^{(+)}][e^-][I_M]^{-1} = K_6 \qquad (4.14)$$

$$[I_t] = [I_M] + [I_M^{(+)}] \qquad (4.15)$$

Because the crystal is to remain electrically neutral,

$$[e^-] + [V_M^{(-)}] = [h^+] + [M_i^{(+)}] + [I_M^{(+)}] \qquad (4.16)$$

Equations 4.9 to 4.16 are eight equations that could be solved for the eight unknowns $[M_i]$, $[M_i^{(+)}]$, $[e^-]$, $[h^+]$, $[V_M]$, $[V_M^{(-)}]$, $[I_M]$, and $[I_M^{(+)}]$ if p_M, $[I]_t$ and the reaction constants K_i were known.

The electrochemical potential of the electrons is the Fermi energy E_F. The electron and hole concentrations are related to it by

$$[e^-] = N_c \exp\left(\frac{E_F - E_c}{kT}\right) \qquad [h^+] = N_v \exp\left(\frac{E_v - E_F}{kT}\right) \qquad (4.17)$$

E_c and E_v are the energies of the conduction and valence band edges. N_c and N_v can be calculated if the densities of states of the conduction and valence bands are known. From (4.17),

$$K_5 = N_c N_v \exp\frac{-E_g}{kT} \qquad (4.18)$$

It can be shown that K_3 is related to E_{D1} by

$$K_3 = N_v \exp\frac{-E_{D1}}{kT} \qquad (4.19)$$

with similar expressions for K_4 and K_6. We have assumed that enthalpy and internal energy changes are equal. In fact, $\Delta H = \Delta E + P \Delta V$ at constant pressure. In precise calculations the volume change associated with defect formation, or even ionization, could be taken into account. If the band structure and the positions of the donor and acceptor levels are known, these equilibrium constants can be evaluated. The constants K_1 and K_2 are more difficult to calculate. But if expressions could be found for all these constants, the defect concentrations could be obtained and the position of the Fermi level could be found.

It is important to note that the foregoing reactions take place at a high temperature. Thus equilibrium is attained quickly. When the crystal is cooled to room temperature or below, the defect state characteristic of the higher temperature may be frozen and observed at the lower temperature if the cooling rate is rapid compared with the speed of defect reaction processes. Usually the cooling is rapid compared with some, but not all, of these processes. Some care must be used in taking into account temperature changes.

The book by Kröger[3] is a complete treatise on the defect chemical method. The interested reader can find there applications to many different crystal systems and graphical techniques for finding approximate solutions to equations like (4.9) to (4.16). Even in the case described above in which so many simplifying assumptions were made that the example may not represent any real system, much information is required to calculate concentrations. Application of the theory is often hampered by a lack of such information. As more experimental data become available and methods for calculating energies of defect formation are improved, quantitative defect chemical calculations will become increasingly practical. Many useful deductions can be made from the theory without complete quantitative data, however. These are helpful in deciding what preparative methods to use to enhance or suppress various defects in crystals.[3] Often one is only interested in the order of magnitude of some concentration over a limited range of a variable like p_M. If certain defect species are known to be of negligible concentration and approximations can be made, the problem can become tractable.

THERMODYNAMIC POTENTIALS

The change in Gibbs free energy due to the formation of a vacancy in the covalent crystals germanium, silicon, and diamond has been calculated by Swalin in the following way.[4] The enthalpy change ΔH is assumed to consist of two parts: a part due to the breaking of covalent bonds and a part due to the relaxation of the neighbors around the vacancy. The enthalpy change resulting from changing the length of a bond between two atoms from an equilibrium value R_0 to R is assumed to be given by a Morse potential,

$$\Delta H = D \left\{ 1 + \exp\left[-2\alpha(R - R_0) \right] - 2\exp\left[-\alpha(R - R_0) \right] \right\} \quad (4.20)$$

D and α are obtained from data on heat of sublimation and interatomic force constants. The configuration of lowest energy is attained if the four electrons from the four broken bonds combine to form two bonds. The two bonds can be formed in several equivalent ways, and these equivalences contribute to the entropy change ΔS. Another contribution to ΔS comes from the change in vibrational properties caused by the vacancy and is given approximately by $k\Sigma_i \ln(\omega_i^0/\omega_i)$. The ω_i and ω_i^0 are the lattice vibrational frequencies in the presence and absence of the vacancy.

The calculation of energies of formation of defects in ionic crystals is on a firmer foundation than for covalent crystals.[5] The basic long-range interionic interaction—Coulombic and van der Waals—are not in doubt. The long-range nature of the forces means that many ions must be included in a calculation, but this is no impediment with modern numerical techniques. The short-range repulsive interaction between ions is less well known; good results have been obtained with a Born–Mayer term. One especially useful way of treating ion-ion interactions is the shell model, in which an ion is considered to be composed of a rigid core of nucleus plus inner electrons and an outer deformable shell containing the other electrons. The shell and core are assumed to be coupled by harmonic forces. The interaction between two ions is then resolved into shell-shell, shell-core, and core-core interactions. The crystal is usually divided into two regions: an inner region containing the defect and perhaps several hundred surrounding ions, and an outer region. Interactions between all pairs of ions in the inner region are treated explicitly. The outer region is treated in a continuum approximation. The interionic spacings are allowed to vary, and the minimum value of the total energy is found. The difference between this energy and the total energy of the perfect crystal is the quantity of interest. Calculations have been made for several types of defect in alkali halides,[6] CaF_2,[7] thallium halides,[8] and other ionic crystals. Since digital computers of large storage capacity became available, improvements in speed and accuracy of these calculations have been considerable.[6]

ASSOCIATED POINT DEFECTS

If two defects A and B have opposite charges and if the temperature of the crystal is high enough that at least one of them is mobile, an associate AB may be formed. The associate may resemble a small molecule embedded in the crystal, the $A - B$ spacing being of the order of the nearest-neighbor spacing. Formation of an associate can be described by the equation

$$A + B \leftrightarrow AB \tag{4.21}$$

If the simple defects and the associates are randomly distributed, the concentrations are related by the mass-action law,

$$[AB][A]^{-1}[B]^{-1} = \exp\frac{-\Delta G}{kT} \qquad (4.22)$$

The entropy change ΔS in ΔG depends on the number of equivalent configurations possible for the associate and the change in vibrational properties caused by association. The enthalpy change ΔH is largely due to the interaction between A and B. Contributions from polarization and rearrangement of nearby atoms should be included also, however.

Let us assume that only one type of associate is formed. An example of a type would be A and B substitutional on nearest-neighbor sites. If the total fractional concentration of the defect A, $[A]_t = [A] + [AB]$, is equal to the total fractional concentration of defect B, then from Equation 4.22 the fraction paired, $f = [AB]/[A]_t$, has the simple form,

$$f = [A]_t \exp\frac{\Delta S}{k} \exp\frac{-\Delta H}{kT} \qquad (4.23)$$

f increases linearly with increasing $[A]_t$ and decreases exponentially with increasing $\Delta H/kT$. The higher the temperature, the smaller the number of pairs when the equilibrium implied in Equation 4.21 has been reached.

There is no reason to expect in advance only a single type of associate to form. If A substitutes on cation sites and B substitutes on anion sites, in addition to the nearest neighbor pairs other pairs might occur with B substitutional in the second anion shell about A, or the third anion shell, the fourth, and so on. For each type i an equation like (4.22) could be written, each with a different ΔG_i. In the approximation that ΔH_i is given by the Coulomb attraction only,

$$\Delta H_i = -\frac{q^2}{\varepsilon R_i} \qquad (4.24)$$

the distribution of B defects about an A defect as a function of distance R_i from A has been found by several workers.[9] The terms q and $-q$ are the effective charges of A and B. The interaction is screened by the dielectric constant of the material ε. In one derivation the crystal is treated as a continuum with R_i a continuous variable.[10] In another the discrete nature of the lattice is taken into account.[11] The results of both calculations are very similar. The distribution has two peaks, one at the distance of closest approach R_0 and a second near the separation R_r expected for a completely random distribution with no interaction between A and B. The peaks are separated by a minimum at the distance R_m where the interaction energy is equal to the thermal energy $2kT$. If the binding energy is greater than $2kT$, then there is a tendency toward association.

The temperature T is the temperature at which equilibrium is attained. This is generally a high temperature to which the material was heated during crystal growth or in some annealing process rather than the low temperatures at which an experiment may be performed to investigate the degree of pairing. Only those pairs AB separated by distances R in the range $R_0 < R < R_m$ should properly be called isolated pairs. The pairs with separation $R \approx R_r$ are not isolated, because a member A of such a pair will on the average have several partners B at comparable distances $R \approx R_r$.

Often in systems of interest the peak at $R = R_0$ is very narrow, so that only the two or three smallest separations need be considered. Almost the whole contribution to pairing comes from these. The fraction paired at separation R_j is much larger than the fraction paired with the larger separation R_l if

$$\left(\frac{q^2}{\epsilon kT} \right)\left(\frac{1}{R_j} - \frac{1}{R_l} \right) \gg 1 \tag{4.25}$$

This preponderance of close association has often been observed.[12] At low concentrations a large fraction can be paired only at low temperatures. As concentration increases, appreciable pairing can occur at higher temperatures. Of course, pairing at low temperatures may occur so slowly that is is not observable. Interstitial atoms are sometimes mobile at very low temperatures, however.[13] Associates involving interstitials should form more easily. In a rapidly cooled drystal, the distribution of pairs may be more nearly characteristic of the higher temperature used in preparation. The time development of the approach to equilibrium in pairing has been treated by Reiss, Fuller, and Morin.[10] Electrically neutral defects may also associate. They may exchange an electron to become charged and then interact electrostatically, or a reduction in strain energy may result from association, which can also be a factor favoring association of neutral species. More complex associates consisting of more than two simple defects are also often observed, and they are sometimes present in greater numbers than pairs, even at low defect concentrations. It is usually impossible to predict what type of associate will form in a given system. Under favorable conditions the atoms present in an associate may be identified by magnetic resonance through the nuclear hyperfine structure or infrared local mode absorption spectra through isotope effects.

REFERENCES

1. F. A. Kröger and H. J. Vink in *Solid State Physics*, Vol. 3, F. Seitz and D. Turnbull, Eds. (Academic, New York, 1956) Ch. 5.

2. L. A. Girifalco, *Statistical Physics of Materials* (Wiley-Interscience, New York, 1973).

3. F. A. Kröger, *The Chemistry of Imperfect Crystals* (North Holland, Amsterdam, 1964).

4. R. A. Swalin, *J. Phys. Chem. Solids* **18**, 290 (1961).

5. L. W. Barr and A. B. Lidiard in *Physical Chemistry, An Advanced Treatise*, Vol. 10, W. Jost, Ed. (Academic, New York, 1970) Ch. 3.

6. A. B. Lidiard and M. J. Norgett in *Computational Solid State Physics*, F. Herman, N. W. Dalton, and T. R. Koehler, Eds. (Plenum, New York, 1972) p. 363.

7. C. R. A. Catlow, *J. Phys.* **C 6**, L64 (1973).

8. A. K. Shukla, S. Ramdas, and C. N. R. Rao, *J. Phys. Chem. Solids* **34**, 761 (1973).

9. F. E. Williams, *Phys. Stat. Sol.* **25**, 493 (1968).

10. H. Reiss, C. S. Fuller, and F. J. Morin, *Bell System Tech. J.* **35**, 535 (1956).

11. J. S. Prener, *J. Chem. Phys.* **25**, 1294 (1956).

12. R. Watts, *J. Mat. Sci.* **8**, 1201 (1973).

13. G. D. Watkins in *Radiation Damage in Semiconductors* (Academic, New York, 1964) p. 97.

Chapter 5

Experimental Methods

Experimental techniques of many different sorts have been designed to reveal the arrangement of atoms in a defect and the nature of the defect electronic states that are of importance in the phenomena of interest. In this chapter a few of the many methods that have been found to be especially fruitful are treated in a general way. Particular applications are described in other chapters in which defects in various host crystals are discussed. Analytical methods that determine only the concentration of defects are not discussed and are outside the scope of this book. Detailed descriptions of apparatus can be found in the references given in this chapter and in those cited in later chapters where experimental results are discussed.

MAGNETIC RESONANCE AND THE SPIN HAMILTONIAN

Often the states observed in a magnetic resonance experiment are described by means of an effective Hamiltonian called the spin Hamiltonian. The eigenvalues of the spin Hamiltonian are the observed energy levels in the magnetic field. These levels are assumed to be well separated from other levels of the defect. Their splitting in the magnetic field and dependence on field orientation exactly reproduce the behavior of the actual states of the system, which may not be known in detail. The spin Hamiltonian is sometimes a convenient way of cataloging experimental data. The idea of an effective Hamiltonian whose parameters may be measured experimentally but cannot be calculated because of the inadequacy or lack of a basic theory is a very useful one. It has already been used in Equations 2.17, 2.30, 3.19, and 3.53.

120

If n is the number of energy levels observed, the Zeeman splitting of these levels can be represented by a spin Hamiltonian operator $\mathcal{H}_S$ acting on the $2S + 1 = n$ states $|S, M\rangle$ corresponding to the fictitious spin S. The $\mathcal{H}_S$ contains spin operators $S_i^r (i = x, y, z)$ with powers $r \leqslant 2S$. Some of the coefficients of these operators are proportional to the magnetic field to represent the Zeeman effect, and others are independent of field to represent any zero-field splitting. It can be shown that the matrix elements of the spin Hamiltonian $\langle S, M | \mathcal{H}_S | S, M' \rangle$ are proportional, all with the same proportionality constant, to the matrix elements $\langle \psi_i | \mathcal{H} | \psi_j \rangle$ of the exact Hamiltonian $\mathcal{H}$ of the system whose eigenstates are the ψ_i, which have no orbital degeneracy. In some cases orbitally degenerate states may also be represented by such a spin Hamiltonian. In the other cases the concept of the spin Hamiltonian can be extended[1] and an effective Hamiltonian can also be constructed. These latter cases are excluded from the discussion.

The form of the spin Hamiltonian is determined by symmetry and can be found if the site symmetry of the defect is known. In many cases the form is obvious and can be written down immediately. In more complex cases symmetry arguments must be carefully applied. The spin Hamiltonian that represents the Zeeman effect of the $^3A_2$ states of Equation 1.62 is easily seen to be

$$\mathcal{H}_S = g\beta \mathbf{H} \cdot \mathbf{S} \quad \text{with} \quad S = 1 \quad (5.1)$$

$\mathcal{H}_S$ contains the one unknown parameter g. In this case g is known from theory and, from Equation 1.65, is given by

$$g = 2.0023 - 4\zeta_{3d}/\Delta \quad (5.2)$$

Often the states of interest are the two components of a Kramers doublet which are separated from other states by an energy difference much larger than the Zeeman splitting of the doublet. Let us derive the form of the spin Hamiltonian for a Kramers doublet if the point group of the defect is C_{4v}. The doublet can be represented by the two states $|S = \frac{1}{2}, M = \frac{1}{2}\rangle$ and $|S = \frac{1}{2}, M = -\frac{1}{2}\rangle$. The spin Hamiltonian may contain terms in S_i^r, with $r = 0$ or 1, whose coefficients are proportional to the magnetic field or independent of it. The $r = 0$ field-independent terms merely shift the energy zero and can be ignored. The $r = 0$ terms proportional to H_i and the field independent terms with $r = 1$ are zero because these change sign under time reversal, the operation of replacing the time variable t by $-t$ and taking the complex conjugate. The spin-Hamiltonian must be invariant to this operation. This leaves terms in $H_i S_j$. These products must transform as A_1 because $\mathcal{H}_S$ is invariant to the symmetry transformations of the point group. The z axis is taken to be the axis of the S_4 rotatory reflections. Both **H** and **S** are axial vectors. The H_z and S_z are bases for the A_2 representation of C_{4v}, whereas H_x, H_y and S_x, S_y form bases for the E representation. From the multiplication table for this group[2] $A_2 \times A_2 = A_1$, $A_2 \times E = E$, and $E \times E = A_1 + A_2 + B_1 + B_2$. Because A_1 appears once in $A_2 \times A_2$ and once in $E \times E$, there are two independent constants in the spin Hamiltonian and it has the form

$$\mathcal{H}_S = g_\parallel \beta H_z S_z + g_\perp \beta \left(H_x S_x + H_y S_y \right) \tag{5.3}$$

The two independent constants are $g_\parallel$ and $g_\perp$. The spin Hamiltonian of a Kramers doublet in any symmetry with a single n fold symmetry axis with $n > 2$ is given by Equation 5.3. In lower symmetries there are three independent constants g_1, g_2, g_3 which can be considered the three principal values of the symmetric tensor **g**,

$$\mathbf{g} = \begin{bmatrix} g_1 & 0 & 0 \\ 0 & g_2 & 0 \\ 0 & 0 & g_3 \end{bmatrix}$$

and $\mathcal{H}_S$ is given by

$$\mathcal{H}_S = \beta \mathbf{H} \cdot \mathbf{g} \cdot \mathbf{S} \tag{5.4}$$

The magnetic moment of the defect $\beta \mathbf{g} \cdot \mathbf{S}$ is, in general, then, not isotropic, but reflects the symmetry of the defect site. The symmetry can usually not be inferred from the form of the g tensor alone. For all orthorhombic, monoclinic, and triclinic symmetries (Table 5.1) the measured tensor has three unequal principal values, for example.

Table 5.1   Point Symmetries of Defect centers at Sites of O_h or T_d Symmetry in the Perfect Crystal

O_h		
	Tetragonal	D_{4h}, D_4, C_{4v}, D_{2d}, C_{4h}, S_4, C_4
	Trigonal	D_{3d}, D_3, C_{3v}, C_{3i}, C_3
	Orthorhombic	D_{2h}, D_2, C_{2v}
	Monoclinic	C_{2h}, C_2, C_s
	Triclinic	C_i, C_1
T_d		
	Tetragonal	D_{2d}, S_4
	Trigonal	C_{3v}, C_3
	Orthorhombic	C_{2v}, D_2
	Monoclinic	C_2, C_s
	Triclinic	C_1

These examples are rather trivial. The form of Equations 5.3 and 5.4 could be guessed without the aid of the formal arguments outlined above, but for a system with more levels the group theoretical method is usually necessary and is most easily applied with the aid of the irreducible tensor operators.[3] An irreducible tensor operator component T_l^p is an operator that transforms under the symmetry operations of the full rotation group in the same way as the spherical harmonic Y_l^p. The spherical harmonics have usually been used as basis functions rather than operators. In Equation 1.37 both uses are seen, however: Y_L^M is an operator and $|lm\rangle$ is a wave function. The operator function $T_l^p(x)$ of the coordinates x, y, z is just proportional to Y_l^p expressed in Cartesian coordinates, as in Table 1.4. For a spin Hamiltonian with $S > \frac{1}{2}$ the reduction of the irreducible tensor operators $T_l^p(S)$ and $T_1^q(H)$ in the point group of interest must be considered. In substituting spin components S_i for Cartesian coordinates to form $T_l^p(S)$ from $T_l^p(x)$, S_x, S_y, and S_z are substituted for x, y, and z, respectively, except in products. For example, xy must be replaced by $\frac{1}{2}(S_xS_y + S_yS_x)$, because x and y commute but S_x and S_y do not. The $T_l^p(S)$ and the products $T_l^p(S)\cdot T_1^q(H)$ are reduced to the irreducible representations of the group of interest in the same way as $T_1^p(S)\cdot T_1^q(H)$ was reduced for C_{4v}. In general, the $T_l^p(S)$ lead to a zero-field splitting. An example for $S=5/2$ in C_S symmetry is worked out in reference 4.

If there is interaction with a nuclear spin I of the impurity, the symmetry group operations must be considered to act on **I** as well. The spin Hamiltonian becomes a function of **I** as well as **S** and the states $|S,M\rangle$ are replaced by the product states $|S,M\rangle|I,m\rangle$. The resulting "hyperfine" contribution to the spin Hamiltonian will have the form $\mathbf{I}\cdot\mathbf{A}\cdot\mathbf{S}$, where **A** is a tensor like **g**, which in orthorhombic symmetry has three independent principal values. In tetragonal or trigonal symmetry two of these are equal $(g_{\parallel},g_{\perp},g_{\perp})$, and in cubic symmetry all three are equal. The electron may interact also with nearby nuclei of the host lattice. If these nuclei j have nuclear spins $I^{(j)}$, another term of the form $\sum_j \mathbf{S}\cdot\mathbf{K}^{(j)}\cdot\mathbf{I}^{(j)}$ must be added to the spin Hamiltonian. The tensor $\mathbf{K}^{(j)}$ reflects the symmetry of the interaction of the electron with nucleus j. This symmetry may not be the same as the symmetry of the point group of the site on which the electron is largely localized. The hyperfine tensor **A** or $\mathbf{K}^{(j)}$ is often written as the sum of an isotropic or scalar part A_s and a tensor $\mathbf{A}_p$ with zero trace,

$$\mathbf{A} = A_s\mathcal{G} + \mathbf{A}_p \qquad (5.5)$$

$\mathcal{G}$ is the unit tensor whose only nonzero elements are ones on the diagonal. The A_s is due to the Fermi contact interaction and is proportional to the electron density at the nucleus. The dipole-dipole interaction between the electronic and nuclear moments gives rise to a nonzero $\mathbf{A}_p$. The general form of the spin Hamiltonian for a Kramers doublet is then

$$\mathcal{H}_S = \beta \mathbf{H} \cdot \mathbf{g} \cdot \mathbf{S} + \mathbf{I} \cdot \mathbf{A} \cdot \mathbf{S} + \sum_j \mathbf{S} \cdot \mathbf{K}^{(j)} \cdot \mathbf{I}^{(j)} \qquad (5.6)$$

A nuclear Zeeman term $g_N \beta_N \mathbf{I} \cdot \mathbf{S}$ may also be included, and if the nuclei have quadrupole moments, quadrupole terms of the form $\mathbf{I} \cdot \mathbf{Q} \cdot \mathbf{I}$ can be added as well.

In an electron paramagnetic resonance (EPR) experiment on a defect with a Kramers doublet ground state a time varying magnetic field of fixed microwave frequency ν induces transitions between the two energy levels $E_1(H)$ and $E_2(H)$ which are the eigenvalues of $\mathcal{H}_S$ and functions of the static field $\mathbf{H}$. When the static field $\mathbf{H}$ is adjusted to the resonance value $\mathbf{H}_r$ given by

$$h\nu = |E_1(\mathbf{H}_r) - E_2(\mathbf{H}_r)| \qquad (5.7)$$

transitions with $\Delta M = \pm 1$, $\Delta m = 0$ are induced, power is absorbed from the microwave field, and a resonance signal is detected. If the defect point symmetry is cubic, the Zeeman interaction reduces to $g\beta \mathbf{H} \cdot \mathbf{S}$ and the resonance field H_r is independent of the direction of $\mathbf{H}$. In lower symmetries the value of H_r depends on the orientation of $\mathbf{H}$ with respect to the axes of $\mathbf{g}$. Thus symmetry information is given by the angular dependence of the spectrum. Figure 5.1 shows the angular dependence of a spectrum due to a defect with C_{3v} symmetry. The symmetry axis is one of the four $\langle 111 \rangle$ directions, and there are for a general orientation of $\mathbf{H}$ four resonances. The spin Hamiltonian is that of Equation 5.3 with $g_\parallel = 2$, $g_\perp = 6$. The frequency is 9.2 GHz. The magnetic field $\mathbf{H}$ is rotated in a plane parallel to a $\{110\}$ plane of the sample. Two of the four resonances always occur at the same value of H_r for $\mathbf{H}$ in this plane. When $\mathbf{H}$ is parallel to a $\langle 100 \rangle$ direction, all four types of defect are magnetically equivalent and their resonances occur at the same value of H_r.

The hyperfine interaction with a nucleus of nuclear spin I splits the resonance line into $2I+1$ lines. If the hyperfine interaction is small compared with the Zeeman interaction, these lines are evenly spaced and nearly symmetrically disposed about the position of the resonance in the absence of hyperfine splitting (or the resonance due to isotopes with zero nuclear spin.) The hyperfine structure provides a "signature" of the nucleus. Figure 5.2 shows a resonance spectrum of Dy^{3+} in ZnS. The derivative of the absorption is plotted versus magnetic field. The isotopes ^{161}Dy and ^{163}Dy have nuclear spin $I = 5/2$. Each produces a hyperfine pattern of six lines. ^{161}Dy is 18.9% abundant, ^{163}Dy is 25% abundant, and the other isotopes have zero nuclear spin. The ratio of each ^{161}Dy line to

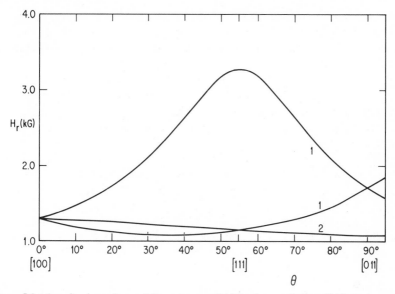

Figure 5.1 Angular dependence of the resonance field for four equivalent $\langle 111 \rangle$ centers with $g_{\parallel} = 2$, $g_{\perp} = 6$. θ is the angle between **H** and the [100] axis. **H** is rotated in the $(01\bar{1})$ plane. The numbers beside the curves represent the degeneracy of the spectra, the curve marked 2 representing twice as many centers as that marked 1.

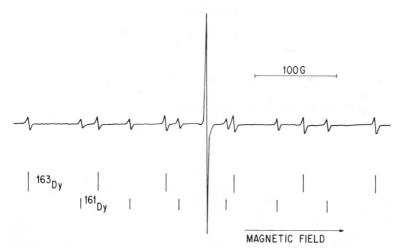

Figure 5.2 EPR spectrum of Dy^{3+} in hexagonal ZnS with the magnetic field along the c axis of the crystal.

the main line due to the 56.1% abundant isotopes with no nuclear spin is
$(1/6)(18.9/56.1) = 0.056$. For the ^{163}Dy lines the ratio is $(1/6)(25/56.1) = 0.074$. In this case the signature is unique because Dy is the only element
with such a distribution of isotopes.

If the hyperfine signature of an element is not unique or if the natural
abundance of an isotope with nonzero nuclear moment is too low for the
hyperfine structure to be apparent, a sample crystal may be prepared with
the element of interest enriched in a particular isotope. The upper part of
Figure 5.3 shows part of the spectrum of an Fe-As pair in ZnSe,[5] described
in Chapter 9. The prominent four-line pattern is due to a hyperfine
interaction with ^{75}As ($I = 3/2$, 100% abundant). Of the iron isotopes only
^{57}Fe ($I = 1/2$) has a nonzero nuclear moment, but the natural abundance is
only 2.2%. The lower trace shows the same part of the spectrum taken with
the ^{57}Fe content enriched to 98%. The ^{57}Fe hyperfine interaction leads to a
doubling of the lines and establishes the presence of one iron nucleus in
the defect center.

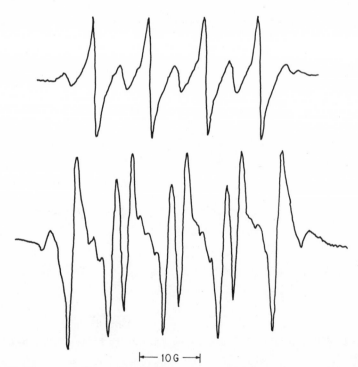

$\vdash$— 10 G —$\dashv$

Figure 5.3 Portion of EPR spectrum of Fe-As pairs in ZnSe (from Watts, reference 5). The
upper trace is from a crystal containing naturally occurring Fe impurities. In the lower trace
Fe is 98% ^{57}Fe.

When hyperfine interactions with more than one nucleus can be observed, more detailed information about the structure of the defect becomes available. The other nuclei may be part of the host lattice or they may be the nuclei of other impurity atoms in a cluster. Often the density of the magnetic electron is small at the positions of these nuclei, and the interactions, sometimes called superhyperfine interactions, with these nuclei are small, leading to small splittings. As an example, let us consider a defect consisting of an impurity at a tetrahedral site coordinated by four nearest neighbors along $\langle 111 \rangle$ directions. We assume that the central atom has no nuclear spin, but the neighbors have nuclear spin $I = 1/2$. The spin Hamiltonian is taken to be

$$\mathcal{H}_S = g\beta \mathbf{H} \cdot \mathbf{S} + \mathbf{S} \cdot \sum_{i=1}^{4} \mathbf{K}^{(i)} \cdot \mathbf{I}^{(i)} \tag{5.8}$$

with spin $S = 1/2$.

Although the symmetry of the center is tetrahedral and the $\mathbf{g}$ tensor is isotropic, $\mathbf{g} = g\mathcal{I}$, each neighbor has trigonal symmetry and $\mathbf{K}^{(i)}$ is given by

$$\mathbf{K} = \begin{bmatrix} K_\perp & 0 & 0 \\ 0 & K_\perp & 0 \\ 0 & 0 & K_\parallel \end{bmatrix}$$

The principal axes are different for each neighbor. The z axis is taken along the $\langle 111 \rangle$ axis joining the neighbor with the central nucleus.

If the field $\mathbf{H}$ is along the $\langle 111 \rangle$ axis, which is the z axis of neighbor $i = 4$, then the other three neighbors are magnetically equivalent. Quantizing the spins along the direction of the magnetic field, we find that the eigenvalues of the spin Hamiltonian are given for small hyperfine interactions approximately by

$$E(M,m) = g\beta HM + K_\parallel m^{(4)} M + M(1/3)\left(K_\parallel^2 + 8K_\perp^2\right)^{\frac{1}{2}} \sum_{i=1}^{3} m^{(i)} \tag{5.9}$$

The transitions $\Delta M = \pm 1$, $\Delta m^{(i)} = 0$ at frequency ν occur at resonance fields H_r which are solutions of

$$h\nu = g\beta H_r + K_\parallel m^{(4)} + \frac{1}{3}\left(K_\parallel^2 + 8K_\perp^2\right)^{\frac{1}{2}} \sum_{i=1}^{3} m^{(i)}$$

$$H_r = \frac{h\nu}{g\beta} + \left(\frac{K_\parallel}{g\beta}\right)m^{(4)} + \frac{\left(K_\parallel^2 + 8K_\perp^2\right)^{\frac{1}{2}}}{3g\beta} \sum_{i=1}^{3} m^{(i)} \tag{5.10}$$

Letting the four $m^{(i)}$ take on their possible values $\pm 1/2$, we see that the spectrum consists of six lines. For convenience we arbitrarily set $K_{\perp} = K_{\parallel}/4$ to display the spectrum schematically in Figure 5.4. There are two groups of three lines with intensity ratios $1:3:3:1$. The factor 3 corresponds to the three permutations with the same energy $m^{(1)} = 1/2$, $m^{(2)} = 1/2$, $m^{(3)} = -1/2$; $m^{(1)} = 1/2$, $m^{(2)} = -1/2$, $m^{(3)} = 1/2$; $m^{(1)} = -1/2$, $m^{(2)} = 1/2$, $m^{(3)} = 1/2$. The splitting of these three lines is $(1/\sqrt{6})(K_{\parallel}/g\beta)$. The centers of the two groups are separated in field by $(K_{\parallel}/g\beta)$. As the magentic field is rotated away from the $\langle 111 \rangle$ direction, the three nuclei cease to be magnetically equivalent, and further splitting occurs. Two identical nuclei which are related by the inversion operation are magnetically equivalent for all orientations of the field. Such is not the case for any two of these four nuclei. In general, then, the hyperfine splittings and their angular dependence show which atoms are present at the defect and yield information about their relative locations.

Usually, the farther from the defect is the nucleus j, the smaller is the magnitude of the corresponding $\mathbf{K}^{(j)}$. Even when this interaction is so small that the splitting it produces is less than the line width and is unresolved, it can sometimes still be studied by the electron-nuclear double resonance or ENDOR method. In this technique the electronic resonance transition $\Delta M = \pm 1$, $\Delta m_j = 0$ is saturated, and the (lower) frequency of a second time varying field is swept through the nuclear resonance $\Delta M = 0$, $\Delta m_j = \pm 1$, changing the electronic level populations and the magnitude of the electronic resonance signal. The value of the nuclear resonance frequency ν_r' and its angular dependence are the data of interest. The atoms of the host crystal, in general, have a distribution of isotopes. If those with nonzero nuclear moments are abundant, the hyperfine interaction of the defect electrons with these moments leads to a broadening of the spectral lines.

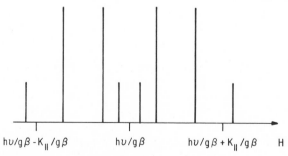

$h\nu/g\beta - K_{\parallel}/g\beta$ $h\nu/g\beta$ $h\nu/g\beta + K_{\parallel}/g\beta$ H

Figure 5.4 Stick figure representing hyperfine structure expected for three equivalent nuclei and one additional nucleus inequivalent magnetically to the others, all with nuclear spins $I = 1/2$. Length of a line is proportional to signal strength.

This broadening may render impossible the observation by EPR of resolved hyperfine structure due to nuclei in the defect center. In such a case the structure might be studied by ENDOR.

The EPR technique is sensitive. The minimum total number of defects in the sample that can be detected is, under favorable circumstances, as low as 10^{11}. For ENDOR the corresponding number is about 10^{15}. Nuclear magnetic resonance (NMR) measurements are much less sensitive and for this reason are not as useful in defect studies; several special schemes have been used to enhance NMR sensitivity. NMR may still be useful when defect concentrations are high. An example is the study of the nuclear resonance of fluorine ions near Yb^{3+} impurities in CaF_2. In this case the Yb concentration was about 0.05% atomic.[6] A discussion of magnetic resonance spectrometers can be found in reference 7. Detailed treatments of the spin Hamiltonian are given in references 8 and 9. Appendix C of reference 9 contains tables that are especially useful in analysing hyperfine spectra arising from more than one nucleus.

OPTICAL ABSORPTION AND LUMINESCENCE

Absorption and luminescence measurements yield the energy levels of the defect and often information about the associated wave functions. Under favorable conditions the defect symmetry can also be obtained from optical experiments in a direct way. At very low temperature only the ground level of the defect is appreciably occupied, and the absorption spectrum gives the energy levels directly. Further information about the states is obtained from the polarization properties of the absorption spectrum. Often, luminescence is excited by absorption of light of wavelength lying in a particular range. This range may correspond to a transition between energy levels of localized states of the defect responsible for the luminescence; a transition between states of another defect, called a sensitizer, which transfers its energy to the luminescent defect;[10] a transition from a localized state to band states of the host crystal; or a transition from valence band states to conduction band states. The luminescence spectrum yields energy level positions also. The minimum detectable defect concentration is usually much lower in luminescence than in absorption measurements, because in luminescence no large background light level is present to reduce the signal-to-noise ratio. A third type of spectrum which is often useful is the luminescence excitation spectrum. This spectrum is a plot of luminescence intensity at a fixed luminescence wavelength versus wavelength of the exciting light. The excitation spectrum, in general, corresponds to a part of the absorption spectrum.

Luminescence may also be excited by electron bombardment of the sample (cathodoluminescence) rather than by light (photoluminescence). Luminescence excited by applying a voltage across the sample is called electroluminescence. The sample may in this case contain a *pn* junction or a surface barrier structure. The most selective excitation is obtained by photoluminescence.

The interaction of an electron of momentum **p** and angular momentum $l + g_0 s$ with a radiation field of vector potential $\mathfrak{A}$ is proportional[11] to $\mathfrak{A} \cdot \mathbf{p}$. When the vector potential is written in field quantized form as a plane wave or sum of plane waves of wavevectors **k**, the dependence of $\mathfrak{A}$ on the coordinate **r** is of the form $\exp(\pm i\mathbf{k} \cdot \mathbf{r})$. Because we are interested in defect states of small extent *r* compared with the wavelength of light λ, these terms can be expanded in powers of the small quantity $\mathbf{k} \cdot \mathbf{r} \sim 2\pi r / \lambda$, and the interaction $\mathfrak{A} \cdot \mathbf{p}$ breaks up into a number of multipolar terms. The most important of these are the electric dipolar interaction, proportional to $\mathbf{e}_E \cdot \Sigma_i e r_i$ and the magnetic dipolar interaction, proportional to $\mathbf{e}_H \cdot \Sigma_i (l_i + g_0 s_i)$, for a system with a number of active electrons *i*. The term **e** is a unit polarization vector of the electric or magnetic field of the light wave. Then the matrix elements of interest for absorption or luminescence transitions are those of the components of the electric and magnetic dipole operators $\mathbf{M}^{(E1)} = \Sigma e r_i$ and or $\mathbf{M}^{(M1)} = \beta \Sigma (l_i + g_0 s_i)$ between the electronic states in question. Electric quadrupole transitions generally are negligibly weak.

In an absorption measurement the measured quantity is the absorption coefficient α, which is recorded as a function of wavelength or frequency ν. For light of polarization **e** incident on the sample in the $+z$ direction the intensity $I(z)$ varies as $dI/dz = -\alpha I$ in the sample. If the intensity is low enough that the upper state ψ_2 is not appreciably populated compared with the lower state ψ_1, the absorption coefficient α and absorption cross section σ are given by

$$\alpha(\nu) = \left(\frac{8\pi^3 \nu}{hc} \right) [\quad] N \left(\frac{g_2}{g_1} \right) |\langle \psi_2 | M_e | \psi_1 \rangle|^2 \hbar(\nu)$$

$$\sigma(\nu) = \frac{\alpha(\nu)}{N} \tag{5.11}$$

The M_e is the component of $\mathbf{M}^{(E1)}$ along $\mathbf{e}_E$ or of $\mathbf{M}^{(M1)}$ along $\mathbf{e}_H$. For the electric dipole transition the square bracket is given by $n^{-1}(E_{loc}/E)^2$; E_{loc}/E is the ratio of the local field at the defect to the average electric field in the crystal. This ratio is not known and is usually set equal to one or to the Lorentz expression $(n^2 + 2)/3$. The *n* is the refractive index of the crystal, and *N* is the density of absorbing centers; g_i is the degeneracy of state i; and $\hbar(\nu)$ is a lineshape function. It is normalized so that $\int \hbar(\nu)\,d\nu = 1$. In a solid the width of the transition, represented by $\hbar(\nu)$, is almost always due to interactions of the electronic system with lattice vibrations. However, for zero-phonon lines at low temperatures $\hbar(\nu)$ may be determined in some cases by the inhomogeneity of the crystal; that is, because of strains or slight disorder the transition frequency depends on the location of the defect in the crystal. This mechanism is called inhomogeneous broadening. For a magnetic dipole transition the bracket $[\ ] = n$. If the light is not polarized, $|\langle M \rangle|^2$ is replaced by the average value $\frac{1}{3}|\langle \mathbf{M} \rangle|^2$ in Equation 5.11.

$$|\langle \mathbf{M} \rangle|^2 = |\langle M_x \rangle|^2 + |\langle M_y \rangle|^2 + |\langle M_z \rangle|^2$$

The absorption oscillator strength is given by

$$f = \frac{8\pi^2 m\nu}{3he^2}\left(\frac{g_2}{g_1}\right)|\langle \mathbf{M} \rangle|^2 \tag{5.12}$$

This dimensionless parameter lies in the range 0.1 to 1 for strongly allowed electric dipole transitions of color centers and in the range 10^{-6} to 10^{-8} for nearly forbidden transitions of rare earth impurities.

The interaction $\mathfrak{A} \cdot \mathbf{p}$ represents both photon absorption and photon emission. The same matrix elements govern emission processes. The probability of spontaneous emission of a photon of any polarization is given by

$$\mathcal{Q} = \frac{64\pi^3 \nu^3}{3hc^3}[\ \ ]n^2|\langle \mathbf{M} \rangle|^2 \tag{5.13}$$

The square bracket has the same meaning as in Equation 5.11. The factor ν^3 alone implies much shorter spontaneous lifetimes $\mathcal{Q}^{-1}$ in the ultraviolet than in the infrared.

The selection rules for optical transitions have been discussed in Chapter 1. The three components of $\mathbf{M}$ may transform according to the same irreducible representation of the impurity point group (cubic symmetry), and in this case the absorption or emission is isotropic. In lower symmetries the components form bases for more than one irreducible representation, leading to polarization of the spectra.

LIFTING OF ELECTRONIC DEGENERACY BY STRESS

For defect centers with high-point symmetry some of the electronic states will generally be degenerate. When the optical spectrum contains sufficiently narrow lines, the degeneracies of the states between which a transition occurs can often be established by studying the splitting of the corresponding spectral line under the application of some external perturbation such as an electric or magnetic field or a uniaxial stress. Let us consider the effects of stress on the optical spectra of centers in cubic crystals. (The effects of applied electric fields or stress on the magnetic resonance spectra are discussed in Chapter 7 of reference 9.) The site symmetry of the defect center is assumed to be the same as the site symmetry of the perfect lattice. When stress is applied, the symmetry is lowered. For a defect with site symmetry O_h the application of stress along $\langle 100 \rangle$, $\langle 111 \rangle$, or $\langle 110 \rangle$ axes lowers the symmetry to D_{4h}, D_{3d}, or D_{2h}, respectively. For T_d defect symmetry the symmetry under these stresses becomes D_{2d}, C_{3v}, and C_{2v}. The applied stress may produce splittings and polarization effects in the spectra, and lines forbidden in the high symmetry may appear in the lower symmetry under stress.

The perturbation $\mathcal{H}$ due to the stress can be considered to be proportional to the stress tensor.

$$\mathcal{H} = \sum_{i,j} V_{ij} \sigma_{ij} \qquad i,j = x,y,z \qquad (5.14)$$

The operators V_{ij} are elements of a second-rank symmetric tensor. The x,y,z axes are $\langle 100 \rangle$ directions. We are interested in the effect of this perturbation on the degenerate states $|\Gamma n\rangle$ belonging to the irreducible representation Γ of O_h or T_d. Some such states are listed in Table 1.4. Two examples are the three T_2 states of T_d (or T_{2u} states of O_h) and the two E states of T_d (or E_g states of O_h).

$$|T_2 x\rangle = x \qquad |Eu\rangle = 3z^2 - r^2$$

$$|T_2 y\rangle = y \qquad |Ev\rangle = \sqrt{3}\,(x^2 - y^2)$$

$$|T_2 z\rangle = z$$

In finding the matrix elements of $\mathcal{H}$ between the T_2 states and between the E states the quantities of interest are $\langle \Gamma n | V_{ij} | \Gamma m \rangle$. Because of symmetry relations between these elements there are never more than three independent elements for any Γ. The splittings are shown in Table 5.2 for stresses T along $\langle 100 \rangle$, $\langle 111 \rangle$, or $\langle 110 \rangle$ directions. The splitting of E is characterized by the single parameter B. The parameter A represents a shift of the center of gravity. The splitting of T_2 is characterized by only two parameters (D and E). The eigenfunctions in the stressed crystal, the components of the split states, are linear combinations of the functions which were degenerate in the absence of stress.

$$|\Gamma a\rangle = \sum_n c_n^{(a)} |\Gamma n\rangle$$

Table 5.2 Splitting of E and T_2 States under Applied Stress

Stress	E Splitting	E Degeneracy	T_2 Splitting	T_2 Degeneracy
$\langle 001 \rangle$	$A - 2B$	1	$C + 2D$	1
$\sigma_{zz} = T$	$A + 2B$	1	$C - D$	2
$\langle 111 \rangle$				
$\sigma_{xx} = \sigma_{yy} = \sigma_{zz} = T/3$ $\quad A$		2	$C + 2E/3$	1
$\sigma_{xy} = \sigma_{yz} = \sigma_{zx} = T/3$			$C - E/3$	2
$\langle 110 \rangle$	$A + B$	1	$C + \frac{1}{2}(D + E)$	1
$\sigma_{xx} = \sigma_{yy} = \sigma_{zz} = T/2$ $\quad A - B$		1	$C + \frac{1}{2}(D - E)$	1
			$C - D$	1

$$A = \tfrac{1}{2}\langle Eu|V_{xx}|Eu\rangle T + \langle Ev|V_{xx}|Ev\rangle T \qquad C = \tfrac{1}{3}\langle T_2 x|(V_{xx} + V_{yy} + V_{zz})|T_2 x\rangle T$$
$$B = (1/\sqrt{3}\,)\langle Eu|V_{xx}|Ev\rangle T \qquad D = \tfrac{1}{3}\langle T_2 x|(V_{xx} - V_{yy})|T_2 x\rangle T$$
$$E = 2\langle T_2 x|V_{xy}|T_2 y\rangle T$$

In a very useful paper Kaplyanskii[12] has tabulated the effects of $\langle 100 \rangle$, $\langle 111 \rangle$, and $\langle 110 \rangle$ stress on the energy of all possible states $|\Gamma n\rangle$ of O_h and T_d symmetry. He has also tabulated the polarization effects that arise under these stresses.

For consideration of polarization of absorption or luminescence spectra it is convenient to define principal axes X, Y, Z with the Z axis parallel to the stress for $\langle 111 \rangle$ or $\langle 100 \rangle$ stress. The polarization properties are determined by the matrix elements of the components of the electric or magnetic dipole operator $\langle \Gamma a|M_\alpha|\Gamma' b\rangle$, where α is X, Y, or Z. These can be expressed in terms of $\langle \Gamma n|M_i|\Gamma' m\rangle$, $i = x, y, z$, which are again expressible in terms of a very few parameters. For transitions $E \leftrightarrow T_2$ all matrix elements $\langle En|M_i|T_2 m\rangle$ are expressible in terms of a single parameter.

The usefulness of Kaplyanskii's results for the experimentalist lies in the small number of free parameters. Splittings and intensities of components of split lines measured under stress applied in the three different directions should stand in simple relationships with each other. These relationships are determined by the symmetry properties of the states and are independent of the particular functions, such as $|T_2 x\rangle = x$, which were chosen to represent the symmetry properties. Accidental degeneracies have been excluded. These are important, for example, for an effective mass excited state of a donor in a semiconductor with several equivalent conduction band minima. The degeneracy due to the several equivalent minima is generally lifted for the ground state, but usually not for the excited p states. The same treatment can be extended to these cases also.[13]

ORIENTATIONAL DEGENERACY

Often the point symmetry of a defect center is lower than the symmetry of the site of the perfect lattice where the defect is located. There are a number of centers that are different only because of their different, but crystallographically equivalent, orientations. The different orientations may be considered "states" of the center. The degeneracy associated with the different orientations is called orientational degeneracy.

The different defect symmetries that could, in principle, arise in a crystal where the symmetry of the site of interest in the perfect lattice has T_d or O_h symmetry are listed in Table 5.1. Tetragonal centers have one fourth-order rotation (C_4) or reflection-rotation axis (S_4). This axis is a $\langle 100 \rangle$ axis of the crystal. Trigonal centers have one three fold rotation axis (C_3), which is a $\langle 111 \rangle$ axis of the crystal. Orthorhombic centers have a twofold rotation axis (C_2), with two other perpendicular twofold rotation axes or two reflection planes (σ_v) containing C_2. All three two fold axes may lie along $\langle 100 \rangle$ directions, or one may lie along a $\langle 100 \rangle$ direction and the other two along $\langle 110 \rangle$ directions. Monoclinic centers have one twofold axis (C_2) or a reflection plane (σ_h). The axis C_2 may lie along a $\langle 100 \rangle$ or $\langle 110 \rangle$ axis. Triclinic centers have no symmetry axes or reflection planes. It is usually not possible to deduce the point group of a defect center from an experiment, but it is often possible to narrow the possibilities to a few point groups.

The orientational degeneracy can be partially or completely lifted, and a splitting of an optical transition line may occur under application of uniaxial stress. For example, there are three types of tetragonal center (or six depending on whether [100] and [$\bar{1}$00] centers can be distinguished), each with axis along one of the three $\langle 100 \rangle$ directions. These remain equivalent under stress applied along a $\langle 111 \rangle$ axis, but are divided into groups of one and two by a $\langle 100 \rangle$ stress. Stress applied along a $\langle 100 \rangle$ or $\langle 111 \rangle$ axis makes the crystal optically uniaxial, and $\langle 110 \rangle$ stress makes the crystal biaxial, so that applied stress can lead to polarized spectra as well. Kaplyanskii[14] has tabulated the splittings and the polarization properties to be expected for tetragonal, trigonal, orthorhombic, monoclinic, and triclinic centers in cubic crystals under stress applied along $\langle 100 \rangle$, $\langle 111 \rangle$, or $\langle 110 \rangle$ axes. In working out the polarization properties each trigonal or tetragonal center was assumed to have an electric or magnetic transition dipole moment parallel or perpendicular to the axis of the center. For the lower symmetries three inequivalent oscillators are necessary. For certain trigonal or tetragonal centers some states may have electronic degeneracy, but this is assumed not to be lifted by the stress. Runciman[15] has considered the lifting of both electronic and orientational degeneracy by stress in these cases.

Even in the absence of applied perturbing fields the luminescence may be partially polarized under polarized excitation. Let us consider a simple example.[16,17] The defect has a symmetry axis called z', and z' has a number of equivalent directions. If the defect consisted of two associated atoms on nearest neighbor sites in the diamond or zincblende lattices, these equivalent directions would be the $\langle 111 \rangle$ axes; in the rock salt lattice for the same type of defect they would be the $\langle 001 \rangle$ axes. It is assumed that an equal number of defects have axes in each of the several equivalent directions. Let the defect have an absorption band in which the intensity of the absorbed light is proportional to $\cos^2\varphi$ and an emission band at longer wavelength with luminescence intensity also proportional to $\cos^2\varphi$. The φ is the angle between the z' axis and the polarization vector **E** of the exciting light in the case of absorption or the emitted light in the case of luminescence. This dependence on φ would result from an electric dipole transition between atomic s and $p_{z'}$ states, for example. If the exciting light is unpolarized, the observed luminescence is unpolarized, but if the exciting light is polarized, the luminescence is also polarized. To see this we consider the case in which the exciting light is incident in the $[\bar{1}00]$ direction with the polarization vector in the (100) plane making an angle θ with the Z or [001] axis, as shown in Figure 5.5. The intensity of the luminescence emitted along the [100] direction with polarization parallel or perpendicular to **E** is called $I_{\parallel}(\theta)$ and $I_{\perp}(\theta)$, respectively. The angular dependencies of these quantities are found by summing the contributions from the several equivalent types of defect. For defects that have axes z' parallel to the four $\langle 111 \rangle$ directions the degree of polarization of the

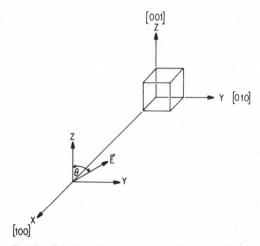

Figure 5.5 Exciting light is incident in the $-X$ direction with electric vector **E**.

luminescence

$$P(\theta) = \frac{I_{\parallel}(\theta) - I_{\perp}(\theta)}{I_{\parallel}(\theta) + I_{\perp}(\theta)} \qquad (5.15)$$

is found to be $P(\theta) = \sin^2(2\theta)$. If the z' directions are the three $\langle 100 \rangle$ directions, $P(\theta) = \cos^2(2\theta)$. From the angular dependence observed the defect axis can be inferred. This method is useful only if the symmetry of the host crystal is high.

When the spectral lines are narrow enough, the Zeeman effect can be used to determine the degeneracy of the energy levels. The symmetry of the defect can be determined from the angular dependence of the Zeeman effect.[18] This method is somewhat analogous to the symmetry determination by magnetic resonance illustrated in Figure 5.1. When the lines are broad, some information may still be obtained by measurement of the magnetic circular dichroism. In this technique the Zeeman splitting of the lower state leads to a difference in absorption of right- and left-hand circularly polarized light. The temperature must be low enough that the populations of the Zeeman levels are sufficiently different.

If the ground state of the defect is paramagnetic, the orientational degeneracy is readily apparent from the EPR spectra. Each differently oriented type of center has a separate spectrum, and from the angular dependence of the spectra the orientations of the centers are found. The orientational degeneracy of the defect with resonance spectra shown in Figure 5.1 is four. There are four different types of trigonal center corresponding to the four $\langle 111 \rangle$ axes.

A defect with orientational degeneracy may have a nonzero static electric dipole moment. Two examples are the $(OH)^-$ molecule ion and the Li^+ ion in KCl. The $(OH)^-$ molecule ion substitutes for a halogen and the dipole is oriented along a $\langle 100 \rangle$ axis. The Li^+ ion occupies a cation site but is displaced from the substitutional position along a $\langle 111 \rangle$ axis. The energy of such a dipole of dipole moment $\mathbf{M}^{(E1)} = q\mathbf{R}$, where $\pm q$ are the charges separated by $\mathbf{R}$, in an applied (local) electric field $\mathbf{E}$ is $\mathbf{M}^{(E1)} \cdot \mathbf{E}$. The orientational degeneracy can be lifted by the electric field in much the same fashion as a magnetic field $\mathbf{H}$ splits apart Zeeman levels of a magnetic dipole $\boldsymbol{\mu} = \beta \mathbf{g} \cdot \mathbf{S}$ according to the interaction $\boldsymbol{\mu} \cdot \mathbf{H}$. A zero-field splitting can also occur in the electric dipole case, as shown schematically in Figure 5.6. The figure shows potential energy curves of a defect with two stable positions A and B (orientational degeneracy two) in zero applied field. Because of the slight overlap of the zero-point vibrational wavefunctions in the two wells the defect can be considered to oscillate or tunnel between positions A and B with frequency δ/h or to have the two eigenstates shown which are separated in energy by δ, the zero-field splitting. An applied field lowers the energy of one well and raises the other, leading to further splitting. If there is no overlap, reorientation may still occur by a thermally activated process.[19] In paraelectric resonance,[19,20,21] the analogue of EPR, transitions between the levels are induced by an oscillating electric field tuned to resonance. In paraelectric resonance the spectra often consist of broad lines and are for this reason difficult to analyse, however. Generally, a defect with lower symmetry produces some distortion of the surrounding lattice. The distortion has been called an "elastic dipole," in analogy with electric and magnetic dipoles.[22] The energy of the elastic dipole is linearly related to an applied stress field σ by an equation like (5.14). The elastic dipole moment V is a tensor rather than a vector. Reorientation of electric dipoles or elastic dipoles is sometimes studied by dielectric loss or acoustic loss measurements.

Another experimental technique used to study reorientation of defects with electric dipole moments is the method of ionic thermocurrents.[23] The dipoles are initially polarized in a strong static electric field at a temperature at which reorientation is possible. Then the crystal is cooled to a temperature at which this orientation of dipoles is frozen in. Tunneling is assumed not to occur. Finally, the crystal is heated with no applied field. The thermally activated reorientation is detected by measuring the depolarization current.

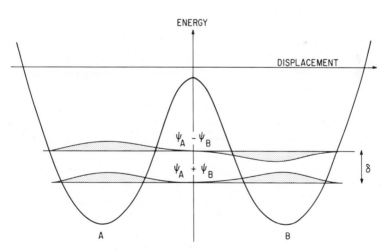

Figure 5.6 Potential energy of a defect with two equivalent orientations, A and B. ψ_A and ψ_B are the states which would be eigenstates if each well were isolated.

RAMAN SCATTERING

In Raman spectroscopy a part of an incident monochromatic light beam of frequency ν_1 linearly polarized in the i ($i = x, y, z$) direction is inelastically scattered by the sample. The Raman scattered photons are in general polarized in a different direction j and have a different frequency $\nu_2 = \nu_1 - \nu_{kl}$. The ν_{kl} is the transition frequency of the system of interest that jumps from the state k with energy E_k to the state l with energy E_l; $h\nu_{kl} = E_k - E_l$. If E_k is larger than E_l, the scattered photon is shifted to a lower frequency relative to the incident photon. If E_l is larger than E_k, ν_2 is larger than ν_1. If the system has only two energy levels, then the Raman spectrum would consist of a line at $\nu_2^+ = \nu_1 + |\nu_{kl}|$ and a line at $\nu_2^- = \nu_1 - |\nu_{kl}|$ in addition to the strong line at ν_1 because of the elastically or Rayleigh scattered light. The ratio of the intensities of the lines at ν_2^+ and ν_2^- depends on the sample temperature through the populations of the states k and l. The fraction of the incident light scattered at an angle θ with respect to the direction of the incident beam, $I_{kl}^{ij}(\theta)$, is

$$I_{kl}^{ij}(\theta) \propto N\nu_2^4 |\langle k|\alpha_{ij}|l\rangle|^2 \sin^2\theta \qquad (5.16)$$

In the usual experimental configuration θ is $90°$; N is the density of the scattering defects, and $\boldsymbol{\alpha}$ is a scattering tensor that has the symmetry of the point group of the defect. The defect symmetry may be inferred from the polarization properties of the scattered light.

The levels E_k and E_l may be electronic[24] or vibrational[15] energies. In the case in which E_k and E_l are the energies of electronic states of the same parity the form of α_{ij} is given in Equation 1.45. The Raman scattering method is complementary to absorption and luminescence because the selection rules are different. Thus transitions that are forbidden and cannot be observed by one method may be allowed and observable by the other.

Because the intensity of the Raman scattered light is many orders of magnitude less than that of the elastically scattered light, great care is necessary to exclude the unwanted light at frequency ν_1 from the detection system, expecially when the Raman shift $\nu_2 - \nu_1$ is small. A double monochromator, in which the scattered light is refracted by the first grating, passes through a slit, and onto a second grating, is commonly used for this filtering. The light emerging from the exit slit of the monochromator is typically detected by a cooled photomultiplier with a very low dark current and a photon counting circuit. The incident beam at frequency ν_1, usually from a laser source, is tightly focused in the sample. The frequency ν_1 should be as large as practical, because the signal is proportional to ν_2^4. The frequency ν_1 should not lie within the excitation band of a luminescence which might interfere with observation of the scattered light. However, if

ν_1 is set near an absorption band, a resonant enhancement of the scattering may occur, as can be seen from Equation 1.45. A tunable dye laser is often a convenient source because of its flexibility. Raman scattering from defects with concentrations as low as 10^{16} cm^{-3} has been observed.

PARTICLE SCATTERING

There is a group of several related experimental techniques[26,27] by which it is often possible to determine the location of impurity atoms in a lattice. No direct information about the electronic state of the impurity is obtained, but ideally the fractional impurity concentrations on the various possible sites may be determined. These techniques are especially applicable to thin crystalline films and are often used to study the effects on the film of ion implantation.

When a collimated beam of mono-energetic charged particles, for example, He$^+$ ions accelerated to a kinetic energy E_0 of 1 MeV or more, is incident on a crystal, the particles are scattered by the atoms of the crystal and atoms present in the crystal as impurities. Some strike atoms on the surface and are scattered backward away from the crystal or forward to collide with other atoms. Others penetrate a bit farther before the first collision, but all forward-scattered particles collide with many atoms in turn, losing kinetic energy in the process. The back-scattered particles may be detected and their energy spectrum analyzed. Those scattered by the atoms of the host crystal and by light impurities have a continuous distribution of energies with some maximum cutoff energy E_M. For simplicity a monatomic lattice is assumed. Particles scattered backward by the surface atoms have energy E_M, whereas particles that have been slowed by forward scattering are scattered backward with energies less than E_M. Particles detected after scattering by impurity atoms localized near the surface and heavier than the atoms of the host crystal suffer a smaller energy loss in the elastic collision and lead to a resolved peak at some energy E_I, where $E_M < E_I < E_0$.

Many perfect lattices have "channels" parallel to special crystallographic directions, along which it is much easier for an incident particle to avoid scattering by the host atoms. Figure 5.7 shows these channels along the $\langle 110 \rangle$ direction of the diamond lattice. Unless an incident particle approaches within a distance of the order of 0.1 Å to 0.2 Å of a lattice atom, it may continue down the channel, penetrating to a depth perhaps ten times greater than for the case in which it is incident at an angle a few degrees away from the channel direction. The number of back-scattered particles detected will be much less. If the crystal contains a heavy substitutional impurity, the number of particles back scattered by the

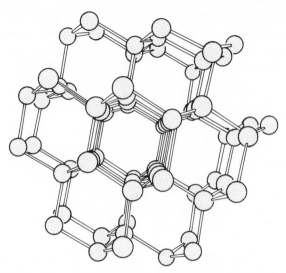

Figure 5.7   View along a $\langle 110 \rangle$ axis of the diamond lattice showing the channels.

impurity will, like the number of particles scattered by the host atoms, dip sharply to a minimum as the crystal is rotated and the angle corresponding to alignment of the incoming beam with the channels is traversed. If, on the other hand, the heavy impurity is at the interstitial site in the channel, the number of particles scattered by these impurities will be unchanged as the orientation is varied through the channeling angle. If, as in the case of the diamond lattice, there is more than one type of interstitial site, it is often possible by observing the presence or absence of such channel effects along several different channel directions, $\langle 110 \rangle$, $\langle 111 \rangle$, $\langle 100 \rangle$, to determine from the relative scattering intensities what fractions of the impurity are located at the various types of site. An interstitial site of a particular type may lie in the $\langle 110 \rangle$ channel, but not in the $\langle 111 \rangle$ channel, for example.

Other schemes for observing such channeling effects are also possible. If the incoming particle reacts with the impurity, a product of the nuclear reaction may be detected rather than the back-scattered particles of the beam. Similarly, if X-ray emission is stimulated by the collision, the X-rays may be detected. If some of the impurity atoms are radioactive isotopes, the emerging reaction products also undergo channeling, and analysis of the channeling effects can yield the location of the impurity. No incident beam is needed in this case. In these schemes the restriction of applicability of the method to heavy impurities is removed.

These methods are not very sensitive. Impurity concentrations of 0.01% atomic or more are required in back-scattering measurements. The technique of the nuclear reaction induced by the incident beam is usually less sensitive. In the case of no incoming beam, sensitivity may be somewhat greater. Most experiments have been performed on silicon because of the importance of ion implanted silicon for integrated circuits. Almost all other experiments have been on germanium.

BIBLIOGRAPHY

Di Bartolo, B., *Optical Interactions in Solids* (Wiley, New York, 1968).

Fowler, W. B., Ed., *Physics of Color Centers* (Academic, New York, 1968).

Pake, G. E., and Estle, T. L., *The Physical Principles of Electron Paramagnetic Resonance* (Benjamin, Reading, Mass., 1973).

Poole, C. P., *Electron Spin Resonance* (Wiley-Interscience, New York, 1967).

REFERENCES

1. S. Sugano, Y. Tanabe, and H. Kamimura, *Multiplets of Transition Metal Ions in Crystals* (Academic, New York, 1970) p. 192.
2. G. F. Koster, J. O. Dimmock, R. G. Wheeler, and H. Statz, *Properties of the Thirty-two Point Groups* (MIT Press, Cambridge, Mass., 1963).
3. B. R. Judd, *Operator Techniques in Atomic Spectroscopy* (McGraw-Hill, New York, 1963) p. 41.
4. W. C. Holton, M. de Wit, T. L. Estle, B. Dischler, and J. Schneider, *Phys. Rev.* **169**, 359 (1968).
5. R. K. Watts, *Phys. Rev.* **B2**, 1239 (1970).
6. J. P. Wolfe and R. S. Markiewicz, *Phys. Rev. Lett.* **30**, 1105 (1973).
7. C. P. Poole, *Electron Spin Resonance* (Wiley-Interscience, New York, 1967).
8. A. Abragam and B. Bleaney, *Electron Paramagnetic Resonance of Transition Ions* (Clarendon, Oxford, 1970).
9. G. E. Pake and T. L. Estle, *The Physical Principles of Electron Paramagnetic Resonance* (Benjamin, Reading, Mass., 1973).
10. R. K. Watts in *Optical Properties of Ions in Solids*, B. Di Bartolo, Ed. (Plenum, New York, 1975) ch. 10.
11. B. Di Bartolo, *Optical Interactions in Solids* (Wiley, New York, 1968) ch. 14.
12. A. A. Kaplyanskii, *Opt. Spectr.* **16**, 557 (1964).
13. C. S. Chen, J. C. Corelli, and G. D. Watkins, *Phys. Rev.* **B5**, 510 (1972).
14. A. A. Kaplyanskii, *Opt. Spectr.* **16**, 329 (1964).
15. W. A. Runciman, *Proc. Phys. Soc.* **86**, 629 (1965).
16. P. P. Feofilov, *J. Phys. Rad.* **17**, 657 (1966).

17. R. J. Elliott, L. G. Matthews, and E. W. J. Mitchell, *Phil. Mag.* **3**, 360 (1958).
18. A. F. Leung, *J. Phys.* **C6**, 2234 (1973).
19. F. Lüty, *J. Phys.* **C9**, 49 (1973).
20. H. B. Shore, *Phys. Rev.* **151**, 570 (1966).
21. T. L. Estle, *Phys. Rev.* **176**, 1056 (1968).
22. A. S. Nowick and B. S. Berry, *Anelastic Relaxation in Crystalline Solids* (Academic, New York, 1972).
23. M. S. Li, M. de Souza, and F. Lüty, *Phys. Rev.* **B7**, 4677 (1973).
24. G. B. Wright and A. Mooradian, *Phys. Rev. Lett.* **18**, 608 (1967).
25. D. T. Hon, W. L. Faust, W. G. Spitzer, and P. F. Williams, *Phys. Rev. Lett.* **25**, 1184 (1970).
26. G. L. Dearnaly in *Reports on Progress in Physics*, Vol. 32, A. C. Strickland, Ed. (Institute of Physics, London, 1969) p. 405.
27. J. W. Mayer, L. Eriksson, and J. A. Davies, *Ion Implantation in Semiconductors* (Academic, New York, 1970) Ch. 4.

Chapter 6

Electrons in Covalent and Ionic Crystals

In Chapters 1 and 2 several models for the states of electrons associated with point defects are briefly discussed. Because the electronic problem is quite complex, involving interactions among many particles, it is always of interest to choose the simplest theoretical model that will adequately describe the defect. Then the mathematical problem becomes tractable, and a physical picture of the defect is more easily formed. Choice of a model depends in large part on the type of host crystal—whether it is a covalent crystal like Si, an ionic crystal like KCl, or a compound with bonding somewhere between these extremes. In the following chapters defects in a variety of crystals are discussed. This short chapter gives a brief discussion of some qualitative aspects of the electron distribution in crystals. It is more useful to classify crystals in a rough way on the basis of the nature of the forces that hold them together than as semiconductors or insulators, for example. Whether a crystal is a semiconductor or an insulator does not depend on intrinsic crystal properties, but on the temperature and on the kinds of defects present. In the last ten years much progress has been made toward relating band structure and intrinsic optical properties to chemical trends in a group of related materials.

Let us consider the series of crystals Ge, GaAs, ZnSe, KBr formed from the elements of the same row of the periodic table. The electronic configurations of the neutral atoms are given below.

K	$\{[Ar]\}(4s)$	As	$\{[Ar](3d)^{10}\}(4s)^2(4p)^3$
Zn	$\{[Ar](3d)^{10}\}(4s)^2$	Se	$\{[Ar](3d)^{10}\}(4s)^2(4p)^4$
Ga	$\{[Ar](3d)^{10}\}(4s)^2(4p)$	Br	$\{[Ar](3d)^{10}\}(4s)^2(4p)^5$
Ge	$\{[Ar](3d)^{10}\}(4s)^2(4p)^2$		

The curly brackets indicate the tightly bound core electrons; the other electrons are the valence electrons. The [Ar] represents the closed shells of the argon configuration, $(1s)^2(2s)^2(2p)^6(3s)^2(3p)^6$.

Ge crystallizes in the diamond structure shown in Figure 7.1 in which each Ge atom is tetrahedrally coordinated by four others. There are two atoms per unit cell, and each has four valence electrons outside the core. Only the valence electrons are used in constructing the energy bands. The lowest four energy bands are the valence bands. The uppermost of these is separated from the lowest conduction band by the forbidden gap. At the lowest temperatures the valence bands are filled with electrons and the conduction bands are empty. The charge density of the electrons in these valence bands is concentrated between the Ge nuclei. (The core electron density is, of course, concentrated at the nuclei.) Qualitatively this is the sort of distribution expected if the four valence electrons are placed in sp^3 hybrid orbitals pointing along the four $\langle 111 \rangle$ directions. The orbital of the central Ge of Figure 1.1 pointing along the [111] or $0-2$ direction is $(1/2)s + (\sqrt{3}/2)(p_x + p_y + p_z)$, and the other three differ only in the signs before the p orbitals. It is advantageous to combine s and p orbitals to form hybrid sp^3 orbitals if the s and p valence electrons have nearly the same energy in the crystal. Generally, some contribution from d orbitals may be included also. The orbitals of nearest-neighbor atoms overlap in the region between the atoms. When the bonding linear combination of these atomic orbitals is filled with electrons, the resulting valence charge density is concentrated in the region between the nuclei, giving the tetrahedral structure its stability. This bonding charge density between the atoms, which results from the sharing of valence electrons, is the essence of covalency.

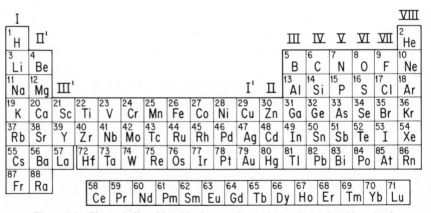

Figure 6.1　The periodic table with elements beyond atomic number 88 omitted.

If enough electrons are excited from the valence bands to fill completely the lowest conduction band, the charge density due to the electrons in this band is, on the other hand, much more evenly spread throughout the crystal, with only a slight concentration near the nuclei. There is no concentration of charge in the region between the nuclei; this distribution is similar to that of a smeared out molecular antibonding orbital.

The case of GaAs is somewhat similar to that of Ge. It has the zincblende structure that is derived from the diamond structure by placing Ga and As alternately on the diamond lattice sites. (Figure 8.1) The two atoms per unit cell, Ga and As, also have eight valence electrons—three from Ga and five from As. The band structures are very similar, although GaAs has a direct gap and Ge an indirect gap (Figures 7.5 and 8.2). If the valence bands are filled with the eight electrons added two at a time, the first two, which fill the lowest band, are much more strongly attracted by the As^{5+} core than by the Ga^{3+} core, and the resulting charge distribution is centered there, giving the As atom an effective charge of $3+$. The contribution to the valence charge distribution coming from the electrons that fill the second, third, and fourth valence bands is largely localized between the nuclei, as in Ge, with a slight shift of the center of the distribution toward the As nucleus. The total valence charge distribution is then located between the As and Ga nuclei, the maximum of the distribution being nearer the As nucleus than the Ga nucleus. The bond is partly covalent and partly ionic.

ZnSe is formed from atoms of the same row but from columns II and VI. The structure is zincblende, and the bonding is more ionic than in GaAs. The larger charge of the Se^{6+} core strongly attracts the valence electrons. The total valence charge density forms a distorted sphere around the Se core with a bulge of maximum density, corresponding to some contribution from sp^3 bonds, along the Zn–Se axis. These valence band electron densities are shown in Figure 6.2, as calculated by Walter and Cohen[1,2] by the pseudopotential method. This method is based on the approximation that the electrons in the solid are much like free electrons and so can be represented as plane waves. Actually, the potentials seen by the electrons are too strong for the approximation to be valid, but the crystal potential can be replaced by a weak pseudopotential that leads to the same band structure.[3] Matrix elements of the potential are obtained by fitting the calculated band structure to experimental optical data. The potential, or pseudopotential, can be split into two parts V_a and V_s. The V_s part is symmetric about the point midway between two nearest neighbors, and V_a is antisymmetric about this point. For a homopolar crystal like Ge or Si, $V_a = 0$, because this is a point of inversion symmetry. The pseudopotential for Ge is nearly equal to V_s for GaAs and ZnSe.

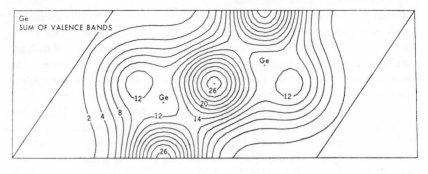

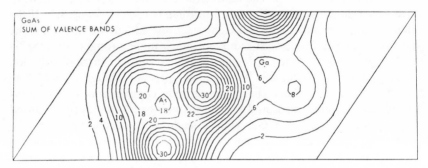

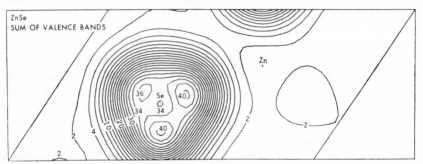

Figure 6.2 Total valence electron charge density for Ge, GaAs, and ZnSe. Contours of constant charge density in a (110) plane are shown (from Cohen, reference 2, copyright 1973 by the American Association for the Advancement of Science).

We have been discussing trends that are evident when we move horizontally along a row of the periodic table. What happens when we move vertically? Generally, as we move down we come to crystals that are semimetals or metals. Tin can exist in two forms: gray Sn, which is a semimetal (no gap between conduction band minimum and valence band maximum) with the diamond structure, and white Sn, a metal with a distorted face centered cubic structure. Lead is a metal with face centered cubic structure; HgSe and HgTe are semimetals. The tendency toward metallic behavior is connected with the diminishing tendency to form covalent sp^3 bonds as the principal quantum number n of the valence electrons increases. The difference in energy of ns and np orbitals increases and hybridization is less favored. For $n = 6$ the $(6s)^2(6p)^2$ configuration has lower energy than the directed hybrid bond orbitals. Pearson[4] has found that for binary compounds MX the metallic character increases with increasing average principal quantum number $\bar{n} = (n_M + n_X)/2$.

Some II-VI compounds that normally have the zincblende structure make a transition under the application of increasing hydrostatic pressure to the NaCl structure, shown in Figure 10.1. The structures differ only in the positions of the cations. KBr has the NaCl structure. The eight valence electrons of the K and Br atoms are very strongly attracted to the Br^{7+} core. They concentrate at the Br nucleus, and an ionic bond results. The crystal can be considered to consist of spheres of small K^+ and large Br^- ions. The structure of lowest total energy for ions of these relative sizes is the NaCl structure in which each ion is sixfold coordinated by nearest neighbors of opposite charge.

If the more covalent compounds we have discussed are centered about column IV of the periodic table in the sense that covalency increases as column IV is approached in the sequence II-VI, III-V, IV, then binary ionic compounds are analogously centered about the noble gas column VIII, as shown in Figure 6.3. The analogy is poor in the sense that noble gas atoms have no charge, and the crystals are held together by polarization forces; but the farther we go from column VIII, the worse the picture of the crystal as a collection of spherical charge distributions becomes.

A tight binding approximation is the natural starting point for describing the band structure of alkali halides, although other methods have also

4 Be	5 B	6 C	7 N	8 O
12 Mg	13 Al	14 Si	15 P	16 S
30 Zn	31 Ga	32 Ge	33 As	34 Se
48 Cd	49 In	50 Sn	51 Sb	52 Te
80 Hg	81 Tl	82 Pb	83 Bi	84 Po

	1 H	2 He	3 Li	
8 O	9 F	10 Ne	11 Na	12 Mg
16 S	17 Cl	18 Ar	19 K	20 Ca
34 Se	35 Br	36 Kr	37 Rb	38 Sr
52 Te	53 I	54 Xe	55 Cs	56 Ba
84 Po	85 At	86 Rn	87 Fr	88 Ra

Figure 6.3 Portions of the periodic table rearranged for covalent compounds at left and ionic compounds at right.

been used. Pantelides[5] has compared the valence bands of a number of ionic compounds with rock salt structure, using as tight binding basis orbitals only three p orbitals on each anion. The main interactions are between orbitals on neighboring anions. The smaller cations act as spacers. This situation constrasts with that for tetrahedrally bonded covalent compounds, where, when a tight binding basis set consisting of sp^3 hybrid orbitals is chosen, the main interactions are between nearest neighbors (cation-anion).[6]

Much progress has been made toward understanding the structures of binary compounds through the work of Phillips and Van Vechten.[7-11] In their phenomenological theory a small number of parameters deduced from optical spectroscopic data on crystals is used to derive results from which several properties of a large class of materials can be accurately predicted.

Phillips and Van Vechten define an average band gap $\overline{E}_g$ in terms of the static dielectric constant ε_s, and the plasma frequency ν_p, the characteristic frequency of oscillation of the valence electrons,

$$\varepsilon_s = 1 + \left(\frac{h\nu_p}{\overline{E}_g} \right)^2$$

$$\nu_p^2 = \frac{Ne^2}{\pi m} \tag{6.1}$$

N is the number of valence electrons per unit volume; $\overline{E}_g$ corresponds to the energy difference between bonding and antibonding states. It is larger than the energy gap E_g between the lowest conduction band and the highest valence band because it represents an average over the bands. The E_g is important for electrical properties of the material, and $\overline{E}_g$ is the important parameter for structural properties. The average energy gap can be separated into a covalent or homopolar contribution E_h and an ionic part C.

$$\overline{E}_g^2 = E_h^2 + C^2 \tag{6.2}$$

The fraction of ionic character of the bond or ionicity f_i is defined by

$$f_i = \frac{C^2}{\overline{E}_g^2} \tag{6.3}$$

For a homopolar crystal such as Si there is no ionic contribution to the binding, and $C = 0$. The term E_h arises from the symmetric part of the potential V_s and C, from V_a. For many $M^n X^{8-n}$ crystals the homopolar energy scales with a power of the bond length or nearest neighbor separation R_{MX} so that E_h can be found from

$$E_h(MX) = E_h(\text{Si})(R_{\text{Si}}/R_{MX})^{2.5} \tag{6.4}$$

The ionic energy parameter is proportional to the electronegativity difference $\mathfrak{X}_M - \mathfrak{X}_X$ of the two constituents. The electronegativity of M is defined to be proportional to the Coulomb interaction between an electron at R_M, the covalent radius, and the ion core charge $|e|Z_M$ screened by the valence electrons. The expression for C is

$$C(MX) = 1.5\left[\frac{e^2 Z_M}{R_M} - \frac{e^2 Z_X}{R_X}\right]\exp\frac{-k_s(R_M + R_X)}{2}$$

$$R_M + R_X = R_{MX} \tag{6.5}$$

The k_s is a Thomas–Fermi screening wave number.

The ionicity defined in Equation 6.3 has proven to be very useful. It is tabulated, along with other data on the compounds examined in Chapters 8, 9, and 10 in Tables 8.1, 9.1, and 10.1. The ionicity $f_i = 0.785$ is a critical value. The $M^n X^{8-n}$ compounds with f_i less than this value have fourfold coordinated structures, and those with f_i greater than 0.785 have the sixfold coordinated rock salt structure. The cohesive energies and heats of formation also show strong dependences on f_i. The force constant for bond bending (but not bond stretching, which depends on central forces) scales with the fraction of covalent character $1 - f_i$.

REFERENCES

1. J. P. Walter and M. L. Cohen, *Phys. Rev.* **B4**, 1877 (1971).
2. M. L. Cohen, *Science* **179**, 1189 (1973).
3. M. L. Cohen and V. Heine in *Solid State Physics*, Vol. 24, H. Ehrenreich, F. Seitz, and D. Turnbull, Eds. (Academic, New York, 1970) Ch. 2.
4. W. B. Pearson in *Treatise on Solid State Chemistry*, Vol. 1, N. B. Hannay, Ed. (Plenum, New York, 1973) Ch. 3.
5. S. T. Pantelides, *Phys. Rev.* **B11**, 5082 (1975).
6. S. T. Pantelides and W. A. Harrison, *Phys. Rev.* **B11**, 3006 (1975).
7. J. C. Phillips and J. A. Van Vechten, *Phys. Rev.* **B2**, 2147 (1970).
8. J. A. Van Vechten and J. C. Phillips, *Phys. Rev.* **B2**, 2160 (1970).
9. J. C. Phillips, *Rev. Mod. Phys.* **42**, 317 (1970).
10. J. C. Phillips, *Bonds and Bands in Semiconductors*, (Academic, New York, 1973).
11. J. C. Phillips, *Covalent Bonding in Crystals, Molecules and Polymers*, (University of Chicago, Chicago, 1969).

Chapter 7

Diamond, Silicon, and Germanium

Diamond, silicon, and germanium crystals have the diamond structure shown in Figure 7.1. The unit cell contains two atoms. Each atom has four nearest neighbors along the $\langle 111 \rangle$ directions, twelve next nearest neighbors along the $\langle 110 \rangle$ axes, twelve third nearest neighbors along $\langle 311 \rangle$ axes, and so on. The spacings of these neighbors are listed in Table 7.1. The tetrahedral interstitial site has four nearest neighbors along $\langle 111 \rangle$ directions, six next nearest neighbors along $\langle 100 \rangle$ directions, and twelve third nearest neighbors along $\langle 311 \rangle$ directions. The distance to the four nearest neighbors is the same as for the substitional site. The lattice can be viewed as two interpenetrating face-centered-cubic lattices displaced from each other by a shift along the body diagonal of the cube. There is a center of inversion symmetry midway between two nearest neighbor atoms, as can be seen from the figure. The presence of the inversion center between the two identical atoms of a unit cell prevents optical absorption because of the excitation of single phonons in the perfect crystal. Two-phonon absorption is allowed, however. A complicating feature of the space group of the diamond lattice is the presence of a screw axis. The infinite crystal is left invariant by a screwlike translational-rotational motion along this axis. This translational motion is not, by itself, a member of the space group. The point group, the group of all the rotation elements of the space group, is O_h.

The Brillouin zone of the diamond lattice is shown in Figure 7.2 with the points of high symmetry labeled in the conventional manner. The electronic configurations of the free carbon, silicon, and germanium atoms are $[He](2s)^2(2p)^2$, $[Ne](3s)^2(3p)^2$, and $[Ar](4s)^2(4p)^2$; [He], [Ne], and [Ar] stand for the closed-shell configurations of the helium, neon, and argon atoms.

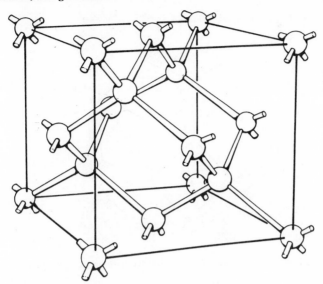

Figure 7.1   Diamond crystal structure.

The four valence electrons, eight per unit cell, in the outer s and p orbitals form the strong covalent bonds on which the crystal structure is based. The band structures are shown in Figures 7.3, 7.4, and 7.5. These have been calculated[1,2] neglecting the spin-orbit interaction. The gap between the highest valence band and the lowest conduction band is small compared with the total width of the valence band. The group of the wave vector at $k = 0$, the point Γ, is the cubic group O_h. The character table for O_h is given in Table 1.5. Because electron spin has been neglected, the energy bands at the point Γ are labeled with the irreducible representations of the single group, rather than the double group. The lowest band, Γ_{1g}, is $s$like, and the upper valence band, Γ_{5g}, is $p$like. At points other than the zone center the wavefunctions are mixtures of s and p orbitals in the language of tight

Table 7.1   Some Properties of Diamond, Silicon, and Germanium

	First, second, and third nearest neighbor spacings (Å)			Band gap (eV)	Melting point (C)	Dielectric constant
C (diamond)	1.54,	2.52,	2.96	5.48	4030	5.5
Si	2.35,	3.84,	4.51	1.17	1420	11.4
Ge	2.45,	4.00,	4.70	0.744	959	15.4

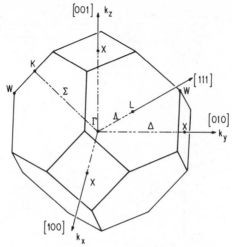

Figure 7.2 Brillouin zone of a crystal with diamond structure.

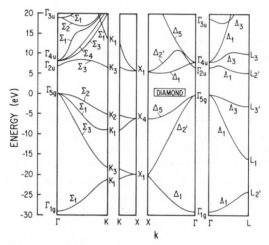

Figure 7.3 Band structure of diamond (from Hemstreet, Fong, and Cohen, reference 1).

binding. The valence band labels have the subscript g, indicating that the wavefunction is even under inversion in the point midway between nearest neighbor atoms. This is the parity we expect for bonding-type orbitals with the electronic charge concentrated between the atoms. The eight available electrons can fill these valence bands. The valence band maximum occurs at Γ for all three crystals.

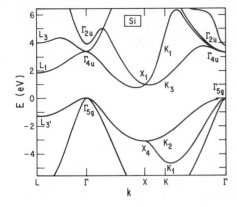

Figure 7.4 Band structure of silicon (from Cohen and Bergstresser, reference 2).

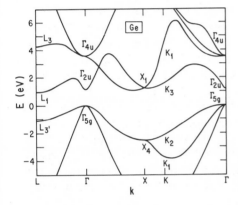

Figure 7.5 Band structure of germanium (from Cohen and Bergstresser, reference 2).

The conduction bands are of antibonding character. The band gaps are said to be indirect because the valence band maximum and conduction band minimum do not occur at the same point in **k** space. For diamond and silicon there are six equivalent conduction band minima along the $\langle 100 \rangle$ directions, and for germanium there are four equivalent minima along the $\langle 111 \rangle$ directions.

Some properties of the three crystals are displayed in Table 7.1. Diamond has the smallest interatomic spacing, the smallest dielectric constant, largest band gap, and highest melting point. The band gap listed is the minimum separation of valence and conduction bands at 0 K—the energy difference between valence band maximum and conduction band minimum. The dielectric constant varies with band gap E_g approximately as[3]

$$\varepsilon = 1 + \left(\frac{\hbar \omega_p}{E_g} \right)^2, \qquad \omega_p^2 = \frac{4\pi N e^2}{m} \tag{7.1}$$

N is the number of electrons per unit volume, and ω_p is the plasma frequency. Because ε appears in the denominator of the expression for the energies of shallow donor and acceptor states (Equation 2.7), we may expect these states to lie deeper or farther from the band edge in crystals with larger band gaps.

SILICON

Much progress has been made toward understanding the behavior of defects in silicon, largely because of the availabilty of good sample crystals. All electronic devices made from silicon depend on impurity-controlled electrical properties. This commercially important material is available as large single crystals of very high quality and purity. Crystals are generally grown by two different methods. In the Czochralski method a single crystal seed contacts the surface of the molten silicon contained in a quartz crucible. The seed is slowly withdrawn, pulling the growing crystal from the melt. Single crystals of mass as large as 10 kg are commonly produced in this way. The crucible is a source of oxygen contamination, and oxygen content may be as high as 10^{18} cm^{-3} in such crystals. No crucible is necessary in the floating-zone method. A narrow disc-shaped molten zone, held in place by surface tension, is produced in a vertically mounted silicon rod by induction heating with a radiofrequency coil. The molten zone is swept slowly along the rod, from a seed crystal at one end, leaving behind a single crystal. The growth chamber is evacuated or filled with an inert gas. Oxygen content may be 10^{16} cm^{-3} or less in crystals grown in this way.

Silicon crystals with resistivity at room temperature greater than 10^4 ohm cm have been produced. This corresponds to a concentration of shallow donors or acceptors of the order of 10^{12} cm^{-3}. The concentration of carbon, an electrically inactive contaminant, may be 10^{17} cm^{-3} or greater. Crystals completely free from dislocations, which are linear lattice defects, are routinely grown. Dislocations may be observed at the point at which they intersect a crystal surface by use of an etch. Etching causes the material at the dislocation to dissolve more rapidly than the surrounding crystal, producing small pits at the dislocations.

The lowest conduction band of silicon has six equivalent minima along the $\langle 100 \rangle$ axes in **k** space. The minimum lies at about $0.85k_{ZB}$, where k_{ZB} is the position of the zone boundary. For the minimum or valley at $\mathbf{k}_j$ the contribution to the impurity wavefunction has the form $F_j(\mathbf{r})\phi_j(\mathbf{r})$, where $F_j(\mathbf{r})$ is a slowly varying envelope function and $\phi_j(\mathbf{r})$ is a Bloch function for an electron in the jth minimum,

$$\phi_j(\mathbf{r}) = u_j(\mathbf{r})\exp(i\mathbf{k}_j\cdot\mathbf{r}) \qquad (7.2)$$

The complete wavefunction of the impurity electron is a linear combination of the contributions from the six valleys,

$$\psi(\mathbf{r}) = \sum_{j=1}^{6} \alpha_j F_j(\mathbf{r})\phi_j(\mathbf{r}) \tag{7.3}$$

The envelope functions and the energy levels E are found by solving the effective mass equations[4]

$$\alpha_j \left[T_j + V(\mathbf{r}) - E \right] F_j(\mathbf{r}) + \sum_{t \neq j} \alpha_t Q_t(\mathbf{r}) F_t(\mathbf{r}) = 0$$

$$Q_t(\mathbf{r}) = \left[T_t + V(\mathbf{r}) - E \right] \exp\left[i(\mathbf{k}_t - \mathbf{k}_j)\cdot\mathbf{r} \right] \tag{7.4}$$

Here T_j stands for the kinetic energy operator for an electron in valley j; $V(\mathbf{r})$ is the potential energy operator; and $Q_t(\mathbf{r})$ is a term that has the effect of mixing the envelope functions corresponding to the six different valleys. The $V(\mathbf{r})$ has the symmetry of the site occupied by the donor; for a simple substitutional donor this is T_d.

For the higher excited states with diffuse wavefunctions, which have smaller amplitude at the impurity site, it is a good approximation to neglect the intervalley mixing by setting Q_t equal to zero. The energies of these states are determined by the asymptotic behavior of the potential at large r, given by $-e^2/\varepsilon r$. For these states Equation 7.4 then simplifies to

$$\left(T_j - \frac{e^2}{\varepsilon r} - E \right) F_j(\mathbf{r}) = 0 \tag{7.5}$$

A separate equation is obtained for each valley, and they are all equivalent. For the [001] valley Equation 7.5 is

$$\left[-\left(\frac{\hbar^2}{2m_\perp^*} \right)\left(\frac{\partial^2}{\partial x^2} + \frac{\partial^2}{\partial y^2} \right) - \left(\frac{\hbar^2}{2m_\parallel^*} \right)\frac{\partial^2}{\partial z^2} - \frac{e^2}{\varepsilon r} - E \right] F(\mathbf{r}) = 0 \tag{7.6}$$

$$m_\parallel^* = 0.1905m, \qquad m_\perp^* = 0.9163m$$

Equation 7.6 is very similar to that solved in Chapter 2 for the shallow donor in CdS. We may also expect the same sort of solution in this case. Because of the axial symmetry of the equation, the solutions can be classified as eigenstates of angular momentum l^2 and l_z quantized along the z axis with quantum numbers l and m. The p states are split into two components corresponding to $m=0$ and $m=\pm 1$. For each eigenvalue the complete wavefunction of Equation 7.3 can be formed by taking the linear combination in several different ways. These ψs form bases for a reducible representation of T_d, the point group of the impurity site. For s states or $p,0$ states this representation reduces to A_1, E, and T_2 and for $p, \pm 1$ states it reduces to two T_1 and two T_2 irreducible representations. These could be split into three levels and four levels, respectively. But we have included in Equation 7.6 nothing to cause such a splitting, so that in our present approximation the degeneracy remains. The energy levels will be just the eigenvalues of Equation 7.6.

Equation 7.6 has been solved by Faulkner[5] by a variational method. Good agreement is found for the upper states, but not for the ground $1s$ levels. This is shown in Figure 7.6, where the calculated energy levels are compared with the observed ones for several donors. For the ground states there are two kinds of discrepancy. The splitting of the $1s$ states into A_1, E, and T_2 sublevels is observed experimentally, and the level positions vary

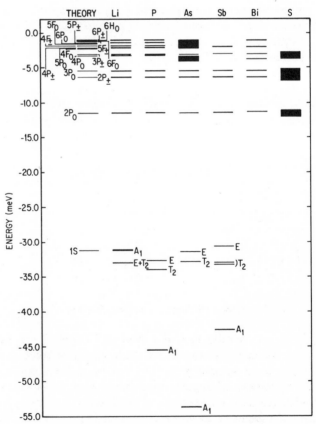

Figure 7.6 Calculated and observed donor levels in silicon (from Faulkner, reference 5).

from donor to donor. Table 7.2 lists the ionization energies of several donors. When the electron spin is taken into account the A_1 and E orbital states become a Γ_6 doublet and a Γ_8 quartet, respectively. The T_2 state leads to a representation reducible to $\Gamma_7 + \Gamma_8$, and therefore a splitting could result. This splitting has never been observed in silicon except for a very small effect in the ground state of the lithium donor.

Table 7.2 Ionization Energies of Some
Donors in Silicon (in meV)

P	45.31	Bi	70.47
As	53.51	Li	32.81
Sb	42.51		

Good agreement with experiment is obtained for the ground state energies by the method of Pantelides and Sah.[4] For phosphorus and sulfur donors they use the general form, Equation 7.4, which includes intervalley mixing. The potential energy $V(\mathbf{r})$ is taken to be

$$V(\mathbf{r}) = -\frac{Ze^2}{\varepsilon r} + U(r) \qquad (7.7)$$

$U(r)$ is a correction term, which is small in the case of phosphorus or sulfur, but is not small for other donors; Z is one for phosphorus and two for sulfur. Phosphorus and sulfur come from the same row of the periodic table as silicon; therefore, the atoms have the same electronic core plus one and two extra electrons, respectively. The Bloch function $\phi_j(\mathbf{r})$ of Equation 7.3 has predominantly $3s$, $3p$, and $3d$ character at the lattice sites in order that it be orthogonal to the $1s$, $2s$, and $2p$ silicon core states. Because phosphorus and sulfur have the same core states, Equation 7.3 with these $\phi_j(\mathbf{r})$ is a good representation of the wavefunction of the extra bound electron. Even though the sulfur level is deep, good agreement with experiment is found, as shown in Table 7.3. The valley was assumed to be spherical, as an approximation, in the calculation. Sulfur is a double donor and may bind one electron (S^+) or two electrons (S^0).

For other donors with electron cores different from that of silicon the Bloch functions $\phi_j(\mathbf{r})$ do not represent well the rapid variation of the donor electron wavefunction at small r. In these cases ground-state energies have also been calculated from first principles by a modification of the effective mass theory wherein ψ is transformed in such a way that orthogonality with the donor core states is obtained.[6] Some results are shown in Table 7.4.

Table 7.3 Experimentally Observed Lower Levels of
Phosphorus and Sulfur Donors in Silicon Compared with
Calculated Values of Pantelides and Sah (units are meV)

Donor	State	Calculated[a]	Experimental
Phosphorus	A_1	−42.4	−45.5
	E	−30.5	−32.6
	T_2	−31.3	−33.9
Sulfur (S⁺)	A_1	−659.3	−613.6
	E		
	T_2		
Sulfur (S⁰)	A_1	−297.1	−302.2
	E	−118.2	
	T_2	−129.2	−188.3

[a] Reference 4.

The effective mass theory of shallow acceptor states is somewhat more complicated because of the degeneracy of the valence band edge at the maximum point Γ. Because of the degeneracy, Bloch states of more than one band must be included in the impurity wavefunction. The upper valence band at $k=0$ is the orbital triplet Γ_{5g}, as shown in Figures 7.4 and 7.5. When the spin-orbit interaction is taken into account, this band splits into an upper quartet Γ_{8g} and a lower doublet Γ_{7g}. The separation in energy of the Γ_{8g} and Γ_{7g} bands at $k=0$ is called Δ_{so}. If the splitting Δ_{so} is negligibly small, we must include contributions from all these valence band

Table 7.4 Experimental and Calculated Donor Levels in
Silicon (units are meV)

Donor	State	Calculated[a]	Experimental
Arsenic	A_1	−53.1	−53.7
	E	−29.6	−31.2
	T_2	−29.8	−32.6
Antimony	A_1	−31.1	−42.7
	E	−28.5	−30.5
	T_2	−27.8	−32.9
Nitrogen (N⁻)	A_1	−52.5	−45.0
	E	−31.0	
	T_2	−32.6	

[a] Reference 6.

states on an equal footing in constructing the impurity wavefunction. If Δ_{so} is very large, on the hand, contributions from Γ_{7g} could be neglected, just as contributions from the higher lying conduction bands are neglected for the donor impurities. The effective mass equations are

$$\sum_{j=1}^{6} \sum_{\alpha,\beta} D_{ij}^{\alpha\beta} \frac{\partial}{\partial x_\alpha} \frac{\partial}{\partial x_\beta} F_j(\mathbf{r}) - \left(\Delta_{so}\delta_j - \frac{e^2}{\varepsilon r} + E \right) F_i(\mathbf{r}) = 0. \qquad (7.8)$$

The index j runs over the six valence band states Γ_{8g} and Γ_{7g}, and x_α and x_β take on the three values x, y, and z. The δ_j is zero for the four Γ_{8g} states and one for the two Γ_{7g} states. The elements $D_{ij}^{\alpha\beta}$ are parameters of the band structure.[7] The potential energy has been approximated by $-e^2/\varepsilon r$. The acceptor wavefunction is a linear combination of the $F_j\phi_j$, where ϕ_j is again a Bloch function.

For silicon the splitting Δ_{so} is comparable with the acceptor ionization energies, and all six bands must be included in the description of each acceptor state. For germanium Δ_{so} is much larger than the acceptor ionization energies. In this case it is a good approximation to consider only the Γ_{8g} band in calculating the acceptor states associated with this band and only the Γ_{7g} band in calculating the acceptor states associated with it. The two "ladders" of acceptor states associated with the different bands are shown schematically in Figure 7.7. They are inverted compared with the donor levels. For the sake of simplicity we discuss the upper ladder for germanium only. The states in silicon are qualitatively similar.

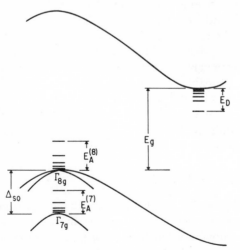

Figure 7.7 Schematic representation of donor and acceptor levels on a band diagram for a crystal with an indirect band gap.

The envelope functions F_j are again nearly hydrogenic functions and can be labeled $1s$, $2s$, $2p$, and so on. The group of the point Γ in the Brillouin zone is O_h. The s functions transform as Γ_{1g} of O_h; the p functions, as Γ_{5u}; and the d functions, as $\Gamma_{3g} + \Gamma_{5g}$. The complete wavefunction $F\phi$ transforms according to the product representation $\Gamma_F \times \Gamma_\phi$; these are for s, p, and d envelope functions.

$$\Gamma_{1g} \times \Gamma_{8g} = \Gamma_{8g} \rightarrow \Gamma_8$$

$$\Gamma_{5u} \times \Gamma_{8g} = \Gamma_{6u} + \Gamma_{7u} + 2\Gamma_{8u} \rightarrow \Gamma_7 + \Gamma_6 + 2\Gamma_8$$

$$(\Gamma_{3g} + \Gamma_{5g}) \times \Gamma_{8g} = 2\Gamma_{6g} + 2\Gamma_{7g} + 3\Gamma_{8g} \rightarrow 2\Gamma_6 + 2\Gamma_7 + 3\Gamma_8$$

The irreducible representations to the right of the arrows are those of T_d, the true symmetry of the substitutional acceptor site and the symmetry of the actual potential seen by the hole. The ground state is a quartet (Γ_8), and the excited states are doublets and quartets. It is conventional to label the acceptor states by the double group labels in the Γ notation and to label the donor states according to the orbital degeneracy with the chemical notation (A_1, E, T_1, T_2). We follow this convention so that the reader will have less difficulty in turning to other literature. The states associated with the Γ_{7g} band are found in the same way. The complete wavefunction of the states with s envelope functions transforms as Γ_7. Those with p envelopes transform as $\Gamma_6 + \Gamma_8$.

The acceptor energy levels and wavefunctions have been calculated by Schechter[8] for silicon and germanium, by Mendelson and Schultz[9] for silicon, and by Mendelson and James[10] for germanium. Agreement between theory and the measured positions of the excited states is fairly good for germanium and poor for silicon. In both cases calculated ionization energies are in worse agreement with experiment than calculated positions of excited levels, probably for the same reasons as in the case of donor levels. Baldereschi and Lipari have treated the problem of acceptor states in a different way, emphasizing more strongly the similarity with a free atom.[11] Their method is described in Chapter 2. Both notations for the acceptor states are shown in Figure 7.18.

GROUP-V DONORS

The atoms from group V of the periodic table, nitrogen, phosphorus, arsenic, antimony, and bismuth, have one more valence electron than silicon and act as donors when they form isolated substitutional defects. Nitrogen has been studied very little. Electrical measurements of Zorin, Pavlov, and Tetelbaum[12] on ion implanted nitrogen indicate a donor level at 45 meV. Annealing of the crystal causes a decrease in the concentration of this defect. Maximum concentration was $8 \times 10^{16} cm^{-3}$. Pantelides and Sah[6] have assigned this level to N^-, substitutional nitrogen with two loosely bound electrons, on the basis of a match with the calculated level position (Table 7.4). The nature of the observed donor center is actually not known. Ion implantation produces much damage and can lead to more complex defect centers.

Phosphorus, arsenic, antimony, and bismuth form simple substitutional donors. These have been studied thoroughly and their properties are well understood. The typical hydrogenlike optical absorption spectrum[13] of phosphorus is shown in Figure 7.8. At the lowest temperatures only transitions from the $1sA_1$ ground state are seen, as in Figure 7.8. As the sample temperature is raised, transitions from the higher lying $1sE$ and $1sT_2$ states are also observed. The nature of the states has been established by Zeeman and stress-induced splitting of the spectral lines.[14] Because of the anisotropy of the conduction band valleys, the Zeeman effect is anisotropic, as described in Equations 2.20.

Uniaxial stress applied to the crystal produces changes in the relative energies of the six conduction band minima and leads to partial removal of energy level degeneracies. For example, under a stress parallel to a $\langle 100 \rangle$ axis the $1sA_1$ state is shifted in energy, the $1sT_2$ state splits into an orbital doublet and a singlet, and the $1sE$ state splits into two singlets. Similarly, the $p,0$ and $p,\pm 1$ levels are each split into two levels. The electric dipole selection rules are different, depending whether the polarization vector of the light is parallel or perpendicular to the stress direction. They may be

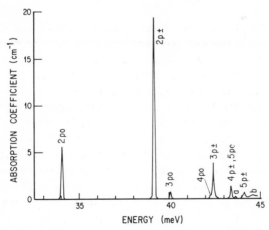

Figure 7.8 Absorption spectrum of phosphorus in silicon (from Aggarwal, Fisher, Mourzine, and Ramdas, reference 13).

worked out in the usual way by consideration of the transformation properties of the states in the reduced symmetry brought about by the stress.

From Figure 7.8 we see that only absorption peaks corresponding to transitions, which occur with a change of parity of the wavefunction $1s \rightarrow np$, are apparently observed in accordance with the electric dipole selection rule. Weaker lines corresponding to transitions with no change of parity have also sometimes been observed[15] when the impurity concentration is high. The group of the impurity site T_d does not contain the inversion operation. For this reason the true wavefunctions need not really have a definite parity. The solutions of Equation 7.5 do have definite parities, of course, but the solutions of Equation 7.4 do not. The matrix elements of the electric dipole operator between states ψ are approximately given by the elements between the corresponding envelope functions F.[7] In Raman scattering experiments the transitions $1sA_1 \rightarrow 1sE$ and $1sA_1 \rightarrow 1sT_2$ have also been seen.

The donor ground state has been extensively studied by means of magnetic resonance techniques. The EPR spectrum of phosphorus or arsenic donors, for example, consists of $2I + 1$ lines due to the nuclear hyperfine interaction where the nuclear spin I is $1/2$ for phosphorus and $3/2$ for arsenic. The magnetic resonance of an electron in a particular valley is described by a spin Hamiltonian with axial symmetry. The principal values of the g tensor, $g_\parallel$, and $g_\perp$, differ very slightly from the free electron g factor 2.0023. However, the resonance of the ground $1sA_1$ state of Equation 2.8 is described by an isotropic g tensor with diagonal elements

$$g = (1/3) g_\parallel + (2/3) g_\perp. \tag{7.9}$$

As the donor concentration is increased from 10^{16} cm^{-3}, interactions between the increasingly closely spaced donors become more important, and weaker lines corresponding to exchange coupled pairs of donors appear. As concentration increases further, the coupling becomes stronger. At high concentrations (10^{18} cm^{-3}) the electrons are not localized on particular donors, the hyperfine splitting is no longer seen, and the resonance consists of a single line, the "free carrier" resonance.[7]

With ENDOR spectroscopy it has been possible to map in great detail the donor wavefunction through the hyperfine interaction of the electron with the nuclear moments of the surrounding silicon atoms. Very many shells of silicon neighbors are observed in this way. The Fermi contact interaction strength is proportional to the squared amplitude of the wavefunction. For the A_1 state of Equation 2.8 this is given by

$$|\psi(\mathbf{r}_q)|^2 = (2/3) u_{k_m}^2 (\mathbf{r}_q) \big[F_x(\mathbf{r}_q)\cos(k_m x_q) + F_y(\mathbf{r}_q)\cos(k_m y_q)$$
$$+ F_z(\mathbf{r}_q)\cos(k_m z_q) \big]^2. \tag{7.10}$$

k_m is the magnitude of the wavevector at a conduction band minimum, $|\mathbf{k}_j| = k_m$. The expression is evaluated at the position $\mathbf{r}_q$ of the qth silicon nucleus. F_x stands for the envelope function F_j corresponding to the [100] valley. In writing Equation 7.10 we use the fact that F_{-x}, the function for the [$\overline{1}$00] valley, is equal to F_x. The cosine terms produce interference effects leading to rapid variation of the interaction with distance. These measurements also reveal that the envelope function departs slightly from symmetry under inversion of coordinates in the donor site.[16]

Dean, Haynes, and Flood[17] have studied the luminescence due to the decay of excitons bound to the neutral donors phosphorus, arsenic, antimony, and bismuth. In the initial state the complex consists of the charged donor with the electron and an exciton bound to it, $D^+e^-e^-h^+$. In the final state only the electron is bound, D^+e^-. Transitions have been observed in which in the final state the bound electron is left in the $1sE$ state or in the $2s$ state, as well as the ground $1sA_1$ state. Kosai and Gershenzon[18] have studied sharp-line spectra which they attribute to the radiative decay of two-exciton and three-exciton complexes bound to the donor. Other donor luminescence transitions have also been observed and are attributed to radiative recombination of a free hole with one of two electrons bound to the donor. The initial state of the donor in this case, $D^+e^-e^-$, is similar to the negative hydrogen ion. After the recombination the electron of the neutral donor, D^+e^-, is left in an excited $2s$ state. From the spectra the binding energy of the second electron is found to be 3, 4, and 5 meV for antimony, phosphorus, and arsenic, respectively. Smaller values in the range 0.5 to 1 meV have been measured for the thermal ionization energy by a different method.[19] At any rate, the second electron is very loosely bound and must have a quite large orbit.

LITHIUM AND MAGNESIUM

These are interstitial donors in silicon. Magnesium is a double donor. By giving up the two $3s$ electrons the magnesium atom attains the closed-shell neon configuration. No valence electrons are needed for binding when the atom is interstitial. The optical absorption spectra of both charge states, $Mg^{2+}e^-$ or Mg^+ and $Mg^{2+}e^-e^-$ or Mg^0, have been investigated by Ho and Ramdas.[20] Both give rise to spectra similar to that of phosphorus, shown in Figure 7.8. The ionization energies are 108 and 256 meV for Mg^0 and Mg^+, respectively. Samples displaying the Mg^+ spectrum are prepared by diffusing magnesium into p-type silicon. Magnesium can then give one electron to the acceptor. Through piezospectroscopic studies it was shown that the symmetry group of the donor site is T_d and that the ground state of both charge states is $1sA_1$.

The $1sA_1$ ground state of Mg^+ has also been studied by EPR.[21] When the crystal is irradiated with light of band gap energy at low temperature, another Mg^+ resonance appears. This resonance has $\langle 111 \rangle$ axial symmetry and might result from association of interstitial Mg^+ with another defect or from Mg^+ in an interstitial site of C_{3v} symmetry with no other defect nearby.

Lithium is also an interstitial donor. If lithium were substitutional, we should expect it to be a triple acceptor. Lithium diffuses rapidly in silicon, even at low temperatures. In otherwise pure material the solubility of lithium is low. In p-type material the solubility may be high, on the other hand, and is controlled by the concentration of acceptors. The absorption spectrum is also similar to that of the other donors.[13] However, the behavior of the spectrum under uniaxial stress is quite different. Additional splittings of spectral lines occur. Detailed analysis leads to the conclusion that the site symmetry is T_d, but the ground state is $1sE$ degenerate in energy with $1sT_2$. The $1sA_1$ level lies only 1.8 meV higher. These conclusions are confirmed by EPR and ENDOR measurements.[22]

The $1sE$ and $1sT_2$ wavefunctions are given by Equation 7.3 with the coefficients α_j,

$$Eu \qquad \alpha_j = \left(1/\sqrt{12} \right)(-1, -1, -1, -1, 2, 2)$$

$$Ev \qquad \alpha_j = (1/2)(1, 1, -1, -1, 0, 0)$$

$$T_2x \qquad \alpha_j = \left(1/\sqrt{2} \right)(1, -1, 0, 0, 0, 0) \qquad\qquad (7.11)$$

$$T_2y \qquad \alpha_j = \left(1/\sqrt{2} \right)(0, 0, 1, -1, 0, 0)$$

$$T_2z \qquad \alpha_j = \left(1/\sqrt{2} \right)(0, 0, 0, 0, 1, -1).$$

In the equations the valleys are arranged in the order $[100], [\bar{1}00]$, $[010], \ldots, [00\bar{1}]$. Each of the three T_2 states gives rise to a resonance of axial symmetry. The components of the doublet are mixed by the magnetic field, and the E resonance has cubic anisotropy like that described by Equation 1.54. Random strain can produce asymmetrical line shapes in the spectrum of the E state. These general features are observed in the rather complicated magnetic resonance spectra. Uniaxial stress applied along a $\langle 100 \rangle$ direction simplifies the spectrum by depressing the energy of one of the T_2 states below those of the other states. The ^7Li hyperfine interaction is temperature dependent because of thermal population of the low-lying singlet. Both the E and T_2 wavefunctions have nodes at the lithium nucleus and, therefore, no Fermi contact lithium hyperfine interaction. Although the ground state is orbitally degenerate, no Jahn–Teller effect is seen. The electron wavefunction spreads over many lattice sites and couples only weakly to the displacement of a single ion.

Lithium and oxygen form an associate, but the structure of the defect has not been established. A local vibrational mode at 1007 cm^{-1} has been ascribed to this center.[23] The frequency of the absorption line does not change when ^{6}Li is substituted for natural lithium (93% ^{7}Li). If the line is due to a lithium-oxygen center, it must correspond to a mode in which the vibration amplitude of the lithium atom is relatively small. The electronic optical absorption spectra have been studied by Aggarwal et al.[13] The center is a donor with ionization energy 39.4 meV. The ground state is $1sA_1$, in contrast with isolated lithium. The results of stress experiments imply that the defect symmetry is T_d. EPR measurements reveal a slight departure of the g tensor from isotropy and $\langle 111 \rangle$ axial symmetry.[24]

SULFUR

Sulfur forms several different donor centers in silicon, both as an isolated impurity and through association with other defects. Isolated substitutional sulfur is a double donor with the three possible charge states S^{2+}, S^+ with one extra bound electron, and S^0 with two bound donor electrons. S^+ has the relatively large ionization energy 614 meV.[15] Because the ionization energy is, on the simple effective mass theory, proportional to the square of the core charge, we expect an ionization energy 4 times larger than for an electron bound to a singly charged donor core such as phosphorus. The hydrogenlike donor absorption spectrum, including some transitions to states of even "parity," has been observed by Kleiner and Krag[15] for S^+. The inferred T_d site symmetry is consistent with the model of an isolated substitutional impurity. Magnetic resonance of the ground state of S^+ has been studied by Ludwig.[25] The resonance is isotropic, and hyperfine interactions with the ^{33}S nucleus and the nuclei of several shells of silicon neighbors are seen. The resonance is seen only if the sample is also doped with acceptors. An acceptor can trap one of the sulfur electrons. If the acceptor concentration is too high, the signal intensity decreases, probably because the concentration of non magnetic S^{2+} increases. In these samples it is possible to increase the concentration of S$^+$ by irradiating the sample with light of energy equal to the band gap energy or greater. Some of the electrons created are trapped at S^{2+} sites to produce S$^+$. Because the level is so deep, agreement between measured hyperfine parameters for the silicon neighbors and simple effective mass theory is worse than in the case of shallow donors.

A resonance due to sulfur pairs has also been observed by Ludwig. From the ratio of the intensity of the ^{33}S hyperfine lines to the intensity of the main line due to isotopes with no nuclear spin, it is found that the defect contains two equivalent sulfur nuclei. Although the symmetry of such a defect must be lower than tetrahedral, no asymmetry is observed in the g tensor or in the hyperfine tensor. When the sample is irradiated by infrared light of energy $h\nu \gtrsim 0.37$ eV, the intensity of the resonance decreases and photoconductivity is observed. The light may release electrons from the center. The ionization energy would then be 0.37 eV. An optical absorption spectrum from this same sulfur center has also been seen.[15]

Two other hydrogenlike absorption spectra have also been assigned to sulfur.[26] The ionization energies are 110 and 188 meV. Five $1s$ levels are seen for the shallower donor rather than the three (A_1, E, T_2) expected in tetrahedral symmetry. Uniaxial stress experiments show that the site symmetry of the other is also lower than T_d. These spectra have been assigned to the two-electron configuration S^0. The low symmetry implies that the sulfur is associated with some other defect. Rosier and Sah[27] have ascribed a level with thermal ionization energy[4] 275 meV to isolated substitutional S^0, but no direct evidence concerning site symmetry is yet available.

GROUP III ACCEPTORS

The atoms of group III of the periodic table, boron, aluminum, gallium, and indium, have one less valence electron than silicon and form substitutional acceptors. (Some may also occupy interstitial positions. This is discussed in a following section.) Their ionization energies, $E_A^{(8)}$ of Figure 7.7, are 45, 57, 65, and 160 meV, respectively. The spin-orbit splitting of the bands Δ_{so} is 44 meV. The positions of the p-like excited states vary somewhat more from one acceptor to another than in the case of donors. Transitions have been observed in optical absorption measurements not only from the ground state to excited states of the Γ_{8g} ladder of levels, but also to p-like excited states of the other ladder associated with the Γ_{7g} band.[28] These observations allow estimation of the ionization energy $E_A^{(7)}$ for this series of states. The values of $E_A^{(7)}$ for these acceptors are 88, 113, 117, and 200 meV, respectively.[29] The degeneracies of the various states have been checked by stress-optical experiments. Under uniaxial stress the Γ_8 states may split into two components. The Kramers degeneracy of the Γ_7 and Γ_6 doublets is not lifted, of course. In particular, it has been established that the ground states of these four acceptors are of the Γ_8 type, as predicted theoretically.[27]

The Zeeman splitting of the Γ_8 ground state should be characterized by a cubic anisotropy like that described by Equation 1.54. No such EPR spectrum has been observed, possibly because the lines are broadened greatly by internal strain. However, a resonance is observed when uniaxial stress is applied to split the quartet into two doublets.[30] If we describe the Zeeman splitting of the Γ_8 state by an effective spin Hamiltonian with spin $S = 3/2$, the observed transition would be $M = 1/2 \leftrightarrow M = -1/2$.

Zinc is a double acceptor in silicon, and copper is a triple acceptor. Gold can be either a donor or an acceptor. The energy levels associated with these and other dopants have been tabulated by Sze and Irvin.[31] We shall discuss some of these impurities in connection with germanium for which spectroscopic data exist.

In Chapter 3 we saw how a light substitutional impurity in the diamond lattice has associated with it a triply degenerate local mode of T_2 (or Γ_5) symmetry. Smith and Angress[32] have studied the local mode due to boron in silicon. Because boron is an acceptor and high concentrations are necessary to see the absorption, the sample must be compensated by the addition of donors. The compensation leaves the boron acceptors in the negative charge state, B^-. Otherwise, the absorption produced by free holes would mask the local mode spectrum. The absorption occurs at 644 cm^{-1} for ^{10}B and 620 cm^{-1} for ^{11}B. The second harmonics were also seen.[33]

BORON ASSOCIATES

Local mode spectra have also been observed from boron paired with phosphorus, arsenic, antimony, or lithium and from boron pairs. All these impurities are presumably substitutional except for lithium. We expect, roughly speaking, that in the case of a light impurity atom paired with a heavy one the triplet mode of the light impurity would be split in the lower symmetry of the pair. For a pair of light impurities the triplet associated with each isolated light atom would split, and there would be twice as many local modes as in the case of the light atom–heavy atom pair. Elliott and Pfeuty[34] have considered the theoretical problem in detail.

If two identical light impurity atoms are substitutional at nearest neighbor sites, the symmetry of the defect is D_{3d}. There is a center of inversion midway between the two atoms. The six degenerate modes of the isolated substitutional impurities split into two singlets Γ_{1g} and Γ_{2u} and two doublets Γ_{3g} and Γ_{3u}. The highest frequency mode is Γ_{1g} and corresponds to vibration along the defect axis, the two atoms vibrating in opposite directions or with a 180° phase difference. The next highest frequency is that of Γ_{3g}, corresponding also to out-of-phase vibration but perpendicular to the axis. Next lower in frequency is Γ_{3u}, an in-phase transverse vibration, and lowest is Γ_{2u}, an in-phase vibration along the axis. If the force constants remain the same as in the perfect crystal, the frequency of the triply degenerate mode of the isolated single impurity would lie between the frequencies of Γ_{3g} and Γ_{3u}. Because the three components of the electric dipole operator transform as Γ_{2u} and Γ_{3u}, electric dipole transitions are allowed from the Γ_{1g} ground state (no vibration) to the Γ_{2u} and Γ_{3u} states only.

Now let us suppose that the two members of the pair are not identical. The inversion operation is no longer an element of the symmetry group of the defect. The symmetry group is C_{3v} rather than D_{3d}. There are still four modes, two of Γ_1 symmetry and two of Γ_3 type. The correlation with the modes of D_{3d} symmetry is

$$\Gamma_1 - \Gamma_{1g} \qquad \Gamma_3 - \Gamma_{3g} \qquad \Gamma_1' - \Gamma_{2u} \qquad \Gamma_3' - \Gamma_{3u}$$

All four modes are infrared active. Elliott and Pfeuty have calculated the frequencies of these modes for the cases of two substitutional impurities of equal charges and also for oppositely charged impurities—the case of the boron–donor pair—as a function of the mass of one impurity with the mass of the other equal to that of ^{10}B or ^{11}B. The lower frequency Γ_1' and Γ_3' modes are only at higher frequency than the lattice modes and observable if the mass of the second impurity is less than about 85% of the mass of the silicon atom. In the special case in which the masses of the two impurities are equal, the absorption of these two modes is zero. For boron paired with the heavy donors phosphorus, arsenic and antimony, then, two modes should be seen in absorption. For boron pairs there are four local modes, but only two are observable in absorption because of the selection rule.

The spectrum of the boron-heavy donor pair consists of four lines,[35] one pair from ^{10}B and one from ^{11}B. They lie in the range 600 to 670 cm^{-1}, the B–P pairs giving the lowest frequencies and B–Sb the highest. Good agreement with the calculations of Pfeuty are obtained[36] if the higher frequency line of each pair is assigned to the Γ_1 mode and lower frequency line to Γ_3. The force constants must be varied to obtain a good fit. In samples containing especially high concentrations of boron, lines assigned to the vibrations of boron pairs have also been seen.[37] When the two members of the pair are of the same isotopic species, the pair symmetry is D_{3d}. The frequencies of the Γ_{2u} and Γ_{3u} modes for ^{11}B–^{11}B pairs are 552 and 615 cm^{-1}. The corresponding calculated values are 559 and 613 cm^{-1}. The frequencies of these modes are independent of the force constant between the two boron atoms, because they vibrate in phase and have no relative displacement.

Local modes of boron-lithium pairs have been studied experimentally by several investigators.[23,38,39] Elliott and Pfeuty[34] have considered the cases of local modes of an isolated interstitial atom and of a pair of impurities, one of which is interstitial, whereas, the other is substitutional. For an isolated atom, the smaller the force constant, the lighter the impurity must be in order that a local mode emerge above the band modes. Because lithium diffuses interstitially quite rapidly, the force constant for the atom in the interstitial site is expected to be small. This may explain why no local mode due to isolated interstitial lithium has been observed. The symmetry of the substitutional-interstitial pair is C_{3v}. Absorption lines from excitation of both Γ_3 and Γ_1 modes have been observed in the region 560 to 680 cm^{-1} for boron-lithium pairs, the Γ_3 frequency being higher. The frequencies are accurately predicted by a calculation of Elliott and Pfeuty. There are also lines in the region near 520 cm^{-1}, very near the maximum phonon frequency of silicon, which show no dependence on the boron isotope. The absorption is proportional to boron concentration, however, and similar lines are observed when aluminum or gallium is used rather than boron as the acceptor.[38,40] Assignment of these lines to particular modes is uncertain.

CARBON

Carbon is a common contaminant of silicon crystals. Because carbon, like silicon, is a group-IV element, we should expect it to occupy a substitutional site and to be electrically neutral. The measured activation energy for diffusion of carbon in silicon is 3.2 eV, only slightly smaller than the values 3.5 to 4.0 eV typical of the substitutional group-III acceptors and group-V donors.[41] Furthermore, in silicon carbide each carbon atom is tetrahedrally bonded to four silicon nearest neighbors.

The absorption spectrum of carbon in silicon consists of a single line near 16 μm, as expected for a substitutional impurity. At 77 K the absorption occurs at 608, 589, and 573 cm^{-1} for the ^{12}C, ^{13}C, and ^{14}C isotopes, respectively. Absorptions due to excitation of the second harmonics have also been observed.[42] Absorption strength is found to be proportional to carbon concentration. The absorption spectrum is similar to that of boron in silicon. For both impurities the frequency of the second harmonic is less than double the frequency of the fundamental by the small amount 3.8 cm^{-1}. This indicates that the anharmonicity of the vibrational potential is similar for the two impurities. The nearest neighbor distance in SiC is 1.89 Å, much smaller than in silicon (2.35 Å). The substitutional carbon then, is similar to carbon in silicon carbide, but with a larger bond length. This expansion might reduce the force constants relative to those of silicon carbide, leading to the longer wavelength position of the mode compared with the reststrahlen wavelength of silicon carbide, 12 μm.

The local mode absorption of carbon in silicon is about 6 times that of boron. This seems surprising, for carbon is expected to be uncharged, but boron has a single negative charge. According to the theory of vibrations of a charged light impurity in a cubic homopolar crystal the total integrated absorption in the local mode and band modes is given by[43]

$$\int \alpha(\omega)d\omega = \frac{2\pi^2 N\eta^2}{ncM'} \tag{7.12}$$

$\alpha(\omega)$ is the absorption coefficient, n the refractive index, N the impurity concentration, and η and M' the charge and mass of the impurity. The ratio of the integrated absorption in the local mode to that in the band modes is 3000. If Equation 7.12 is applied to the case of carbon, from measurement of $\int \alpha(\omega)d\omega$ for the local mode and N, it is found that the effective charge is $|\eta| = 2.5|e|$. This may be compared with $|\eta| \approx |e|$ for boron. Leigh and Szigeti[44] have treated the problem of the effective charge associated with impurity vibrations in covalent crystals. They conclude that there is no simple relationship between the effective charge η and the valence of the impurity. The η is largely determined by differences in the bonding to neighboring atoms between the impurity and the host atom it replaces.

When silicon crystals containing appreciable concentrations of carbon are heated to 1000 C or higher, the carbon coalesces to form silicon carbide particles. This behavior has been monitored by observing the decrease of absorption in the local mode of isolated substitutional carbon and a corresponding rise in the absorption in a band near 12 μm, ascribed to silicon carbide. Bean and Newman conclude that most of the carbon present in silicon is either dissolved and occupies substitutional sites or is present as silicon carbide precipitates.[45]

Isolated interstitial carbon has been identified in silicon by EPR.[46] From the ^{13}C hyperfine interaction in a sample containing carbon enriched to 60% ^{13}C it could be determined that a single carbon nucleus was present in the defect. Rather than occupying the tetrahedral interstitial position, however, the carbon is displaced along a $\langle 100 \rangle$ direction toward a nearby substitutional silicon. This silicon atom is thought to be pushed away from the substitutional site along the $\langle 100 \rangle$ axis, so that it and the carbon atom lie opposite each other with the substitutional site between them. As sample temperature is raised, the carbon atom migrates away to pair with another atom, the two again sharing a substitutional site.

VACANCY

There has been much investigation of the effects of various kinds of radiation on semiconductor materials, especially silicon. Not only is it of interest to determine how electronic components made from such materials would be affected by proximity to either a nuclear reactor or a nuclear explosion or by a space flight, but the techniques of ion implantation and electron beam lithography are becoming more common in semiconductor device processing, and they also damage the material. Much of our knowledge of the defect centers created by electron bombardment of sililcon is due to the investigations of G. D. Watkins, J. W. Corbett, and co-workers.

A beam of electrons of high energy—1 or 2 MeV—may penetrate a solid to a distance of several millimeters. Although there is a great difference between the masses of electron and nucleus, the electron may transfer enough energy to the nucleus to knock it from the lattice site into an interstitial position, creating two defects: a vacancy and an interstitial atom. The interstitial atom may fall back into the vacancy, it may migrate away and eventually, perhaps, associate with some other defect, or it may remain near the vacancy. Similarly, the vacancy may move away. What happens depends on many factors, one of the most important being the temperature of the crystal.

When *p*-type silicon is bombarded with high-energy electrons at temperatures below 20 K an EPR spectrum ascribed to the isolated vacancy is observed.[47] The interstitial silicon atom apparently migrates far enough away, even at these low temperatures, that it does not perturb the electronic properties of the vacancy. The resonance is described by the spin Hamilitonian of a Kramers doublet,

$$\mathcal{H} = \beta g_{\|} S_z H_z + \beta g_{\perp} \left(S_x H_x + S_y H_y \right) + \sum_{i=1}^{4} \mathbf{I}_i \cdot \mathbf{A}_i \cdot \mathbf{S}, \qquad (7.13)$$

with effective spin $S = 1/2$. The g tensor is axially symmetric about a $\langle 100 \rangle$ direction. There are thus three spectra corresponding to the three $\langle 100 \rangle$ directions. For a general direction of the magnetic field (H_x, H_y, H_z) these spectra are separated, occurring at different values of the field. The tensor $\mathbf{A}_i$ is axially symmetric, the axis lying very nearly along a $\langle 111 \rangle$ direction. The term $\Sigma \mathbf{I}_i \cdot \mathbf{A}_i \cdot \mathbf{S}$ describes a hyperfine interaction with four silicon nuclei. The ^{29}Si isotope is 4.7% abundant and has nuclear spin $I = 1/2$. The measured parameters are

$$g_{\|} = 2.0087 \qquad A_{\|} = 43.9 \times 10^{-4} \text{ cm}^{-1}$$
$$g_{\perp} = 1.9989 \qquad A_{\perp} = 29.8 \times 10^{-4} \text{ cm}^{-1}.$$

When a uniaxial stress is applied to the crystal along a $\langle 100 \rangle$ axis at 2 K the relative intensities of the three spectra change rapidly because of reorientation of some centers. That is, some centers may change from a [100] symmetry axis to [010], for example. The rapid reorientation at such low temperature probably implies that the lower symmetry is not due to association of simple defects. The observations are consistent with the following model.

Coulson and Kearsley[48] first proposed for the vacancy in the diamond lattice a simple model in which the only orbitals considered are the four σ orbitals extending into the vacancy from the four nearest neighbors. These are shown in Figure 7.9. These might, for example, be approximated by sp^3

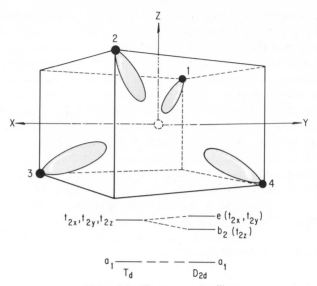

Figure 7.9 The vacancy in silicon.

hybrid orbitals, as in a small molecule. It is not at all obvious that such a model is approriate. The proper basis states for the perfect crystal are not tight-binding states, but almost-free-electron Bloch waves. However the observed spin is in a rather well-localized orbital. If the σ orbitals are assumed to be composed of the silicon $3s$ and $3p$ functions, then the magnitudes of the observed ^{29}Si hyperfine parameters imply that about 60% of the total electron density is localized in these four σ orbitals. That is, the orbital at site i is written

$$\sigma_i = \alpha_i \psi_s^{(i)} + \left(1 - \alpha_i^2\right)^{1/2} \psi_p^{(i)} \tag{7.14}$$

and the wavefunction of the magnetic electron is

$$\Psi = \sum_i \eta_i \sigma_i \tag{7.15}$$

Then the hyperfine interaction with the ith nucleus is given by

$$A_{\|_i} = a_i + 2b_i, \qquad A_{\perp_i} = a_i - b_i$$

$$a_i = \frac{16\pi}{3} \left(g_N^i \beta_N \right) \beta \alpha_i^2 \eta_i^2 |\psi_s^{(i)}(0)|^2 \tag{7.16}$$

$$b_i = \frac{4}{5} \left(g_N^i \beta_N \right) \beta \left(1 - \alpha_i^2\right) \eta_i^2 \langle r_p^{-3} \rangle_i$$

$g_N^i \beta_N \mathbf{I}_i$ is the magnetic moment of nucleus i and $\langle r_p^{-3} \rangle_i$ is the expected value of r^{-3} for an electron in the p orbital. The contribution of a is due to the Fermi contact interaction between the nonzero electron density in the s orbital $|\psi_s(0)|^2$ at the nucleus and the nuclear moment; b is due to the dipolar interaction of the nuclear moment with the electron in the p orbital. Overlap of orbitals has been neglected. If $\langle r_p^{-3} \rangle_i$ and $|\psi_s(0)|_i^2$ are known, α_i^2 and η_i^2 can be determined from measurement of a_i and b_i. The total density in the orbitals considered is $\Sigma \eta_i^2$.

Let us consider the four σ orbitals of Figure 7.9. In Chapter 1, Equations 1.67 and 1.72, we found that these four orbitals form bases for the irreducible representations a_1 and t_2 of T_d, where

$$\phi(a_1) = \tfrac{1}{2}(\sigma_1 + \sigma_2 + \sigma_3 + \sigma_4)$$

$$\psi(t_2 x) = \tfrac{1}{2}(\sigma_2 - \sigma_1 + \sigma_3 - \sigma_4)$$

$$\psi(t_2 y) = \tfrac{1}{2}(\sigma_2 - \sigma_1 + \sigma_4 - \sigma_3) \tag{7.17}$$

$$\psi(t_2 z) = \tfrac{1}{2}(\sigma_2 - \sigma_4 + \sigma_1 - \sigma_3).$$

Thus in T_d symmetry there would be two single-particle energy levels corresponding to the a_1 and t_2 states. We suppose that the a_1 level lies lower, as shown in Figure 7.9. The electrically neutral vacancy would have four electrons distributed in these orbitals. Because the resonance is seen only in p-type material, the singly positive charge state is assumed. The first two electrons can go into the a_1 orbital with opposite spin. The third may occupy any of the three t_2 orbitals. This orbital degeneracy can lead to a Jahn–Teller effect. A distortion along the z or [001] axis that reduces the symmetry to D_{2d} will split the t_2 states into a doublet e of D_{2d} and a singlet b_2. If b_2 lies lower, the third electron will occupy the b_2 orbital. Thus the center will have an [001] symmetry axis, and there will be a hyperfine interaction with four equivalent silicon nuclei, as observed. The three $\langle 100 \rangle$ symmetry axes are equally likely if the strains in the crystal are random.

We saw in Chapter 3 that an electronic state transforming as T_2 in T_d symmetry can interact not only with an E-type distortion, which leads to $\langle 100 \rangle$ symmetry axes, but also with a T_2 distortion, which leads to $\langle 111 \rangle$ symmetry axes, and with the symmetric A_1 distortion. Messmer and Watkins[49] have carried out a molecular orbital calculation involving the orbitals of a large number of atoms surrounding the vacancy. For the neutral state of the vacancy they predict a Jahn–Teller energy of about 0.5 eV for diamond. The calculation is not sufficiently accurate to predict whether the trigonal or tetragonal distortion leads to a lower energy.

The neutral vacancy V^0 would not be expected to be paramagnetic because the additional electron could pair with the magnetic electron of the positive vacancy V^+. It is not observed by magnetic resonance. The singly negative vacancy V^- is observed. Of the two extra electrons, compared with V^+, one can pair with the electron in the b_2 orbital of Figure 7.9 and the other then must go into one of the two e orbitals. A further reduction of the symmetry to C_{2v} lifts this degeneracy, splitting the e state of D_{2d} into b_1 and b_2 singlet states of C_{2v}. In one of these only the orbitals σ_1 and σ_2 appear and in the other, only σ_3 and σ_4, so that the electron is localized on two silicon atoms in this case rather than four. The observed hyperfine splitting indicates interaction with two silicon nuclei. Uniaxial stress at low temperature can make the electron hop from the b_1 to the b_2 orbital, or vice versa, and this inversion occurs more readily than a change of orientation—a flip of the axis of the defect from one $\langle 100 \rangle$ direction to one of the other two $\langle 100 \rangle$ directions. The ground state energies of these charge states of the vacancy all lie within the forbidden gap with the V^+ level near the valence band, the V^0 level above this, and the V^- level highest. In this simple picture of filling one electron orbitals we have completely neglected electron-electron interactions. These may not be negligible in comparison with the Jahn–Teller stabilization energy and should be included in an accurate calculation.

After electron bombardment of a silicon sample at room temperature the resonance of the isolated vacancy is not observed. Instead, more complex defects consisting of a vacancy associated with other simple defects are generally produced. At elevated temperatures the vacancy is able to migrate through the lattice and is often trapped by another defect. This behavior can be monitored by magnetic resonance. During isothermal anneals at 140 K the amplitude of the V^+ resonance signal decays exponentially with time, and the decay time τ has the temperature dependence

$$\tau \propto \exp \frac{U}{kT} \qquad (7.18)$$

U, the activation energy for motion of the vacancy is 0.33 eV. At much lower temperatures there may be a thermally activated electronic reorientation of the defect. The much lower activation energy which represents the barrier between equivalent distortions is only 0.02 eV for V^+. As the temperature increases from 0 K, then, the defect may first reorient, and at higher temperatures it may begin to migrate.

VACANCY—GERMANIUM ASSOCIATE

Germanium impurities can trap vacancies. EPR spectra due to two charge states of the complex have been studied.[50] In the singly negative state the spectrum is very similar to that of isolated V^- except that the hyperfine structure shows that the electron is localized in an orbital shared between a silicon atom and a germanium atom rather than two silicon atoms. The spectrum of $(GeV)^+$ is also similar to that of V^+ One of the four silicon atoms is replaced by a germanium atom, and hyperfine interactions with three silicon nuclei and one germanium nucleus are seen.

VACANCY—GROUP-V DONOR ASSOCIATE

When silicon grown in vacuum by the floating-zone method and containing phosphorus is irradiated by 1.5 MeV electrons at room temperature, the dominant defect center observed by magnetic resonance is the vacancy-phosphorus associate.[51] The resonance is observed only after the Fermi level has receded more than 0.4 eV below the conduction band edge as a result of the irradiation. The resonance is associated with this empty acceptor level. The spectra are described by the spin Hamiltonian

$$\mathcal{H} = \beta \mathbf{H} \cdot \mathbf{g} \cdot \mathbf{S} + \mathbf{I_P} \cdot \mathbf{A_P} \cdot \mathbf{S} + \sum_i \mathbf{I}_{Si}^i \cdot \mathbf{A}_{Si}^i \cdot \mathbf{S} \qquad (7.19)$$

$$\mathbf{g} = \begin{bmatrix} g_1 & 0 & 0 \\ 0 & g_2 & 0 \\ 0 & 0 & g_3 \end{bmatrix} \qquad \begin{array}{l} g_1 = 2.0005, \qquad g_2 = 2.0112 \\ g_3 = 2.0096 \end{array}$$

with spin $S = 1/2$. The g tensor has orthorhombic symmetry, one principal axis (2) being parallel to a $\langle 110 \rangle$ axis, and another (1), only 3° from a $\langle 111 \rangle$ axis. There is a hyperfine interaction with one phosphorus nucleus. The ^{31}P isotope is 100% abundant and has nuclear spin $I_P = 1/2$. A strong hyperfine interaction with one ^{29}Si nucleus, and weaker interactions with other silicon nuclei are also observed. There are 24 different orientations of the defect. But half of these differ from the others only by a 180° rotation about the principal axis of g_2, and so there are 12 magnetically inequivalent centers or resonance spectra. The strong ^{29}Si hyperfine interaction is characterized by an axially symmetric tensor, the symmetry axis being the same $\langle 111 \rangle$ axis which is nearly parallel to axis 2 of the g tensor. The smaller ^{31}P hyperfine interaction departs only slightly from axial symmetry about another $\langle 111 \rangle$ axis.

The model for the defect center is shown in Figure 7.10. We shall start as in the case of the isolated vacancy of Figure 7.9 with the four broken bonds that remain after a silicon atom is removed to form the vacancy. In this case atom 1 is phosphorus rather than silicon. The extra positive nuclear charge at site 1 causes the σ_1 orbital to lie lower in energy than the other three orbitals σ_2, σ_3, and σ_4, which form bases for a doubly degenerate and a singly degenerate irreducible representation of C_{3v}. The phosphorus atom supplies an extra electron, so that in the neutral charge state there are five electrons to be distributed in the available orbitals. Two can be paired in σ_1. Two can go into the apparently lower lying singlet, and the last into the doublet. This degeneracy can be removed by a distortion which causes the energy of the bonding combination of σ_3 and σ_4 to decrease. Then two electrons can fill this bonding orbital, shown schematically in Figure 7.10, and the third can occupy the σ_2 orbital. Thus the magnetic electron is largely localized on the single silicon atom 2, and this accounts for the large ^{29}Si hyperfine interaction. The magnitudes of the hyperfine interactions imply that about 60% of the electron density is localized in this orbital, which is found to be almost entirely of p character. Because this orbital is nonbonding, silicon atom 2 may be pulled away from the vacancy and tend toward a planar configuration with its three remaining nearest neighbors. In a planar arrangement the bonding orbitals are sp^2, and the dangling σ_2 orbital would be a pure p orbital. Another 6% of the electron density resides on silicon atoms 3 and 4, and about 1% is in the phosphorus orbital σ_1. The σ_1 orbital is tilted somewhat away from the $\langle 111 \rangle$ direction. This accounts for 70% of the density. The rest is distributed over more distant atoms.

Figure 7.10 shows only one type of center. Two others could be drawn by taking the bond between atoms 2 and 3 or 2 and 4, rather than 3 and 4. Then by changing the position of the phosphorus atom to 2, 3, or 4 we could generate additional equivalent centers. First, however, let us concentrate on the three equivalent centers possible for a fixed position of the phosphorus atom. If the defect has one of these configurations, there is a certain energy barrier that must be surmounted in order for it to switch to one of the other configurations. As the temperature increases, thermally activated switching from one distortion to another becomes more rapid.

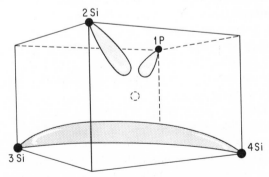

Figure 7.10   Vacancy–phosphorus associate.

Because the g tensor is anisotropic, this switching causes the magnetic field at which resonance occurs to switch also. But in a magnetic resonance experiment the field is held constant, tuned to a particular resonance. The defect then switches rapidly in and out of resonance as it reorients. As the switching becomes more rapid, the resonance lines first broaden and then disappear. New lines appear at the motionally averaged positions and become narrower as the reorientation rate increases. There are two different spectra: a low-temperature spectrum and a high-temperature spectrum which appears near 150 K. The positions of the spectral lines at high temperature are predictable from the positions of the lines in the low-temperature spectrum. From study of the change of linewidth as a function of temperature the activation energy or height of the energy barrier between equivalent orientations is found to be 0.06 eV.

In the high-temperature spectrum the rapid hopping of the magnetic electron from atom 2 to atom 3 to atom 4 makes these atoms equivalent. The spectrum reveals a hyperfine interaction with three equivalent silicon nuclei just as if there were no motion and the electron were in an orbital on these three atoms. This averaging of the hyperfine interaction indicates that the motion is electronic in nature rather than involving atomic rearrangements. At low temperatures a uniaxial stress can render the three distortions inequivalent and bring about a redistribution of populations of centers among the three distortions. For example, a stress along the $\langle 110 \rangle$ axis connecting silicon atoms 3 and 4 of Figure 7.10 reduces the energy of the configuration shown in the figure with respect to the other two distortions corresponding to 3—2 or 2—4 bonds.

At higher temperatures the vacancy can change its position with respect to the phosphorus atom. Uniaxial stress applied at room temperature leads to preferential orientation of the phosphorus-vacancy complex. When the return to an equilibrium distribution of orientations is monitored by isochronal annealing at various temperatures, it is found that an activation energy of 0.93 eV is associated with this atomic motion. To lead to an observable reorientation of the defect the vacancy must make four jumps —two away from the phosphorus atom, followed by two toward it. By successive steps of phosphorus-vacancy reorientation and phosphorus-vacancy interchange the defect can move through the lattice. If the energy barrier against interchange is small compared with that against reorientation, then 0.93 eV is also about equal to the activation energy for diffusion of the defect. This is 0.6 eV greater than the activation energy for diffusion of the vacancy. The phosphorus-vacancy binding energy is somewhat larger than 0.6 eV, then.

Similar arsenic-vacancy and antimony-vacancy associates have also been observed.[52] In these cases also about 70% of the electron density is localized near the vacancy. At low temperatures electronic bond switching is also observed. The activation energies are 0.06 eV in the case of arsenic and 0.07 eV for antimony—nearly the same as for phosphorus (0.06 eV). At higher temperatures atomic reorientation occurs with activation energies 1.07 and 1.29 eV for the arsenic and antimony centers, respectively. The larger activation energies reflect the greater difficulty of movement of these larger atoms through the lattice.

From strain measurements Elkin and Watkins have measured c, the electronic-vibrational linear coupling parameter. (See Equation 3.55.) From a calculation of the spring constant A they are able to estimate the Jahn–Teller energy $E_{JT} = c^2/2A$ and the magnitude of the distortion $|c|/A$. They find for the three defects the values given in Table 7.5. These energies are large and are comparable with the band gap of silicon.

Table 7.5 Jahn–Teller Energies and Distortions for Three Associates

	$c^2/2A$	$|c|/A$
$V_{Si} - P_{Si}$	1.36 eV[a]	0.61 Å[a]
$V_{Si} - As_{Si}$	0.57 eV[b]	0.40 Å[b]
$V_{Si} - Sb_{Si}$	0.66 eV[b]	0.43 Å[b]

[a] Reference 49.
[b] Reference 50.

ALUMINUM CENTERS

During electron bombardment interstitial silicon atoms are produced as well as vacancies. No resonance has been identified with the isolated interstitial silicon atom. In crystals containing aluminum, however, the aluminum interstitial has been observed.[53] There is a nearly one-to-one correlation between the number of these and the number of vacancies produced by the irradiation. Watkins has suggested that this may imply that the silicon atom knocked from its substitutional site in the electron bombardment migrates, even at very low temperatures, to a substitutional aluminum atom and changes places with it, producing interstitial aluminum.[54] The resonance ascribed to aluminum is characterized by an isotropic g tensor and a large isotropic hyperfine interaction with the ^{27}Al nucleus. The g factor is 2.0019. The aluminum atom has the configuration $[Ne](3s)^2(3p)^1$. The resonance is due to the single electron in the $^2S_{1/2}$ state of Al^{2+}. Smaller hyperfine interactions with the nearest-neighbor and next-nearest-neighbor ^{29}Si nuclei indicate by their symmetry that the aluminum ion is interstitial, not substitutional.[55] Interstitial boron and gallium may also be similarly produced by electron bombardment.[54] Interstitial arsenic is produced by the ion implantation process.[56] At higher temperatures this spectrum disappears and spectra due to interstitial aluminum paired with substitutional aluminum appear.[51] In this case the magnetic electron is largely on the interstitial ion, but there is a smaller hyperfine interaction with the nucleus of the substitutional aluminum atom. The symmetry of the defect is C_{3v}. Two such spectra are seen. They are quite similar, but the numerical values of g factors and hyperfine interactions are slightly different. Therefore, at least one spectrum must be that of a more complex defect. Similar interstitial-substitutional gallium pairs have also been observed.

Another magnetic resonance spectrum observed in electron-bombarded aluminum-doped silicon has been attributed to an excited state of an aluminum-vacancy complex.[57] The resonance appears after room-temperature bombardment or after bombardment at low temperatures followed by annealing at higher temperatures. Its appearance is correlated with the disappearance of the resonances due to the isolated vacancy, but during these changes the concentration of interstitial Al^{2+} does not change. This suggests that substitutional aluminum may be trapping vacancies. The excited state is not populated thermally. The sample must be irradiated with infrared light to produce the signal.

The spectra are rather complex and are described by the spin Hamiltonian

$$\mathcal{K} = \beta g_{\parallel} H_z S_z + \beta g_{\perp} (H_x S_x + H_y S_y) + D\left[S_z^2 - \tfrac{1}{3} S(S+1) \right]$$

$$+ \sum_i \mathbf{I}_i \cdot \mathbf{A}_i \cdot \mathbf{S} + \sum_i \mathbf{I}_i \cdot \mathbf{Q}_i \cdot \mathbf{I}_i \tag{7.20}$$

with spin $S = 1$. The z axis is a $\langle 111 \rangle$ direction. The dominant hyperfine interaction is with a single ^{27}Al nucleus, and there is a quadrupole interaction with this nucleus as well. Both $\mathbf{A}_{Al}$ and $\mathbf{Q}_{Al}$ are axially symmetric about the z axis. Smaller hyperfine interactions with neighboring silicon nuclei are also evident.

The model for this defect center is similar to that for the vacancy-phosphorus complex: a substitutional aluminum atom beside a vacancy. There is no Jahn–Teller effect. The symmetry is C_{3v}. The number of electrons is even since $S = 1$. Starting with the same four orbitals of Figure 7.10, we see that if the phosphorus atom is replaced by aluminum, the σ_1 orbital will be highest in this case. The other three orbitals σ_2, σ_3, and σ_4, again form bases for a_1 and e representations of C_{3v}. The charge state of the defect is not known. The magnitudes of the hyperfine interactions are best explained as those of the $^3A_2$ state of the four-electron configuration obtained by pairing two electrons in a_1 and one in each of the e states. The effective charge of the defect would then be -1. The small magnitude of the quadrupole interaction, a measure of the electric field gradient at the aluminum nucleus, may imply that the aluminum atom is displaced slightly along the $\langle 111 \rangle$ axis away from the vacancy. In fact, it is not possible to conclude from the data where the aluminum atom lies, except that it is somewhere along the $\langle 111 \rangle$ axis not so very far from the vacancy.

INTERSTITIAL OXYGEN

There has been much study of defects containing oxygen. Oxygen is a major contaminant of crystals pulled from quartz crucibles. The oxygen content may be of the order of 10^{18} cm^{-3} in such crystals and is introduced by reduction of the crucible during the growth process. It can also be introduced in semiconductor device processing if a silicon wafer covered in part by an oxide layer is subjected to high temperatures. The behavior of oxygen in silicon is rather complex. As a crystal containing a large concentration of oxygen is cooled, oxygen agglomerates may be formed. Oxygen may also form associates with other defects. Although heat treatment may lead to donor levels which are believed to be due to some type of oxygen defect, isolated oxygen forms a defect center that is electrically neutral. The model for the center is now well established, largely through detailed analysis of the local mode spectra.

The model is shown in Figure 7.11. It consists of a single interstitial oxygen atom bonded to two nearest-neighbor silicon atoms. X-ray measurements show that dissolved oxygen leads to an increase in the lattice parameter, suggesting the interstitial location.[58] There are several vibrational absorption peaks in the infrared that have been correlated with this center. For analysis of the spectra it has been sufficient to disregard coupling to the rest of the lattice and to consider only a small "molecule" consisting of the oxygen atom and the nearest silicon atoms. The Si_2O molecule, Si1—O—Si2 of Figure 7.11, has three normal modes, all of

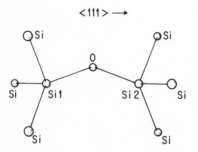

Figure 7.11 Interstitial oxygen in silicon.

which are infrared active. These are shown schematically in Figure 7.12. The infrared absorption spectra consist of bands at 9 μm (1136 cm^{-1}), 8 μm (1203 cm^{-1}), 19 μm (517 cm^{-1}), and in the far infrared, four lines in the region 200 to 340 μm (29.3, 37.8, 43.3, and 49.0 cm^{-1}). The isotope effect was observed in samples containing oxygen enriched in ^{18}O for the lines at 1203, 1136, 37.8, and 29.3 cm^{-1}, showing that these lines are due to vibrations of a defect containing one oxygen atom.[59]

If a uniaxial stress is applied to the crystal with the sample at a high enough temperature that the oxygen atom can move about, then when the crystal is cooled with the stress still applied to a low enough temperature that the oxygen is no longer mobile, the oxygen tends to assume preferentially the orientation that has lowest energy in the strain field. This produces a dichroism in the 9-μm band. Stress along a $\langle 100 \rangle$ direction has no effect. This implies that the band is due to complexes oriented along the $\langle 111 \rangle$ directions, because these remain equivalent under $\langle 100 \rangle$ stress. Stress effects have also been observed for the 1203 and 29.3 cm^{-1} lines.[59]

Table 7.6 shows the expected effects of stress on a spectral line associated with defect centers oriented along the $\langle 111 \rangle$ directions. The oscillation may be either of the π type (oscillation along the symmetry axis) or the σ type (oscillation in the plane perpendicular to the axis). Under [100] stress no degeneracy is lifted, and the absorption intensity for light polarized along [100], $I_{\parallel}$, is the same as for light polarized perpendicular to the stress direction. Stress along [111] splits the line into two components, one 3 times as intense as the other. For stress along [110] the crystal becomes biaxial. The degeneracy raised by the stress is an orientational degeneracy. This effect should be distinguished from the splitting of a degenerate state of a defect by application of stress or an electric field to lower the symmetry. Both effects could occur simultaneously, of course, but in the case of interstitial oxygen only the former is observed.

Table 7.6 Stress-Induced Splitting of a Spectral Line Associated with $\langle 111 \rangle$ Defect Centers (from Bosomworth, Hayes, Spray, and Watkings, reference 59)

Stress Direction	Degeneracy	π	σ
		$I_{\parallel}:I_{\perp}$	
[100]	4	1:1	1:1
[111]	1	3:0	0:3
	3	1:4	8:5
		$I_{110}:I_{001}:I_{1\bar{1}0}$	
[110]	2	2:1:0	1:2:3
	2	0:1:2	3:2:1

The line at 29 cm^{-1} is very sharp at low temperatures. It does not split under $\langle 100 \rangle$ stress. Under $\langle 111 \rangle$ or $\langle 110 \rangle$ stress it splits into two components with polarization properties identical to those predicted by Table 7.6 for a σ oscillator. Thus it may represent vibration of the ω_2 type, shown in Figure 7.12. Similar stress experiments on the 8 and 9 μm lines indicate that these are due to π oscillators oriented along the $\langle 111 \rangle$ axes, suggesting a vibration of the ω_3 type.

The lines at 37.8, 43.3, and 49.0 cm^{-1} appear when the crystal temperature is raised to 35 K, suggesting that they represent transitions from an excited vibrational state to higher-lying states. In addition to the 1136 cm^{-1} line there are two other lines in the 9-μm region at 1121.5 and 1127.9 cm^{-1} which appear only at higher temperatures. There is also a fine structure due to the isotope effects associated with the nuclei Si1 and Si2 of Figure 7.11; that is the molecules ^{28}Si—O—^{28}Si, ^{28}Si—O—^{29}Si, ^{28}Si—O—^{30}Si, and so on have slightly different frequencies and the ratios of the corresponding intensities are given by the isotopic abundances. This is an important confirmation of the model. Bosomworth et al.[59] explain the various lines in the 8 to 9 μm region as combination bands involving ω_2 and ω_3 vibrations. We see below how this occurs.

The Si—O—Si molecule, if it were free rather than trapped in a solid, could rotate about the Si—Si axis. In the solid this might also occur, or the oxygen atom might be constrained at one of several equilibrium positions and execute a vibrational motion about this position. Which possibility actually occurs depends on the height of the energy barriers separating the positions of minimum energy. At any rate, there are two degrees of

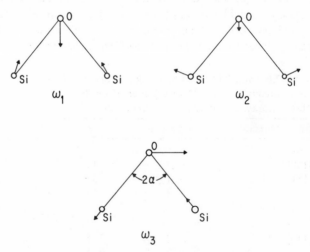

Figure 7.12 Normal modes of the Si1-0-Si2 "molecule" of Figure 7.11.

freedom for motion in the plane perpendicular to the Si—Si axis, rather than the single one implied in Figure 7.12, corresponding to the mode ω_2. The vibrational levels associated with this motion have been calculated from the Hamiltonian for the two-dimensional oscillator

$$\mathcal{H} = (1/2M)P^2 + (1/2)M\omega^2 R^2 + A\exp(-bR^2) \qquad (7.21)$$

R is the radial distance from the Si—Si axis. Because there is no azimuthal dependence in the potential, the oxygen atom can rotate freely. The anharmonic term $A\exp(-bR^2)$ leads to splitting of the harmonic oscillator levels and to a mixing of states. This phenomenological Hamiltonian closely predicts the observed positions of the lines at 29.3, 37.8, 43.3, and 49.0 cm^{-1} and the oxygen isotope effect, if M is the mass of the oxygen atom (^{16}O or ^{18}O) and ω, A, and b are chosen to fit the data. The energy levels and observed transitions are shown in Figure 7.13. The states are

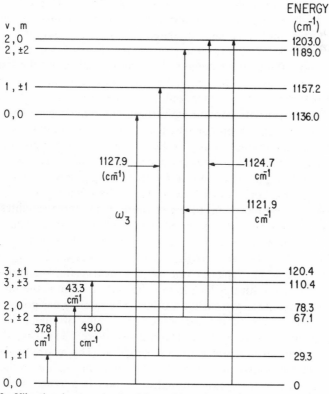

Figure 7.13 Vibrational energy levels of interstitial oxygen with observed transitions indicated by arrows (from Bosomworth, Hayes, Spray, and Watkins, reference 58).

labeled $|v, m\rangle$, where v is the vibrational quantum number and m is the angular momentum quantum number corresponding to azimuthal rotation.

From the fit of the eigenvalues of the Hamiltonian to the experimental energy levels it is found that the Si—O—Si bond angle 2α is 162°. The ω_2 excitations are like those of a free molecule. The D_{3d} symmetry of the molecule in the lattice could lead to a splitting of the v, $m = 3$, ± 3 state, but none is observed. The line at 517 cm^{-1} is not explained by the model. It is very near the maximum vibrational frequency of silicon (523 cm^{-1}) and a peak in the lattice phonon density of states (500 cm^{-1}). If it corresponds to a vibration of the ω_1 type, it may not be describable on the basis of an isolated molecule, but may rather be described as a resonance mode.

CARBON—OXYGEN ASSOCIATE

When both carbon and oxygen are present in sufficiently high concentration in a crystal, weak satellite lines are observed near the 9-μm absorption of interstitial oxygen and near the 16-μm line of isolated substitutional carbon. These lines have been assigned to a carbon—oxygen associate.[42] There are three satellites near 9 μm, one much stronger than the other two. These are not affected by changing the carbon isotope, but the strength of the two stronger lines is proportional to carbon concentration. This implies that the lines are due to a center containing one carbon atom. The lines are not present in crystals containing carbon but no oxygen. They do not decrease in strength or broaden with increasing temperature as does the 9-μm oxygen line corresponding to the transition from the ground state in Figure 7.13. Three satellite lines for each carbon isotope are seen near the carbon line at 16 μm. These are not seen in oxygen-free samples, and the strength is proportional to carbon content. The intensity of the strongest 9-μm satellite is proportional to the strength of the three satellites near 16 μm as carbon concentration is varied. These four lines then seem to be due to the same defect center which contains carbon and oxygen.

The model for this center is the same as for interstitial oxygen, except that a carbon atom replaces one of the next-nearest-neighbor silicon atoms. Consistent with this location of the carbon atom is the insensitivity of the 9-μm satellite to the carbon isotope. The satellite has poorly resolved structure that is due to the various combinations of silicon isotopes as nearest neighbors (silicon atoms 1 and 2 of Figure 7.11). This also shows that the carbon atom is not a nearest neighbor. The nearby carbon atom quenches the transverse motion of the oxygen atom, leading to the temperature-independent absorption strength. Carbon is a much smaller atom than silicon. The oxygen atom might prefer to occupy the position nearest the carbon. A carbon atom in such a site would be at a position of such low symmetry that the threefold degeneracy of the local mode of isolated carbon would be completely lifted. This could account for the three satellites near 16 μm.

SUBSTITUTIONAL OXYGEN

Interstitial oxygen may trap a vacancy and then fill the vacancy to form the substitutional oxygen defect often called the silicon "A center." An energy level 0.17 eV below the conduction band edge is associated with the defect. When an electron is trapped in the level a characteristic electron resonance signal can be observed.[60] For this reason the resonance is studied in n-type silicon containing phosphorus donors which supply the necessary electrons, the phosphorus level being nearer the band edge. The resonance spectrum is described by a spin-Hamiltonian

$$\mathcal{H} = \beta \mathbf{S} \cdot \mathbf{g} \cdot \mathbf{H} + \sum_i \mathbf{I}_i \cdot \mathbf{A}_i \cdot \mathbf{S},$$

$$\mathbf{g} = \begin{bmatrix} g_1 & 0 & 0 \\ 0 & g_2 & 0 \\ 0 & 0 & g_3 \end{bmatrix} \tag{7.22}$$

with $S = 1/2$. The principal axes 1 and 2 of the g tensor are $\langle 110 \rangle$ directions and the third axis is $\langle 100 \rangle$. The three g factors are very near 2. A strong hyperfine interaction is seen with two equivalent silicon nuclei along $\langle 111 \rangle$ axes. Weak hyperfine interactions are seen with other silicon nuclei. All these hyperfine interactions have axial symmetry about $\langle 111 \rangle$ axes. No hyperfine interaction is seen with the ^{17}O nucleus (nuclear spin $I = 5/2$), although a sample in which the oxygen had been enriched in this isotope was studied. However, the presence of oxygen is established by infrared absorption measurements, described below.

The structure of the defect is indicated in Figure 7.14. Let us start the discussion of the model by referring again to Figure 7.9, neglecting the oxygen atom at first. The vacancy is surrounded by the four broken bonds σ_1, σ_2, σ_3, and σ_4 of the nearby silicon atoms. The oxygen atom forms bonds with silicon atoms 1 and 2. The orbitals on the two silicon atoms 3 and 4 can be combined to form a bond between these atoms. This bond will be filled with two electrons of opposite spin. But if we take overlap of σ_3 and σ_4 into account, we find that in addition to the bonding combination of σ_3 and σ_4 there will also be an antibonding combination at higher energy. If S is the overlap integral of σ_3 and σ_4, the equation describing

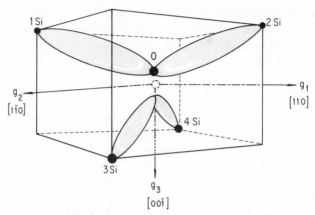

Figure 7.14   Substitutional oxygen in silicon. Axes of the g tensor are shown.

this splitting is similar to Equation 1.76:

$$\begin{vmatrix} \mathcal{H}'_{33} - E & \mathcal{H}'_{34} - E \mathcal{S} \\ \mathcal{H}'_{43} - E \mathcal{S} & \mathcal{H}'_{44} - E \end{vmatrix} = 0 \qquad (7.23)$$

$\mathcal{H}'_{34}$ is the matrix element of an Hamiltonian like that of Equation 1.75 between σ_3 and σ_4. $\mathcal{H}'_{33} = \mathcal{H}'_{44}$. The two solutions, the bonding and antibonding orbitals ψ_a and ψ_b are given by,

$$\psi_a = (2 - 2\mathcal{S})^{-\frac{1}{2}} (\sigma_3 - \sigma_4)$$

$$\psi_b = (2 + 2\mathcal{S})^{-\frac{1}{2}} (\sigma_3 + \sigma_4)$$

The spin resonance is believed to be due to a single electron in the antibonding orbital. This orbital is schematically indicated in Figure 7.14. The oxygen atom lies in a nodal plane of ψ_a. This accounts for the absence of hyperfine structure from the oxygen nucleus. In a perfect crystal the antibonding orbitals would be conduction band states, the bonding orbitals corresponding to valence band states. Because the state ψ_b represents a weaker bond than that of the normal lattice and ψ_a is also less antibonding than the conduction band states, the level E_a lies 0.17 eV below the edge of the conduction band.

From the magnitude of the silicon hyperfine interactions it can be estimated that about 70% of the electron density is localized in the antibonding orbital ψ_a. The principal axes of the g tensor are also shown in Figure 7.14. The axes of the two strong silicon hyperfine interactions are the $\langle 111 \rangle$ axes connecting atoms 3 and 4 with the central site. The oxygen atom is thought to be pulled away from the substitutional site, shown in the figure as a vacancy, by the bonds to atoms 1 and 2.

The oxygen atom may form bonds with any two of the four silicon atoms around the vacancy. These six equivalent defect centers are seen as six separate resonance spectra. A uniaxial stress applied to the crystal can destroy this equivalence, and the energy of the electron of a particular center may increase or decrease depending on the orientation of the axis of the center with respect to the stress. If the electrons are able to redistribute themselves among the centers, the relative intensities can change. This effect is observed when the number of A centers is much greater than the number of available electrons (the number of phosphorus donors). But when the number of donors is larger than the number of A centers, each A center contains an electron in the antibonding orbital, and no change of relative amplitudes is observed.[60] The rate of return of the intensities to an equilibrium distribution after the removal of stress, when studied as a function of temperature, reveals an activation energy for the redistribution which is 0.2 eV, nearly equal to the trap depth, 0.17 eV.

At higher temperatures the oxygen atom can move, allowing the defect to reorient under applied stress. The activation energy for this process is 0.38 eV. This is the activation energy for reorientation of the defect with no electron in the antibonding orbital, because the concentration of the phosphorus donors was 10% of the A-center concentration in this study. Thus a particular A center contains an electron only 10% of the time on the average. This reorientation can occur by the breaking of a silicon-oxygen bond and pivoting of the oxygen atom about the other bond.

Because the magnetic resonance spectra contain no direct evidence for the presence of the oxygen atom in the A center, Corbett, Watkins, Chrenko, and McDonald[61] have carefully correlated the resonance with a characteristic optical absorption band. The presence of oxygen can be directly inferred from the isotope effect in the optical band. The electron bombardment of silicon containing oxygen produces an absorption band at 12 μm (828 cm^{-1}). When the oxygen in the sample is enriched in ^{18}O, another band appears at 791 cm^{-1}, indicating that the band corresponds to a localized mode of a defect containing an oxygen atom. The band is thought to be analogous to the 9-μm band of interstitial oxygen. Indeed, the isotope shifts of the two bands are identical:

$$\frac{\omega(^{16}\text{O})}{\omega(^{18}\text{O})} = 1.047.$$

The application of uniaxial stress affects the optical spectrum also. When the stress is applied at sufficiently high temperature that the Si—O—Si molecule can reorient, the vibrational dipole moment, which is parallel to the Si—Si axis, reflects the new alignment. The result is that the absorption of plane polarized light will depend on the orientation of the polarization direction with respect to the crystal axes. The rate at which the induced dichroism disappears after release of the stress was studied as a function of temperature. The result is shown in Figure 7.15, together with the resonance data. The good agreement indicates that the same defect is observed in both experiments. Table 7.7 shows a correlation of the degree of stress-induced alignment measured optically and by magnetic resonance. The R is the ratio of the peak absorption coefficient measured under stress to that measured with zero stress or the value of this ratio predicted from the magnetic resonance results. The polarization indicated is that of the light with respect to the stress axis. Again the agreement is very good. Further evidence for the identification of the optical band with the A center is provided by the detailed measurements of polarized absorption by Bosomworth, Hayes, Spray, and Watkins.[59] These show that the defect responsible for the optical absorption has the same orthorhombic symmetry as the resonance spectrum.

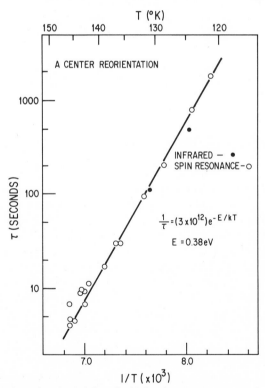

Figure 7.15 Reorientation time of the substitutional oxygen center as a function of temperature (from Corbett, Watkins, Chrenko, and McDonald, reference 61).

Table 7.7 Stress-Induced Alignment of *A* Centers (from Corbett, Watkins, Chrenko, and McDonald, reference 61)

Stress axis	Viewing axis	Polarization	R (resonance)	R (optical)
$\langle 100 \rangle$	$\langle 100 \rangle$	$\parallel$	1.23	1.24
		$\perp$	0.92	0.92
$\langle 110 \rangle$	$\langle 100 \rangle$	$\parallel$	1.11	1.06
		$\perp$	1.08	1.03
$\langle 110 \rangle$	$\langle 110 \rangle$	$\parallel$	1.07	1.06
		$\perp$	0.85	0.80
$\langle 111 \rangle$	$\langle 110 \rangle$	$\parallel$	1.00	0.98
		$\perp$	1.00	1.00

A resonance attributed to an excited electronic state of the neutral charge state of this defect has been studied by Brower.[62] The excited state is populated by optical pumping, as in the case of the aluminum-vacancy complex. The resonance spectrum is described by a spin-Hamiltonian with spin $S = 1$. In the excited state one electron is in the bonding orbital ψ_b and one is in the antibonding orbital ψ_a. An observed hyperfine interaction with the ^{17}O nucleus proves that a single oxygen atom is present. The stress-induced alignment is affected by annealing in the same way for both resonances, implying that both resonances are connected with the same defect.

DIVACANCY

The divacancy is a complex defect consisting of two vacancies at nearest-neighbor sites. Two charge states of the defect have been observed in magnetic resonance experiments,[63] and several absorption bands in the infrared have also been ascribed to the divacancy.[64] The divacancy is another form of radiation damage produced by electron bombardment. After bombardment at low temperature followed by annealing at higher temperatures the concentration of isolated vacancies decreases and the number of divacancies increases as the vacancies associate. If the crystal contains oxygen, however, this association is much less probable, because trapping of a vacancy by oxygen is a more efficient process than divacancy formation.

Resonance of the singly positive charge state of the divacancy $(VV)^+$ is observed in low resistivity p-type silicon. Optical bleaching of the resonance with monochromatic light indicates that a level at about 0.25 eV above the valence band edge is associated with this charge state. The resonance is observed when this level contains an electron. The $(VV)^-$ resonance is seen only in high-resistivity n-type silicon when the associated energy level at 0.4 eV below the conduction band edge does not contain an electron. The two spectra are similar, but quantitatively different. Both resonances are described by a spin Hamiltonian like that of Equation 7.22 with g factors near two and a strong hyperfine interaction with two equivalent silicon nuclei along a $\langle 111 \rangle$ axis.

Figure 7.16 shows the model for the divacancy. When the two silicon atoms at the adjacent sites c and c' are removed to form the divacancy, six broken or dangling bonds remain. Atoms a and d move toward each other to form a bent pair bond, and atoms a' and d' similarly form a bond. The resonance arises from unpaired electrons in the extended bond between atoms b and b'. In the charge state $(VV)^+$ there is one electron in the bb' bonding orbital. In the other charge state, $(VV)^-$, there are two electrons in the bb' bonding orbital and one in the antibonding bb' orbital. The bonding can also be described in terms of a distortion of the Jahn–Teller type which causes the symmetry to decrease from D_{3d} to C_{2h}. The strong silicon hyperfine interaction is due to the two equivalent nuclei of atoms b and b'.

Analysis of the hyperfine interactions reveals that about 60% of the electron density is concentrated in the bb' orbital. For $(VV)^+$ approximately another 20% is concentrated on the four nearby atoms a, a', d, and d', and for $(VV)^-$ about 10% is accounted for there. A principal axis of the g tensor lies near the $\langle 111 \rangle$ axis connecting sites b and c' or c and b' but is tilted toward the $b - b'$ axis.

As in the cases of many of the other damage centers, motional effects are also observed for the divacancy. At low temperatures thermally activated electronic reorientation or bond switching can occur, and at higher temperatures the vacancy-vacancy axis can change orientation. From Figure 7.16 we see that two other equivalent electronic arrangements are bonding of a to b and a' to b' with the unpaired electron in a dd' orbital and bonding of b to d and b' to d' with the unpaired electron in an aa' orbital. The activation energies for this bond switching are 0.073 and 0.056 eV for the $(VV)^+$ and $(VV)^-$ centers, respectively. The activation energy for reorientation of the vacancy-vacancy axis is found to be 1.3 eV. When the axis changes orientation, the complex moves one lattice spacing. Therefore, 1.3 eV is also the activation energy for diffusion of the divacancy. During reorientation the vacancies in an intermediate step become separated by an additional lattice spacing, as for example, when atom b hops to site c' and then to c. Thus the vacancy-vacancy binding energy is also of the order of 1 eV. From annealing studies Watkins and Corbett estimate that this binding energy is at least 1.6 eV. The fact that

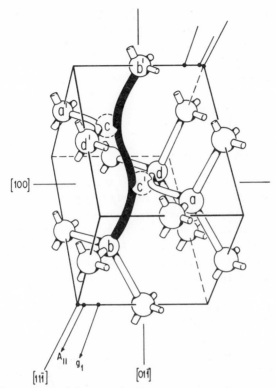

Figure 7.16 Model of the divacancy in silicon (from Watkins and Corbett, reference 63).

the high-temperature stress and annealing studies yield the same values for activation and binding energies regardless of which spectrum is monitored implies that both resonances are indeed due to the same defect. Probably the neutral charge state is that to which these energies refer.

An interesting effect is observed when the influence of the orientation of the incoming beam of bombarding electrons is studied. If the beam is incident on the crystal along the [111] direction, it is found that more divacancies are produced with the vacancy-vacancy axis along this direction than along the other $\langle 111 \rangle$ directions. The effect becomes more pronounced at lower electron energies. When an incoming electron collides with a lattice atom, imparting recoil energy to it, this atom must, to create a divacancy, collide with a neighbor atom and knock it from the lattice site also. The first atom must have enough energy remaining that it does not take the place of the second atom at the substitutional site, but goes on to an interstitial site. The first atom can recoil with most energy along the beam direction. As beam energy decreases, a point is finally reached at which there is enough energy to produce a divacancy oriented along the beam direction, but not along the other $\langle 111 \rangle$ directions. Thus divacancies can be produced in a primary damage event as well as forming at a high temperature as a result of vacancy motion and association. The divacancy has also been produced as a result of ion implantation.[65] Other spectra have been assigned to three-vacancy, four-vacancy, and even five-vacancy clusters.[66,67] It is, of course, more difficult to prove the uniqueness of these more complicated models.

d^n TRANSITION METAL IONS

The transition metal ions act as donor or acceptor impurities. For the most part they occupy interstitial sites, although some can be substitutional as well. They form complex defects by associating with each other or with other acceptors. No optical spectra have been reported, probably because of the low maximum concentrations attainable. Electrical properties have been studied, and donor and acceptor levels have been assigned in many cases. The most detailed information has come from EPR and ENDOR measurements.[68]

The magnetic resonance data are explained by the following model. A substitutional ion uses four electrons for bonding with the silicon neighbors, and the other electrons are localized on the ion in the d shell. For example, the configuration of the neutral chromium atom is $[\text{Ar}](3d)^5(4s)^1$. If four electrons are used in bonding, the configuration of the other electrons is $(3d)^2$. The ion with a singly negative effective charge would have the $(3d)^3$ configuration, and so on. The splitting of the d^n configuration is that expected for a weak field of T_d symmetry. An interstitial atom uses no electrons for bonding. The $4s$ electrons are transferred instead to the d shell. The chromium atom would have the $(3d)^6$ configuration if it were interstitial. The symmetry of the interstitial site is also T_d. However, the splitting for an atom or ion in this site is reversed compared with the substitutional site, the t_2 states lying below the e states in the strong field language. This is the type of splitting expected in sixfold octahedral coordination. The interstitial atom is sixfold coordinated with respect to the second nearest neighbors, which are only 12% more distant than the four nearest neighbors. Again the field is weak so that the t_2 and e states are filled so as to give maximum total spin in the ground configuration. The resulting ground state terms together with the observed species are listed in Table 7.8.

The resonances of ions with singlet ground states, $^3A_2$, $^4A_2$, and $^6A_1$, are described by a spin Hamiltonian with effective spin 1, 3/2, or 5/2, respectively. There is a small splitting of the $^6A_1$ state in zero magnetic field as a result of high-order effects, because the six spin functions $|5/2, M_s\rangle$

Table 7.8 Ground States of Iron-group Transition Metal Ions in Weak Sixfold Octahedral or Tetrahedral Fields

	Interstitial		Substitutional	
d^1	$^2T_2$		^2E	
d^2	$^3T_1$		$^3A_2$	Cr^0_{Si}, Mn^+_{Si}
d^3	$^4A_2$	V^{2+}_i	$^4T_1$	
d^4	^5E		$^5T_2$	
d^5	$^6A_1$	Cr^+_i, Mn^{2+}_i	$^6A_1$	Mn^{2-}_{Si}
d^6	$^5T_2$	Cr^0_i, Mn^+_i	^5E	Pd^-_{Si}, Pt^-_{Si}
d^7	$^4T_1$	Mn^0_i, Fe^+_i	$^4A_2$	
d^8	$^3A_2$	Mn^-_i, Fe^0_i	$^3T_1$	
d^9	^2E	Ni^+_i	$^2T_2$	

form bases for a twofold and a fourfold irreducible representation of T_d. In n-type silicon containing phosphorus the two species interstitial manganese Mn_i and substitutional manganese Mn^{2-}_{Si} are observed simultaneously, indicating that two different types of site are occupied by manganese. In p-type silicon the substitutional manganese is in the form Mn^+_{Si}. The species Mn^+_{Si} and Mn^{2-}_{Si} are stable, but the interstitial forms Mn^-_i and Mn^{2+}_i diffuse at room temperature to form more complex defect centers or to become substitutional through interaction with a vacancy. In the case of substitutional neutral chromium, Cr^0_{Si}, the hyperfine interaction with the twelve next-nearest-neighbor ^{29}Si nuclei has been observed. This is definite proof that the atom is substitutional rather than interstitial.

For the d^6 and d^7 configurations the ground term is an orbital triplet $^5T_2$ or $^4T_1$. Taking the product of orbital and spin representations, we see that spin-orbit coupling and higher-order electrostatic effects split the $^5T_2$ term into a singlet, a doublet, and four triplets. The ground state of Cr^0_i and Mn^+_i is a triplet. Resonance from the excited states has also been observed, indicating that the spin-orbit splitting is much reduced compared with the free ion values. For d^7 the $^4T_1$ term splits into two Kramers doublets and two quartets, a doublet lying lowest. Excited state resonance has also been observed for Mn^0_i. The orbital contributions to the g factors are also greatly reduced for these ions. Reduction of spin-orbit splittings and g factors in these cases are ascribed by Ham to the dynamical Jahn–Teller effect of an orbital triplet.[69] The spectra of Ni^+_i, Pd^-_{Si}, and Pt^-_{Si} are anisotropic, several equivalent centers of low-symmetry being present. For Ni^+_i and Pd^-_{Si} reorientation effects indicate that a Jahn–Teller distortion may be responsible for the low symmetry.

The transition metals have a great tendency toward association with other defects. At room temperature interstitial manganese diffuses to form four-atom clusters. This is the dominant defect in crystals containing only manganese impurities and cooled slowly from high temperatures.[70] The resonance spectrum of this defect indicates that all four atoms are equivalent and are arranged in a tetrahedron, probably occupying interstitial sites. The transition metals also associate with the acceptors boron, gallium, aluminum, or indium. The interstitial transition metal pairs with the substitutional acceptor. The ground state of the transition metal is not greatly different from that of the isolated interstitial atom except for the lower symmetry caused by the acceptor. Usually the symmetry is trigonal corresponding to closest association. The iron-indium pair is aligned along the $\langle 100 \rangle$ directions. In this case the iron may occupy the next-nearest interstitial position. The manganese-gold pair is more complicated. Two charge states have been observed, one resonance having effective spin 5/2 and the other, 3/2. Most pairs are electrically active.

GERMANIUM

Germanium, like silicon, is an important semiconductor material, and crystals of high quality and purity are routinely produced. Much less is known about defect structures in germanium than in silicon. The conduction band minima occur at the zone boundary along the $\langle 111 \rangle$ directions in reciprocal space. There are four equivalent minima, because the minima related by the inversion in the origin are not really different because their positions differ by a reciprocal lattice vector. The effective mass tensor for an electron in such a valley is more anisotropic than in silicon:

$$m_{\parallel}^* = 1.588\,m, \qquad m_{\perp}^* = 0.0815\,m$$

GROUP-V DONORS AND GROUP-III ACCEPTORS

The dielectric constant is larger than in silicon, and impurity states are correspondingly shallower. The fourfold degeneracy associated with the conduction band minima leads to the splitting of an s-like donor state into A_1 and T_2 components. For the group-V donors the ground state splitting into A_1 and T_2 states varies from 0.32 meV for antimony to 4.23 meV for arsenic. As in the case of donors in silicon, the positions of the excited states agree well with predictions of the simple effective mass theory.[5] But ionization energies and ground state splitting are not in agreement. The pseudopotential method of Pantelides and Sah[6] has not been applied to donor states in germanium. Figure 7.17 compares the calculations of Faulkner[5] with typical experimental spectra. Table 7.9 lists measured

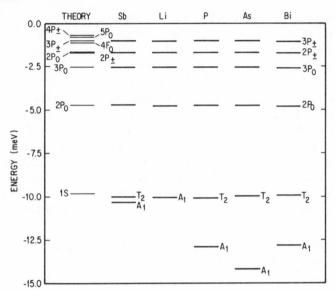

Figure 7.17 Calculated and observed donor levels in germanium (from Faulkner, reference 5).

Table 7.9 Ionization Energies of Some Donors in Germanium (in meV)

P	12.76	Bi	12.68
As	14.04	Li	9.89
Sb	10.19		

ionization energies of some donors. Two-electron transitions have also been observed in the luminescence accompanying the decay of an exciton bound to a neutral donor in germanium. The remaining electron is excited to the $1sT_2$ level in the transition.[17] The group-V donors are also substitutional in germanium. As in silicon, they have rather high solubilities and diffuse slowly, whereas interstitial impurities such as lithium and some transition metal ions have lower solubilities and diffuse more rapidly.

Electron paramagnetic resonance spectra of the group-V donors are qualitatively similar to the spectra seen in silicon. The g tensor is isotropic, as expected for the A_1 state. At low concentration the impurity hyperfine splitting is observed.[7] As in silicon, the T_2 states have nodes at the impurity nucleus. The departure of the g factor from the free electron value is larger. This g shift is proportional to the magnitude of the spin-orbit coupling of the host lattice, which is much larger for germanium than for silicon. Linewidths are generally greater in germanium than in silicon, and less magnetic resonance data are available.

The theory of the acceptor states in germanium has been briefly discussed in the previous section. The situation is simpler than in silicon because the separation of the Γ_{8g} and Γ_{7g} valence bands in germanium is 0.3 eV, and this is much larger than the acceptor ionization energies. Some of these are listed in Table 7.10 for neutral substitutional acceptors. The positions of the excited states of these are in rough agreement with the calculations of Mendelson and James.[10] Some of their calculated level positions are displayed in Figure 7.18. On the left-hand side the levels are labeled in their notation. The first number following the Γ specifies the number of nodes in the radial part of the envelope function. Because there are two Γ_8 states associated with each p state, a second index is added to distinguish them. The predicted ionization energy is 9.28 meV. The envelope functions used in the calculation have definite parities. This is

Table 7.10   Ionization Energies of Neutral Acceptors in Germanium (units are meV)

Boron	10.47	Beryllium	24.45
Aluminum	10.80	Zinc	32.63
Gallium	10.97	Cadmium	54.61
Indium	11.61	Mercury	91.5
Thallium	13.10	Copper	42.8

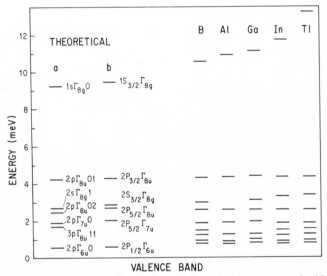

Figure 7.18   Observed energy levels of acceptors in germanium compared with calculated values of (*a*) Mendelson and James, reference 10, and (*b*) Balderschi and Lipari, reference 11.

indicated by the subscripts g and u. As in the case of silicon, the spectral lines have been identified by the raising of degeneracies under application of uniaxial stress or magnetic field.[14]

In the second column from the left in Figure 7.18 the levels calculated by Baldereschi and Lipari[11] are also shown with the states labeled in their notation. Their calculation is in slightly better agreement with the experimental results. Calculated ionization energy is 9.73 eV. Note the large variation from acceptor to acceptor of the position of the $1S_{3/2}$ ground level. The variation of $2S_{3/2}$ is greater than that of $2P_{3/2}$, probably also because of the central cell effects.

Local mode vibrational spectra have been observed for isolated boron, as in silicon, and also for boron-lithium pairs,[71] for isolated phosphorus and for phosphorus-gallium pairs.[72] Substitutional silicon[73] produces a local mode absorption at 389 cm^{-1}.

DOUBLE AND TRIPLE ACCEPTORS

The group-II elements beryllium, zinc, cadmium, and mercury are double acceptors in germanium. Copper, having one less electron than zinc, is a triple acceptor. The absorption spectra of all of these in the neutral charge state have been studied[74] and are found to be similar to those of the group-III acceptors. In the neutral charge state the group-II acceptors have two bound holes, $A^{2-}h^+h^+$. Copper has three bound holes, $Cu^{3-}h^+h^+h^+$. Copper-doped germanium is often used as an infrared detector, the infrared photons ionizing the neutral copper. Current due to the liberated holes is detected. The ionization energies are shown in the second column of Table 7.10. The excitation spectrum of the singly ionized zinc acceptor, $Zn^{2-}h^+$, or Zn^-, has been observed.[75] The spacing of the spectral lines is about 4 times as great as for the neutral acceptors because of the larger effective charge. Ionization energy of Zn^- is 85.8 meV.

Although the spacings of the excited states of the neutral double acceptors are nearly equal to those of the (neutral) group-III acceptors, the degeneracies of the states are larger, because two holes are bound. Consequently, the splitting of the spectral lines under applied stress is more complex. The similarity of the spectra are due to the weakness of the electrostatic interaction between the two holes. The observed excited states are states in which one hole is excited and the other remains in the Γ_8 ground state. Such a state might be written $|\Gamma_8, \Gamma_i\rangle$, for example, to specify that one hole is in a Γ_8 state and the other is in some other state Γ_i. When Γ_i takes on the values of Figure 7.18, the "ladder" of energy levels is similar to that for a group-III acceptor. There is one observable effect of the coupling between holes: a small splitting of the ground state of the neutral mercury acceptor.[76] In the ground state Γ_i is also Γ_8. The product wavefunction is antisymmetric under interchange of the coordinates of the two holes in accordance with the Pauli principle. The product wavefunctions transform as

$$\{\Gamma_8 \times \Gamma_8\}_A = \Gamma_1 + \Gamma_3 + \Gamma_5$$

$\{\}_A$ is the antisymmetric product. The two spins $j = 3/2$ of the holes couple to form states of total $J = 0(\Gamma_1)$ and $J = 2(\Gamma_3 + \Gamma_5)$. The electrostatic interaction splits these by the small amount 0.7 meV. The splitting is not observed for the other acceptors. The others listed in Table 7.10 have smaller ionization energies. The more diffuse wavefunctions would lead to smaller splittings.

TRANSITION METALS

Germanium containing various transition metal impurities has been the subject of many studies of electrical properties. No optical spectra have been reported. Manganese[77] and nickel[78] have been studied by EPR. Both are double acceptors, and both are thought to be substitutional. Manganese is observed as Mn_{Ge}^{2-}. The nickel spectrum is attributed to Ni_{Ge}^{-}. This would probably have the configuration $(3d)^8$ with a $^3T_1$ ground state. Six equivalent centers are seen corresponding to distortions along the $\langle 100 \rangle$ axes. The g tensor has orthorhombic symmetry. Thermally activated switching of the defect from one orientation to another is seen as the sample temperature is raised.

OXYGEN

The main infrared absorption band in germanium containing oxygen occurs[79] at 855 cm^{-1} (11.7 μm). The absorption is proportional to oxygen concentration, and a corresponding band at 818 cm^{-1} is seen for ^{18}O. A weaker line is also seen at 1260 cm^{-1}. The 855 cm^{-1} line may correspond to the ω_3 vibration of interstitial oxygen in silicon. A strong absorption is seen in GeO_2 at 11.5 μm. GeO_2 has the same structure as SiO_2 in which the line occurs near 9 μm. In germanium the 11.7-μm line is narrower than the 9-μm line in silicon and does not show the satellite structure. When samples are annealed between 650 and 810 C, a broad line appears at 11.5 μm, suggesting the formation of a precipitated GeO_2 phase. Thus the behavior of oxygen in germanium is similar to that in silicon, but no detailed model has yet been established.

In n-type germanium containing oxygen an EPR spectrum similar to that of the silicon A center (substitutional oxygen) is produced by electron bombardment.[80] The center has effective spin $1/2$ and the same g tensor principal axes as the silicon A center. The line width is greater in germanium, and no hyperfine structure is seen. There is a rough correlation between the resonance signal and infrared absorption bands at 766, 807, and 847 cm^{-1}.

DIAMOND

Diamond, because of a remarkable combination of physical properties, finds application as a gemstone and as an industrial abrasive. Of course, it has the diamond structure. There is also an hexagonal modification. Carbon may also crystallize in the graphite structure which consists of parallel sheets of contiguous hexagonal six-atom rings, the sheets being widely separated compared with the smallest interatomic spacing within a sheet. Here we are concerned only with the diamond structure. Compared with other crystals of large band gap, diamond has a large refractive index and a high dispersion of the index. These properties account for its brilliance as a gem stone. As the stone is rotated slightly, the light refracted into the viewer's eye changes color. The extreme hardness, several times that of other hard materials, makes diamond superior to other highly brilliant, but more easily scratched gem stones. The hardness makes diamond a cheaper abrasive than other cutting materials with much shorter useful lifetimes. The thermal conductivity of diamond is very high —about 5 times that of copper at room temperature. For this reason diamond heat sinks are sometimes employed in electronic devices.

The color, hardness, and thermal conductivity of diamond are changed by the incorporation of impurities. Generally, large impurity concentrations are necessary in order to produce the desired changes. The main interest in high purity diamond is academic. Diamond is very difficult to grow. Crystals are grown[81,82] in a liquid transition metal solution by transport of carbon from a hot region to a cooler region of the solution where crystallization occurs. For growth, temperatures and pressures in the ranges 1500 to 2400 C and 4×10^4 to 10^5 atm are required. Crystals grown in this way are generally 140 mg in mass or smaller. Research on defects in diamond has been hampered by lack of sample crystals with controlled impurity content. Often natural diamonds have been used as samples. As in the cases of other relatively "dirty" materials, much of the research on defects has consisted of attempts to correlate certain observed properties—intensity of an absorption band or concentration of a particular impurity, for example—with other observed properties. It has not yet been possible to understand the structures of many defects, but because of the simplicity of the crystal structure and of the carbon electronic configuration, much theoretical work has been done on defect states in diamond.

Diamonds have been roughly classified into two types according to optical spectra.[83] Diamonds of type II have negligible absorption in the near ultraviolet, visible, or infrared spectral regions except for two-phonon bands from 2 to 6 μm and the fundamental absorption associated with the band gap. Type I diamonds are less pure and have other absorption bands in the infrared and near ultraviolet.

NITROGEN

Nitrogen is a common impurity in diamond. Nitrogen concentrations as high as 0.2% atomic have been measured. The absorption strengths of two bands, one at 3065 Å and one at 7.8 μm, have been shown to be proportional to nitrogen content, but the bands have not been assigned to definite defect structures.[84] From electron microscopic examination of diamonds with high nitrogen concentrations it has been found that much of the nitrogen may be present in the form of platelets aligned along {100} planes.[85] The linear dimensions of the platelets range from 600 to 1000 Å. Presumably the platelets form during growth when rapid cooling occurs, leading to decreased solubility of nitrogen, which precipitates into the platelets. The nitrogen occupies substitutional positions in the platelets. The presence of vacancy loops near the platelets has been interpreted as implying that nitrogen diffuses by a vacancy mechanism. If interstitial diffusion occurred, the vacancies would disappear when the nitrogen became substitutional. The concentration of platelet nitrogen has been shown to be correlated with the strength of an infrared absorption line at 1368 cm^{-1} (7.3 μm).[86]

Isolated substitutional nitrogen has been identified by magnetic resonance.[87,88] The spectrum is characterized by an isotropic g tensor and hyperfine interactions with the ^{14}N nucleus and with a ^{13}C nucleus. The hyperfine interactions have axial symmetry about the same $\langle 111 \rangle$ axis. There are four equivalent centers, one for each $\langle 111 \rangle$ axis. The magnetic electron is localized in an orbital concentrated on the nitrogen and a nearest-neighbor carbon atom, with most of the density on the carbon. Weaker hyperfine interactions with the three other nearest neighbors of the nitrogen atom have also been identified. At higher temperatures the switching of the electron among the four possible orbitals is evident. The activation energy for this reorientation is 0.76 eV.[89] This large barrier implies a large Jahn–Teller energy. In photoconductivity and Hall effect measurements on diamonds containing isolated substitutional nitrogen a deep donor level 4 eV below the conduction band edge was observed and attributed to the substitutional nitrogen donor.[90] In a molecular orbital calculation for the center Messmer and Watkins[49] predict a trigonal distortion and a very large Jahn–Teller effect, in accord with experiment.

Although we might expect the nitrogen donor in diamond to be similar to the phosphorus donor in silicon, this is not the case. Nitrogen has a deep donor level because of the Jahn–Teller effect.

A very similar EPR spectrum results from the nearest-neighbor nitrogen pair.[91] The g factor is the same as for isolated nitrogen, $g = 2.0024$. The axial hyperfine interaction with one nitrogen nucleus is also nearly the same. In addition, there is a weaker hyperfine interaction with a second nitrogen nucleus. The pair of nitrogen atoms would have two extra electrons not required for bonding. One is apparently removed, because the resonance is characterized by effective spin $1/2$.

ACCEPTORS

Some acceptor ground states in diamond are shallow enough to make certain crystals p-type semiconductors. Hall measurements indicate a ground state 0.368 eV above the valence band.[92] The semiconducting diamonds are also characterized by infrared absorption bands that are due to transitions from the acceptor ground state to a number of excited states, as in silicon and germanium. Boron seems to be the dominant acceptor, although most diamonds contain other acceptor impurities as well.[93] The concentration of acceptors producing the 0.368-eV level has been found to be proportional to the acceptor infrared absorption strength,[92] which has in turn been correlated with boron content.[93] Diamonds heavily doped with boron are blue as a result of the red photoionization continuum of the infrared acceptor absorption.

The infrared acceptor absorption spectrum is complex. The spin-orbit splitting of the conduction band of diamond is only about 6 meV, so that both Γ_{7g} and Γ_{8g} bands contribute and can be considered degenerate as a first approximation. The acceptor ionization energy is large, and the effective Bohr radius is small. The effeccts of applied stress and of a magnetic field on the spectrum have been studied.[94,95] In labeling the energy levels Crowther, Dean, and Sherman[94] have used a different coupling scheme, more appropriate to the condition of very small spin-orbit splitting. The Γ_{5g} (no spin) valence band states are first coupled to the s or p envelope function, and spin is added last. This leads to a notation for the levels different from that of Mendelson and James. The ground $1s$ levels are a Γ_7 doublet and a Γ_8 quartet. The first identified excited level is at 0.304 eV.

The "N3" luminescence in diamond consists of a narrow zero-phonon line at 4152 Å followed by a broad band at longer wavelengths. This luminescence has been ascribed by Dean[96] to decay of an exciton bound to an ionized donor-acceptor pair, based on variation of the luminescence decay time with emission wavelength and the dependence of the shape of the spectrum on excitation intensity. The zero-phonon line and associated phonon replicas are due to nearest neighbor pairs. The polarization of this line and its splitting under applied stress show that it is due to a transition from an E (of C_{3v}) state to an A state of a defect with a $\langle 111 \rangle$ symmetry axis.[97,98] Dean assigns the spectrum to nitrogen-aluminum donor-acceptor pairs. Smith, Gelles, and Sorokin[99] have reported an EPR spectrum which they assign to a hole bound to an aluminum acceptor. Shcherbakova et al.[100] have observed additional structure in the spectrum and attribute it to an electron bound to a nearest-neighbor nitrogen-aluminum pair. They find that the intensity of the resonance is proportional to that of the $N3$ optical band. The g tensor has $\langle 111 \rangle$ axial symmetry, and hyperfine interactions with aluminum and nitrogen nuclei are observed.

VACANCY

Baldwin[101] has observed in a type-II diamond bombarded with electrons an EPR spectrum which he assigns to the singly ionized vacancy V^+. The g factor is equal to the free electron value. The anisotropic linewidth may indicate a very small anisotropy with a $\langle 100 \rangle$ symmetry axis. Hyperfine interactions with four equivalent nearest-neighbor and twelve next-nearest carbon nuclei are seen. The center is similar to the positive vacancy in silicon. The resonance is due to a single unpaired electron in a t_2 orbital, which interacts with an E vibrational mode to produce a $\langle 100 \rangle$ Jahn–Teller distortion. There have been many theoretical calculations of the states of the vacancy in diamond. Reference 49 is the latest, and references to earlier work may be found there. As yet no calculation has predicted that the tetragonal distortion should have lowest energy. A number of absorption lines observed in electron irradiated diamond have been assigned to the neutral vacancy, but the assignment is not certain.[102]

REFERENCES

1. L. A. Hemstreet, C. Y. Fong, and M. L. Cohen, *Phys. Rev.* **B2**, 2054 (1970).
2. M. L. Cohen and T. K. Bergstresser, *Phys. Rev.* **141**, 789 (1966).
3. D. Pines, *Elementary Excitations in Solids* (Benjamin, New York, 1963) p. 174.
4. S. T. Pantelides and C. T. Sah, *Phys. Rev.* **B10**, 621 (1974).
5. R. A. Faulkner, *Phys. Rev.* **184**, 713 (1969).
6. S. T. Pantelides and C. T. Sah, *Phys. Rev.* **B10**, 638 (1974).
7. W. Kohn in *Solid State Physics*, Vol. 5, F. Seitz and D. Turnbull, Eds., (Academic, New York, 1962) p. 258.
8. D. Schechter, *J. Phys. Chem. Solids* **23**, 237 (1962).
9. K. S. Mendelson and D. R. Schultz, *Phys. Stat. Sol.* **31**, 59 (1969).

10. K. S. Mendelson and H. M. James, *J. Phys. Chem. Solids* **25**, 729 (1964).

11. A. Baldereschi and N. O. Lipari, *Phys. Rev.* **B8**, 2697 (1973).

12. E. I. Zorin, P. V. Pavlov, and D. I. Tetelbaum, *Sov. Phys. Semicond.* **2**, 111 (1968).

13. R. L. Aggarwal, P. Fisher, V. Mourzine, and A. K. Ramdas, *Phys. Rev.* **138A**, 882 (1965).

14. P. Fisher and A. K. Ramdas in *Physics of the Solid State*, S. Balakrishna, M. Krishnamurthi and B. R. Rao, Eds., (Academic, New York, 1969) p. 149.

15. W. H. Kleiner and W. E. Krag, *Phys. Rev. Lett.* **25**, 1490 (1970).

16. E. B. Hale and R. L. Mieher, *Phys. Rev.* **184**, 739 (1969); ibid., 751 (1969).

17. P. J. Dean, J. R. Haynes, and W. F. Flood, *Phys. Rev.* **161**, 711 (1967).

18. K. Kosai and M. Gershenzon, *Phys. Rev.* **B9**, 723 (1974).

19. D. D. Thornton and A. Honig, *Phys. Rev. Lett.* **30**, 909 (1973).

20. L. T. Ho and A. K. Ramdas, *Phys. Rev.* **B5**, 462 (1972).

21. J. E. Baxter and G. Ascarelli, *Phys. Rev.* **B7**, 2630 (1973).

22. G. D. Watkins and F. S. Ham, *Phys. Rev.* **B1**, 4071 (1970).

23. R. M. Chrenko, R. S. McDonald, and E. M. Pell, *Phys. Rev.* **138A**, 1775 (1965).

24. G. Feher, *Phys. Rev.* **114**, 1219 (1959).

25. G. W. Ludwig, *Phys. Rev.* **137A**, 1520 (1965).

26. W. E. Krag, W. H. Kleiner, H. J. Zeiger, and S. Fischler, *J. Phys. Soc. Japan* **21** Suppl., 230 (1966).

27. L. L. Rosier and C. T. Sah, *Solid State Electron.* **14**, 41 (1971).

28. S. Zwerdling, K. J. Button, B. Lax, and L. M. Roth, *Phys. Rev. Lett.* **4**, 173 (1960).

29. A. Onton, P. Fisher, and A. K. Ramdas, *Phys. Rev.* **163**, 686 (1967).

30. G. Feher, J. C. Hensel, and E. A. Gere, *Phys. Rev. Lett.* **5**, 309 (1960).

31. S. M. Sze and J. C. Irvin, *Solid-State Electron.* **11**, 599 (1968).

32. R. S. Smith and J. F. Angress, *Phys. Lett.* **6**, 131 (1963).

33. J. F. Angress, A. R. Goodwin, and S. D. Smith, *Proc. Roy. Soc.* **287A**, 64 (1965).

34. R. J. Elliott and P. Pfeuty, *J. Phys. Chem. Solids* **28**, 1789 (1967).

35. R. C. Newman and R. S. Smith in *Localized Excitations in Solids*, R. F. Wallis, Ed., (Plenum, New York, 1968) p. 177.

36. R. C. Newman, *Infrared Studies of Crystal Defects* (Taylor and Francis, London, 1973) p. 103.

37. R. C. Newman and R. S. Smith, *Phys. Lett.* **24A**, 671 (1967).

38. W. G. Spitzer and M. Waldner, *J. Appl. Phys.* **36**, 2450 (1965).

39. M. Balkanski and W. Nazarewicz, *J. Phys. Chem. Solids* **27**, 671 (1966).

40. S. D. Devine and R. C. Newman, *J. Phys. Chem. Solids* **31**, 685 (1970).

41. R. C. Newman and J. B. Willis, *J. Phys. Chem. Solids* **26**, 373 (1965).

42. R. C. Newman and R. S. Smith, *J. Phys. Chem. Solids* **30**, 1493 (1969).

43. P. G. Dawber and R. J. Elliott, *Proc. Roy. Soc.* **A273**, 222 (1963).

44. R. S. Leigh and B. Szigeti in *Localized Excitations in Solids*, R. F. Wallis, Ed., (Plenum, New York, 1968) p. 159.

45. A. R. Bean and R. C. Newman, *J. Phys. Chem. Solids* **32**, 1211 (1971).

46. G. D. Watkins and K. L. Brower, *Phys. Rev. Lett.* **36**, 1329 (1976).

47. G. D. Watkins, *J. Phys. Soc. Japan* **18** Suppl., 22 (1962).

48. C. A. Coulson and M. J. Kearsley, *Proc. Roy. Soc.* **A241**, 433 (1957).

49. R. P. Messmer and G. D. Watkins, *Phys. Rev.* **B7**, 2568 (1973).

50. G. D. Watkins, *IEEE Trans. Nucl. Sci.* **NS-16**, 13 (1969).

51. G. D. Watkins and J. W. Corbett, *Phys. Rev.* **134A**, 1359 (1964).

52. E. L. Elkin and G. D. Watkins, *Phys. Rev.* **174**, 881 (1968).

53. G. D. Watkins in *Radiation Damage in Semiconductors*, P. Baruch, Ed., (Academic, New York, 1964) p. 97.

54. G. D. Watkins in *Radiation Effects in Semiconductors*, F. L. Vook, Ed., (Plenum, New York, 1968) p. 67.

55. K. L. Brower, *Phys. Rev.* **B1**, 1908 (1970).

56. F. H. Eisen in *Ion Implantation in Semiconductors and Other Materials*, B. L. Crowder, Ed., (Plenum, New York, 1973) p. 99.

57. G. D. Watkins, *Phys. Rev.* **155**, 802 (1967).

58. W. L. Bond and W. Kaiser, *J. Phys. Chem. Solids* **16**, 44 (1960).

59. D. R. Bosomworth, W. Hayes, A. R. L. Spray and G. D. Watkins, *Proc. Roy. Soc.* **A317**, 133 (1970).

60. G. D. Watkins and J. W. Corbett, *Phys. Rev.* **121**, 1001 (1961).

61. J. W. Corbett, G. D. Watkins, R. M. Chrenko, and R. S. McDonald, *Phys. Rev.* **121**, 1015 (1961).

62. K. L. Brower, *Phys. Rev.* **B4**, 1968 (1971).

63. G. D. Watkins and J. W. Corbett, *Phys. Rev.* **138A**, 543 (1965); ibid., 555 (1965).

64. L. J. Cheng, J. C. Corelli, J. W. Corbett, and G. D. Watkins, *Phys. Rev.* **152**, 761 (1966).

65. H. J. Stein, F. L. Vook, and J. A. Borders, *Appl. Phys. Lett.* **14**, 328 (1969).

66. K. L. Brower, *Radiation Effects* **8**, 213 (1971).

67. Y.-H. Lee and J. W. Corbett, *Phys. Rev.* **B9**, 4351 (1974).

68. G. W. Ludwig and H. H. Woodbury in *Solid State Physics*, Vol. 13, F. Seitz and D. Turnbull, Eds., (Academic, New York, 1962) p. 223.

69. F. S. Ham, *Phys. Rev.* **138A**, 1727 (1965).

70. H. H. Woodbury and G. W. Ludwig, *J. Phys. Chem. Solids* **117**, 102 (1960).

71. W. Nazarewicz and J. Jurkowski, *Phys. Stat. Sol.* **31**, 237 (1969).

72. A. E. Cosand and W. G. Spitzer, *Appl. Phys. Lett.* **11**, 279 (1967).

73. A. E. Cosand and W. G. Spitzer, *J. Appl. Phys.* **42**, 5241 (1971).

74. W. J. Moore, *J. Phys. Chem. Solids* **32**, 93 (1971).

75. F. Barra, P. Fisher, and S. Rodriguez, *Phys. Rev.* **B7**, 5285 (1973).

76. R. A. Chapman, W. G. Hutchinson, and T. L. Estle, *Phys. Rev. Lett.* **17**, 132 (1966).

77. G. W. Ludwig, H. H. Woodbury, and R. O. Carlson, *J. Phys. Chem. Solids* **8**, 490 (1959).

78. G. W. Ludwig and H. H. Woodbury, *Phys. Rev.* **113**, 1014 (1959).

79. W. Kaiser and C. D. Thurmond, *J. Appl. Phys.* **32**, 115 (1961).

80. J. A. Baldwin, Jr., *J. Appl. Phys.* **36**, 793 (1965).

81. R. H. Wentorf, *J. Phys. Chem.* **75**, 1833 (1971).

82. H. M. Strong and R. M. Chrenko, *J. Phys. Chem.* **75**, 1838 (1971).

83. C. D. Clark in *Physical Properties of Diamonds*, R. Berman, Ed., (Clarendon, Oxford, 1965) p. 295.

84. W. Kaiser and W. L. Bond, *Phys. Rev.* **115**, 857 (1959).

85. T. Evans and C. Phaal, *Proc. Roy. Soc.* **A270**, 538 (1967).

86. E. V. Sobolev, V. I. Lisoivan, and S. V. Lenskaya, *Sov. Phys. Doklady* **12**, 665 (1968).

87. W. V. Smith, P. P. Sorokin, I. L. Gelles, and G. J. Lasher, *Phys. Rev.* **115**, 1546 (1959).

88. J. H. N. Loubser and L. Du Preez, *Brit. J. Appl. Phys.* **16**, 457 (1965).

89. J. H. N. Loubser and W. P. van Ryneveld, *Brit. J. Appl. Phys.* **18**, 1029 (1967).

90. P. Denham, E. C. Lightowlers, and P. J. Dean, *Phys. Rev.* **161**, 762 (1967).

91. M. Y. Shcherbakova, E. V. Sobolev, N. D. Samsonenko, and V. K. Aksenov, *Sov. Phys. Solid State* **11**, 1104 (1969).

92. A. T. Collins and A. W. S. Williams, *J. Phys.* **C4**, 1789 (1971).

93. R. M. Chrenko, *Phys. Rev.* **B7**, 4560 (1973).

94. P. A. Crowther, P. J. Dean, and W. F. Sherman, *Phys. Rev.* **154**, 772 (1967).

95. D. M. S. Bagguley, G. Vella-Coleiro, S. D. Smith, and C. J. Summers, *J. Phys. Soc. Japan* **21** Suppl. 244 (1966).

96. P. J. Dean, *Phys. Rev.* **139A**, 588 (1965).

97. W. A. Runciman, *Proc. Phys. Soc.* **86**, 629 (1965).

98. C. D. Clark, G. W. Maycraft, and E. W. J. Mitchell, *J. Appl. Phys.* **33**, 378 (1962).

99. W. V. Smith, I. L. Gelles, and P. P. Sorokin, *Phys. Rev. Lett.* **2**, 39 (1959).

100. M. Y. Shcherbakova, E. V. Soboleva, N. D. Samsonenko, V. A. Nadolinnyi, P. V. Schastnev, and A. G. Semenov, *Sov. Phys. Solid State* **13**, 281 (1971).

101. J. A. Baldwin, Jr., *Phys. Rev. Lett.* **10**, 220 (1963).

102. J. Walker, L. A. Vermeulen, and C. D. Clark, *Proc. Roy. Soc.* **A341**, 253 (1974).

Chapter *8*

III-V Compounds

In discussing III-V compounds we restrict ourselves to the nitrides, phosphides, arsenides, and antimonides of aluminum, gallium, and indium. Of these, the nitrides have been little studied. Much more is known about point defects in the gallium and indium compounds than in the less stable aluminum compounds. Table 8.1 contains some properties of interest. All the compounds listed have the zincblende structure shown in Figure 8.1 except for AlN and GaN, which have the hexagonal wurtzite structure. For a given cation the band gap decreases with increasing anion atomic weight, and a similar law holds for the cation sequence Al, Ga, In for any particular anion. Most have direct band gaps. The four compounds with indirect gaps, AlP, AlAs, AlSb, and GaP, have conduction band minima at the X points of the zone or near the X points. All have the valence band maximum at the zone center Γ. Figures 8.2 and 8.3 show band structures[1] of two representative types. The Brillouin zone is the same as for the diamond structure (Figure 7.2).

The combination of controllable semiconduction properties together with a wide range of band gaps has led to many applications of these compounds. Although the vapor pressure at the melting point is very high for the phosphides, the problem of growth of large single crystals from the melt has been solved, and growth of epitaxial crystal layers in gas phase or liquid phase reactions is also well developed for many of the crystals. It is possible to form alloys or mixed crystals of most of the pairs of compounds, perhaps because the lattice spacings are all rather similar. In a mixed crystal, like $Ga_{1-x}In_xAs$, for example, the properties of the system change continuously from those of InAs to those of GaAs as the composition x is varied. The compounds with small band gaps, InAs, $Ga_{1-x}In_xAs$, and InSb, find application as infrared detectors. Those with larger band

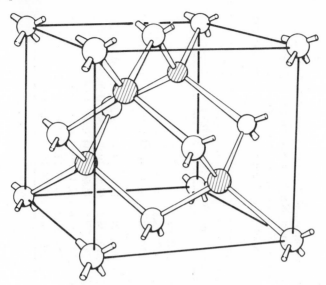

Figure 8.1   Zincblende lattice structure.

gaps are used in diode emitters of visible or near infrared light (GaP, $GaAs_xP_{1-x}$, GaAs, $Ga_xAl_{1-x}As$) and diode lasers (GaAs, $Ga_xAl_{1-x}As$). Other applications include photocathode materials for photomultiplier tubes, Gunn-effect microwave oscillators, and field effect transistors.

The presence of two different types of atom in the lattice leads to a more complicated defect structure than is possible in a single component lattice such as diamond. Impurities from column II of the periodic table may form simple acceptors when they replace cations, and those from column VI may substitute as simple donors at anion sites. Column IV atoms are amphoteric. They behave as donors on cation sites and acceptors on anion sites; conditions of crystal preparation determine which dominates. The two types of vacancy can be expected to have very dissimilar electronic properties, although little is yet known in this regard. Interstitial atoms are less affected by the structure difference. However in place of the single tetrahedral interstitial site of the diamond lattice there are two in the zincblende lattice: one has four cations as nearest neighbors along $\langle 111 \rangle$ directions and six anions as next nearest neighbors along $\langle 100 \rangle$ directions. In the other type the roles of cation and anion are interchanged. EPR measurements have been less useful in elucidating defect structures in the III-V compounds than in silicon. The atoms of the host lattices have nuclear spins, leading to broadening of spectral lines by the unresolved hyperfine structure.

Table 8.1 Some Properties of III–V Compounds. The Parameters Are, from Left to Right, Band Gap, Phillips Ionicity, Nearest Neighbor Separation, Melting Point, Longitudinal Optical Phonon Energy at Zone Center Γ, Static Dielectric Constant, Optical Dielectric Constant, Spin-orbit Splitting of Valence Band at Γ, and Electron Effective Mass.

	E_g (4K) (eV)	f_i	R_{nn} (Å)	T_m (C)	$\hbar\omega_{lo}$ (cm^{-1}, meV)	ε_s	ε_o	Δ_{so} (eV)	m_e^*/m
AlN^a	6	0.449	2.04	>2400					
AlP	2.52(X)	0.307	2.35	>2000	500, 62.0				
AlAs	2.22(X)	0.274	2.43	1740	402, 49.8	10.06	8.16	0.29	
AlSb	1.70(X)	0.250	2.66	1080	340, 42.2	14.4	10.24	0.75	1.5($\parallel$), 0.214($\perp$)
GaN^a	3.47	0.500	2.16	~1500		4			
GaP	2.34(X)	0.327	2.36	1457	402, 49.9	11.0	9.04	0.082	1.5($\parallel$), 0.18($\perp$)
GaAs	1.52	0.310	2.43	1238	292, 36.2	12.5	10.9	0.34	0.0665
GaSb	0.81	0.261	2.65	712	240, 29.8	15.69	14.44	0.80	0.049
InP	1.42	0.421	2.54	1070	345, 42.8	12.3	9.61	0.11	0.078
InAs	0.41	0.357	2.62	943	244, 30.2	14.55	11.8	0.41	0.022
InSb	0.235	0.321	2.80	525	200, 24.8	17.88	15.68	0.82	0.014

^{a}wurtzite structure

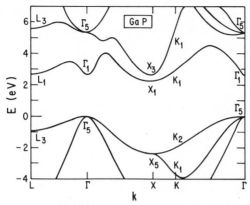

Figure 8.2 Band structure of GaP (from Cohen and Bergstresser, reference 1).

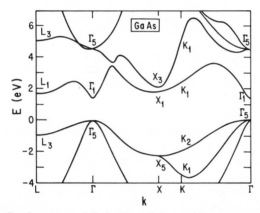

Figure 8.3 Band structure of GaAs (from Cohen and Bergstresser, reference 1).

There is no center of inversion symmetry in the zincblende lattice. The screw operation of the diamond lattice is not an element of the space group. The point group is T_d rather than O_h. One result of the lowered symmetry is the reduced degeneracy of the conduction bands at the point X, evident from comparison of Figures 8.2 and 8.3 with Figures 7.3, 7.4, and 7.5. The splitting results from the difference in electronegativity of anion and cation. The degeneracy of the transverse optical and longitudinal optical phonon branches at the zone center is also lifted. The two frequencies are related by

$$\frac{\omega_{lo}}{\omega_{to}} = \left(\frac{\varepsilon_S}{\varepsilon_o}\right)^{\frac{1}{2}} \tag{8.1}$$

ε_s and ε_o are the static and optical dielectric constants. Another result is the allowed optical absorption by excitation of single phonons. The inequality of ε_s and ε_o in a polar crystal leads to some ambiguity in calculation of impurity energy levels on the effective mass hydrogen-like model. But on the basis of the discussion of Chapter 2, we expect polaron effects to be small.

The small effective mass and large dielectric constant of some of the compounds of Table 8.1 imply that the orbit of an electron bound to a shallow donor is very large in these materials. From $a_D = (\varepsilon_s m / m_e^*) a_0$ we find values of 104, 350, and 676 Å for the ground state radius in GaAs, InAs, and InSb, respectively. For excited states the radii are, of course, even larger. These large values mean that the donor concentration must be very low if properties characteristic of isolated donors are to be measured. Overlapping of orbits leading to impurity "bands" or to hopping of the charge carrier from one impurity to the next can occur already at small concentrations. Because, generally, the concentrations of other impurities than the one to be studied must be even lower, very pure crystals are usually required when spectra of isolated donors are of interest. Similarly bound excitons cannot exist in high concentrations.

GALLIUM PHOSPHIDE

Although GaP has an indirect bandgap and consequently exhibits inefficient radiative recombination for intrinsic luminescence, the material is used to make efficient diode emitters of red or green light. Impurity dopants, especially donors or isoelectronic impurities on phosphorus sites, can produce efficient extrinsic emission. This and the large band gap account for the commercial importance of GaP. At the melting point the phosphorus vapour pressure is 35 atm, making crystal growth a difficult problem. Crystals are commonly grown by the Czochralski method in a high-pressure crystal puller. The melt is covered by an inert liquid such as B_2O_3 to prevent evaporation, and the crystal is pulled through this liquid. The chamber is pressurized with an inert gas. The method is called the liquid encapsulation technique.[2]

The band structure is shown in Figure 8.2. The valence band maximum is at Γ. The conduction band minima are at the zone boundaries $|k_j| = k_0$ in the six $\langle 100 \rangle$ directions in k space, these X point minima lying slightly lower in energy than the Γ point or the L point minima. As the points $\mathbf{k}_j$ and $-\mathbf{k}_j$ differ by a reciprocal lattice vector, there are only three inequivalent minima. In Figure 8.1 the minima are labled X_1, and the band 0.36 eV higher is labeled X_3, where the origin of coordinates is taken at a phosphorus site. The group of the wavevector $\mathbf{k}_j$ is D_{2d}. The X_1 and X_3 correspond to the representations $\Gamma_1(\text{or}\,A_1)$ and $\Gamma_4(\text{or}\,B_2)$ of D_{2d}, respectively. The labels depend on the choice of origin and are interchanged if the origin is taken at a gallium site instead.[3] Let us see why this is so.

In forming the Bloch functions for the X points we must take linear combinations of the factors $\exp(i\mathbf{k}_j \cdot \mathbf{r})$ and $\exp(-i\mathbf{k}_j \cdot \mathbf{r})$ to represent the points $\mathbf{k}_j$ and $-\mathbf{k}_j$ on an equal footing. We take a Ga site as origin. In this way six functions $\phi_x^+, \phi_y^+, \phi_z^+$ and $\phi_x^-, \phi_y^-, \phi_z^-$ are obtained corresponding to the sum and difference of the exponential factors, where j is x, y, or z. The functions $\phi_x^+, \phi_y^+, \phi_z^+$ transform as $\cos(k_0 x)$, $\cos(k_0 y)$, and $\cos(k_0 z)$ and are of X_1 type. The others transform as the sines or as x, y, z and are of X_3 type. The X_1 functions have large amplitude at the gallium sites and nodes at the phosphorus sites. The opposite is true of the X_3 functions. Phosphorus has a core charge of $5|e|$, which is more attractive for electrons than the core charge $3|e|$ of gallium. Thus the X_3 functions concentrated on the phosphorus sites have lower energy and correspond to the three minima. If now we take the origin at a phosphorus site, the X_1 functions have large amplitude at the phosphorus sites and nodes at the gallium sites and correspond to the three minima.

The symmetry group of a substitutional impurity is T_d. For a shallow donor the electronic states are of the form $F\phi$, and the ground states are $F_{1s}\phi$. The origin of coordinates is taken at the impurity. For a donor on a gallium site the functions of interest are $F_{1sx}\phi_x^-$, $F_{1sy}\phi_y^-$, and $F_{1sz}\phi_z^-$. These form a basis for the the T_2 irreducible representation of T_d. For a donor on a phosphorus site the X_1 Bloch functions are used. They form bases for the A_1 and E representations of T_d.

$$\psi_{1s}(A_1) = \left(F_{1sx}\phi_x^+ + F_{1sy}\phi_y^+ + F_{1sz}\phi_z^+ \right)/\sqrt{3} \qquad (8.2)$$

$$\psi_{1s}(Eu) = \left(2F_{1sz}\phi_z^+ - F_{1sx}\phi_x^+ - F_{1sy}\phi_y^+ \right)/\sqrt{6}$$

$$\psi_{1s}(Ev) = \left(F_{1sx}\phi_x^+ - F_{1sy}\phi_y^+ \right)/\sqrt{2}$$

There are several consequences of these different symmetry properties. The ground state of a donor on a gallium site is triply degenerate, but for a phosphorus site donor there may be a splitting of the ground state giving two levels, A_1 and E. The A_1 state has large amplitude at the donor site and is more affected by the central cell potential than the E and T_2 states which have nodes there. The impurity potential can mix the Bloch function corresponding to the $k=0$ Γ_1 conduction band minimum with the A_1 state. In this way donors on phosphorus sites can increase the interband transition probability. The different state symmetries of donors on the two sites also lead to different selection rules for phonon-assisted transitions.[4] Similar effects on the symmetry labels of branches of phonon dispersion curves occur with different choice of origin.

DONORS AND ACCEPTORS

Most shallow donors and acceptors in GaP are usually studied by means of the donor-acceptor pair spectra and the spectra resulting from decay of bound excitons in the visible rather than by the infrared absorption spectra. The bound exciton spectra are discussed in Chapter 2. However, Onton[5,6] and Onton and Taylor[7] have observed the infrared absorption due to S, Si, and Te donors. In the zincblende lattice one-phonon lattice absorption is allowed, leading to spectra cluttered with phonon peaks in contrast to the clean spectra obtained in silicon and shown in Figure 7.8. The S and Te substitute for phosphorus and have A_1 ground states; Si substitutes for gallium and has a T_2 ground state. The positions of the excited p states were found to agree with the predictions of Faulkner's theory with the effective masses

$$m_\parallel^* = 1.5m, \qquad m_\perp^* = 0.180m$$

In the spectral region near 450 meV optical absorption occurs[6] corresponding to transitions to the $2p_\pm$ state associated with the higher lying conduction band minimum and to this band edge, X_3 for S and Te, X_1 for Si. This transition is analogous to the acceptor transitions seen in silicon. It has allowed precise measurement of the separation of the two bands: 355 ± 3 meV.

The two-electron transitions observed in the decay of excitons bound to neutral S, Se, and Te donors have been interpreted as transitions that leave the donor in excited ns or nd states with small splittings of the ns excited states into A_1 and E components.[7] In the case of Te these assignments are verified by stress-induced splitting of the spectral lines.

As expected for an A_1 ground state, the EPR spectrum[8] for S, Se, and Te donors consists of a single isotropic line with g factor slightly less than 2. The $1sA_1 \rightarrow 1sE$ transition is seen by Raman scattering, but apparently is not seen in the two electron exciton transitions. The $1sA_1 - 1sE$ splittings are quite large, amounting to half the binding energy. The values are 54.0, 53.4, and 40.5 meV for S, Se, and Te, respectively.[9] The nature of the transitions was verified by the polarization properties and by stress-induced splitting of the line.

When electron spin is included in the description of the donor states, the double group representations $A_1 \rightarrow \Gamma_6$, $E \rightarrow \Gamma_8$, $T_2 \rightarrow \Gamma_7 + \Gamma_8$ must be used. A small 2-meV splitting of the T_2 ground state of the Sn donor, which substitutes for Ga, has been seen in the luminescence spectrum resulting from decay of an exciton bound to the neutral donor.[10] The EPR spectrum is somewhat similar to that of the lithium donor in silicon, except for the much larger $\Gamma_7 - \Gamma_8$ splitting for Sn. The resonance appears when uniaxial stress is applied, and the line narrows with increasing stress.[11] The axially symmetric resonance spectrum corresponding to a single pair of $\langle 100 \rangle$ valleys is characterized by

$$g_{\parallel} = 1.991, \qquad g_{\perp} = 1.997$$

A large number of combinations of donors and acceptors have been studied by means of the pair spectra. From Equation 2.35 the energy of the emitted recombination radiation at very large donor-acceptor separation R approaches

$$h\nu_{\infty} = E_g - E_D - E_A \qquad (8.4)$$

From analysis of the spectra according to Equation 2.35 the value of $h\nu_{\infty}$ is obtained. If E_g and E_A are known, for example, then E_D can be found. If E_A is known for some acceptor $A^{(1)}$, but not for some other acceptor $A^{(2)}$, then from analysis of the two donor-acceptor spectra $D - A^{(1)}$ and $D - A^{(2)}$, where the donor is the same in both cases, the ionization energy $E_A^{(2)}$ of $A^{(2)}$ can be found.

$$h\nu_{\infty}^{(1)} - h\nu_{\infty}^{(2)} = -E_A^{(1)} + E_A^{(2)}$$

Some such donor-acceptor systems that have been studied are Si—C[12], S—C[12], Si—Zn[12], Te—Si[12], S—Si[12], Si—Si[12], S—Mg[13], Te—Mg[13], S—Cd, S —Zn[14], Se—Si[14], S—Be[15], Te—Be[15], Sn—C[10], Sn—Zn[10], O—Zn[16], and O—Cd[16].

From analysis of the pair spectra one learns whether the donors and acceptors occupy the same type of site or different types. When both are substitutional on the same type of site, the spectrum is commonly called type I. In type-II spectra one occupies a Ga site and the other, a P site. Wiley and Seman[17] have enumerated the separations and numbers of the neighbors at the various separations for the zincblende lattice and for some other lattices of interest. Several corrections to Equation 2.35 have been proposed to improve the fit to experimental data. A van der Waals term, proportional to R^{-6}, is often added to the right-hand side to account for the interaction between the induced dipole moments of the electron and hole charge distributions on the neutral donor and the neutral acceptor. In a theoretical treatment of Merkam and Williams[18] some admixture of p functions to the s states of donor and acceptor is allowed, forcing the axial

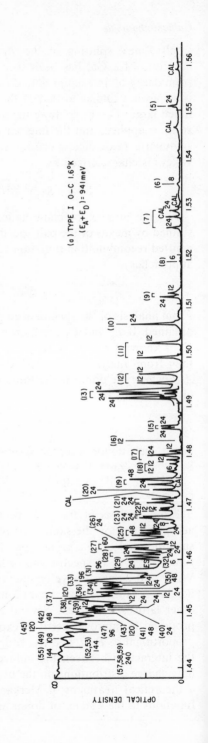

(a) TYPE I 0—C 1.6°K
$(E_A + E_D) = 941$ meV

OPTICAL DENSITY

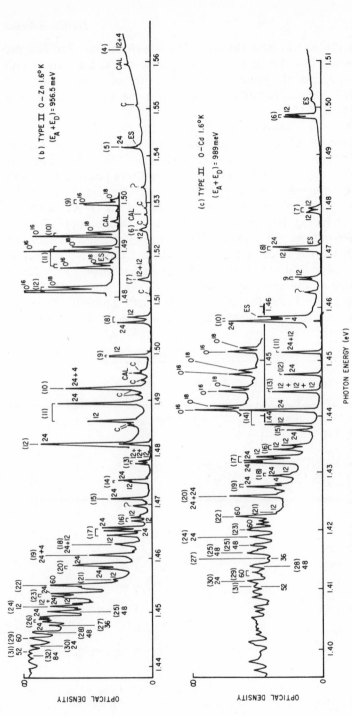

Figure 8.4 Donor-acceptor pair luminescence spectra in GaP. The numbers in parentheses denote the shells. The other numbers denote the number of pairs within a shell. The lines marked C are due to carbon impurities. Oxygen isotope shifts are shown in the inserts (from Dean, Henry, and Frosch, reference 16).

215

symmetry of the pair to be reflected in the wavefunctions. The van der Waals interaction can also be viewed as a mixing of excited p states into the spherically symmetric ground states. In the limit of R large compared with donor and acceptor radii the leading term in the interaction energy is proportional to R^{-6}.

Figures 8.4 and 8.5, from a paper by Dean, Henry, and Frosch[16], show pair spectra for O—C, O—Zn, and O—Cd. A good fit of the data to Equation 2.35 is obtained down to very small separations. The electron is bound so tightly to the oxygen donor that it is not very polarizable. In this sense these spectra are anomolous. Vink, van der Heyden, and van der Does de Bye[19] have made a very careful analysis of the O—C, and S—C pair spectra, including data to $R = 70$ Å, that is to shell number 330. For these large separations the resolution is poor, and the spectra must be analysed with the aid of computor generated simulations. No correction terms are needed to fit the O—C data to Equation 2.35 for $R > 17$ Å. From the fit at large R the two parameters $E_g - E_D - E_A$ and $\varepsilon_s = 11.02$ are accurately obtained. From the nature of the small deviations from the Coulomb law observed at smaller R they conclude that no exisiting theory

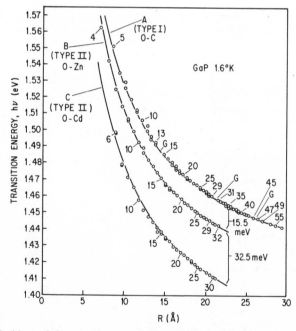

Figure 8.5   Positions of the zero-phonon pair lines of Figure 8.4 plotted versus pair separation. The curves are fits to Equation 2.35 (from Dean, Henry, and Frosch, reference 16).

adequately accounts for these deviations in detail.

Typically no pair spectra are observed for pairs with separation less than some minimum value. The minimum separation is smaller when one of the particles is more tightly bound. Theoretical predictions of this minimum separation have been made, based on simple band extrema.[20] The nearest neighbor O—Zn and O—Cd pairs have been observed and are described in a following section.

Superimposed on the gross dispersion of the pair spectrum by the Coulomb term into a large number of discrete lines with characteristic spacings is often a fine splitting or structure. Several explanations have been proposed for this structure. The acceptors in the nth shell about a donor may for some n be separated into a small number of inequivalent sites. Because of the aspherical charge distributions of the ionized donor and acceptor the interaction will be slightly different for each type of site, leading to the splitting.[21] The core charge distribution is aspherical because of the bonds with the surrounding atoms. For the initial state in which donor and acceptor are neutral, the interaction with the donor may lead to a splitting of the Γ_8 hole state.[22] The interaction might be due to lattice strain caused by the donor. A j—j coupling between electron and hole can also lead to a splitting.

The ionization energies of substitutional donors and acceptors are listed in Table 8.2. As in Si, Li apparently is an interstitial donor. From the spectra due to decay of an exciton bound to the neutral donor and from the Li—Zn donor-acceptor pair spectra Dean[23] concludes that two different types of interstitial Li donor occur corresponding to the two types of tetrahedral interstitial site. Li in the site with four Ga nearest neighbors produces a spectrum with relatively strong zero-phonon lines, as in the case of a donor on P site. This interstitial donor is apparently associated

Table 8.2 Ionization Energies of Isolated Donors and Acceptors in GaP (in meV)

P site donors		Ga site donors	
O	896	Si	82.1
S	104.2	Ge	201.5
Se	102.6	Sn	69.0
Te	89.8		
P site acceptors		Ga site acceptors	
C	46.4	Be	48.7
Si	202	Mg	52.0
Ge	257	Zn	61.7
		Cd	94.3

with another defect, because the local symmetry is trigonal. Ionization energy is 86 meV. Li in the other interstitial site, which because of the smaller amplitude of the band edge wavefunction there is more like a Ga site, has ionization energy 56 meV.

The phonon structure of the distant-pair bands into which the discrete line spectra merge at large R can be understood from consideration of the symmetries of the electronic states. We have seen how the symmetry of the conduction band minima leads for donors on phosphorus sites to an impurity-induced transition probability between conduction and valence bands. For donors on gallium sites the symmetry of the minima is such that coupling to the lowest conduction band at the Γ point cannot occur. Generally the zero-phonon peak in the distant-pair band is much weaker for donors on gallium sites. Transitions assisted by X point phonons of the proper symmetry are allowed for both types of donor. For Ga site impurities these are the longitudinal acoustic, transverse acoustic, and transverse optical modes, and for the P site, the longitudinal acoustic mode.[4]

The excited states of acceptors in GaP have not been extensively investigated. Raman scattering experiments on Zn and Mg acceptors with uniaxial stress applied to the sample have shown that the ground state has Γ_8 symmetry, as expected. Several excited states were also observed.[24]

The localization energy Q of the exciton bound to the neutral donor or acceptor obeys a modified form of Haynes' rule. In contrast with silicon, where $Q/E_A \approx Q/E_D \approx 0.1$, for GaP Q is not proportional to E_D or E_A, although the two quantities are still linearly related. The approximate relations are[23]

$$Q = (0.26)E_D - 7\text{meV}$$

$$Q = (0.056)E_A + 3\text{meV}$$

for donors and acceptors, whether on Ga, P, or interstitial sites.

OXYGEN

Substitutional oxygen is a deep donor with ionization energy 0.89 eV. The donor-acceptor pair spectra for C, Zn, or Cd acceptors are shown in Figure 8.4. The spectra occur in the near infrared region rather than the visible because the large value of E_D shifts the transition energy far from the band gap. The nearest neighbor pairs O—Zn and O—Cd produce the efficient red luminescence used in GaP light emitting diodes. At low temperatures the luminescence consists of a broad band with peak at 1.83 or 1.86 eV for O—Cd or O—Zn, respectively. In the case of O—Cd two sharp zero-phonon lines, one strong (A) and one very weak (B), and several resolved phonon replica peaks are seen on the high energy side. The isotope effect observed on one of these peaks, due to a local resonance mode, on substituting ^{114}Cd for ^{110}Cd confirms the presence of the cadmium atom in the luminescence center.[25] The substitution of ^{18}O for ^{16}O leads to a small shift of the A line to higher energy. A shift of the same magnitude is similarly produced in the infrared spectra of Figure 8.4, as shown in the insert.

From the angular dependence of the Zeeman splitting of the zero-phonon lines it has been shown that the center has $\langle 111 \rangle$ axial symmetry.[25] The luminescence results from decay of an exciton bound to the nearest-neighbor pair. The $j = 1/2$ electron and $j' = 3/2$ hole of the excition form states of total $J = 2$ and $J = 1$. These are split in the trigonal symmetry of the pair. The A line is due to a transition from the $|J = 2, M = \pm 1\rangle$ state to the ground state and the B line, to a transition from the state $|2, \pm 2\rangle$. The $|2, \pm 1\rangle$ state is not pure but is mixed with the $|1, \pm 1\rangle$ state by the axial field. The strong coupling to the lattice is apparently a result of the close localization of the electron.

Superimposed on the luminescence band due to the bound exciton decay is another band corresponding to another luminescence mechanism. The peak of this second band occurs at slightly lower energy. The luminescence of this band decays nonexponentially in the manner typical of luminescence due to the radiation from distant donor-acceptor pairs. This band has been ascribed to the radiative recombination of an electron bound to the nearest-neighbor pair with a hole bound to a distant isolated zinc or cadmium acceptor. Because the oxygen donor level is so deep, an electron may be bound to the O—Zn or O—Cd pair. The pair is similar to an isoelectronic impurity, having no net charge in the absence of the bound electron. The emission frequency would be roughly given by Equation 2.35 with no Coulomb term. The term E_D would represent the binding of the electron to the nearest-neighbor pair. At room temperature a third mechanism may also contribute to the luminescence: the recombination of free holes with electrons bound at the nearest-neighbor pairs.[25] The free holes would be attracted to the net negative charge of the electron bound to the pair. There is evidence that Mg and O form a similar nearest-neighbor complex to which an exciton may be bound.[15] The presence of oxygen is established by observation of the isotope effect. The emission is yellow-orange. Magnesium must be introduced by diffusion, because MgO particles are formed when both impurities are present in the melt.

Another infrared luminescence spectrum has been assigned to oxygen.[27] A zero-phonon line at 0.84 eV accompanied by phonon replicas at lower energy has been shown to be due to oxygen by observation of an isotope shift upon replacement of ^{16}O by ^{18}O. The luminescence is thought to result from a radiative transition of an electron from an excited bound state of the isolated oxygen donor to the ground state. If the effective mass formalism is appropriate for such a deep state, the excited state might be $1sE$ and the ground state $1sA_1$. The zero-phonon line is very weak compared with the phonon replicas. The excited state lies 51.5 meV below the conduction band edge. Transitions to other unidentified higher lying excited states were observed in the luminescence excitation spectrum. Photocapacitance measurements indicate that the oxygen donor can apparently bind a second electron also.[28] The second electron is trapped into a state 0.45 eV below the band edge. Lattice relaxation occurs, lowering this level by the large amount 1.5 eV.

Oxygen can also occupy interstitial sites.[29] An absorption peak at 1002 cm^{-1} has been assigned to interstitial oxygen on the basis of chemical evidence and by analogy with interstitial oxygen in silicon and germanium. The oxygen atom is bound to a nearby gallium and a phosphorus atom as in the oxide $GaPO_4$. The Ga—O—P "molecule" is similar in form to the Si —O—Si and Ge—O—Ge arrangements of interstitial oxygen in silicon and germanium. The absorption is presumably also in this case due to the ω_3 vibration. Crystals grown by the liquid encapsulation Czochralski method are generally contaminated by oxygen from the B_2O_3 encapsulant, and much of the oxygen is apparently in the interstitial form. Other local modes near 1000 cm^{-1} have also been observed by Barker, Berman, and Verleur[29] and have been assigned from comparison with results of a calculation based on a linear chain model to complexes of interstitial oxygen with substitutional carbon and boron impurities.

When lithium is diffused into GaP containing oxygen a red luminescence at low temperature due to decay of an exciton bound to a complex lithium-oxygen associate can be produced.[30] The luminescence is not seen in crystals quenched from 1300 C before diffusion, but does occur in crystals heated to 950 to 1100 C before diffusion at 400 C. The associate is perhaps produced when interstitially diffusing lithium fills a gallium vacancy next to substitutional oxygen atom. The vacancy-oxygen associate, $V_{Ga}-O_P$, may be formed at temperatures in the range 900 to 1000 C by vacancy motion and dissociate at temperatures above about 1200 C.

The spectrum of the lithium-oxygen center consists of four zero-phonon lines near 2.09 eV with phonon replicas at lower energy in emission and higher energy in absorption. In addition to lattice modes, local modes are seen also. From isotopic substitution experiments the presence in the complex of two inequivalent lithium nuclei and one oxygen nucleus is established. The Zeeman effect of the zero-phonon lines has a $\langle 111 \rangle$ axial anisotropy. Dean's model for this defect is the following. An oxygen atom substitutional on a phosphorus site is associated with two lithium atoms.

One of these occupies a nearest-neighbor gallium site, and the other, the nearest interstitial site to the substitutional lithium on the same $\langle 111 \rangle$ axis. The complex requires no charge compensation, having as many valence electrons (one from each lithium atom and six from the oxygen atom) as the gallium-phosphorus pair it replaces. In this respect the defect is similar to the oxygen-zinc and oxygen-cadmium nearest-neighbor pairs. In this case zero-phonon lines from all of the initial states except $|J=2, M=0\rangle$ are observed. This transition, as well as that from $|J=2, M=\pm 2\rangle$ should be forbidden on the simple model. The transition from $|J=2, M=\pm 1\rangle$ becomes allowed through mixing with $|J=1, M=\pm 1\rangle$.

BISMUTH AND NITROGEN

Bismuth and nitrogen are isoelectronic impurities,[31] coming from the same column of the periodic table as phosphorus, the atom they replace. On the basis of the electronegativities listed in Table 2.4 we expect bismuth to be an isoelectronic donor and nitrogen to be an isoelectronic acceptor. A characteristic luminescence spectrum is produced in crystals containing bismuth and one of the donors sulfur, selenium, or tellurium. The spectrum is very similar to the distant pair band in donor-acceptor pair spectra. There is a zero-phonon line accompanied by strong replicas. The shape of the spectrum depends on excitation intensity, and the luminesence decays nonexponentially after pulsed excitation in the same way as the distant pair spectra. Although individual pair transitions are not resolved, the direction of shift of the spectral peak with changing excitation intensity implies that emission frequency increases with increasing pair separation, in contrast with donor-acceptor pair spectra. The spectral positions of the lines depend on the identity of the donor.[32] The spectra result from recombination of holes bound to the bismuth atoms with electrons on the donors. (An isoelectronic donor gives up an electron. It binds a hole.) Because the bismuth atom has no net charge after recombination, the emission energy of the zero-phonon line is given by

$$h\nu = E_g - E_h - E_D \qquad (8.5)$$

E_h is the ionization energy of the hole. The absence of the Coulomb term of donor-acceptor pair spectra accounts for the lack of discrete pair lines. From the measured peak positions $h\nu$ and the known donor ionization energies E_D, the same value of the hole ionization energy, $E_h = 40$ meV, is obtained for each donor.[32]

Generally the optical spectra produced by isoelectronic impurities are due to excitons bound to them, not to single bound particles. The bismuth-donor spectra are unusual. They show that the concept of the isoelectronic donor is really meaningful. Once a hole has been bound by short-range forces, an electron can be bound by Coulomb attraction. Radiative decay of the exciton leads to a luminescence spectrum consisting of two zero-phonon lines, corresponding to transitions from the upper $J = 1$ level and from the $J = 2$ level 2.7 meV below it, accompanied by strong phonon replicas. The $J = 2$ state is split by the small amount 0.25 meV. From analysis of stress-induced splittings of the lines Onton and Morgan[33] find that the $1sE$ state of the electron lies about 30 meV above the $1sA_1$ ground state. The binding energy of the exciton, $E_g - h\nu = 109$ meV, where $h\nu$ is the average emission energy. Because the hole binding energy is 40 meV, the electron and the hole are bound by comparable amounts.

The smaller nitrogen atom is better able to fit into the GaP lattice than the bismuth atom. Nitrogen concentrations as high as 10^{19} cm^{-3} are attainable, but bismuth concentration seems to be limited to about 10^{17} cm^{-3}. This, together with the smaller exciton binding energy and efficient room-temperature luminescence, has led to the commercial importance of GaP:N green light emitters. Figure 2.3 shows the energy levels of the exciton and exciton molecule bound to nitrogen, determined at low concentration. Binding energy of the exciton is only 21 meV. The second exciton is bound by almost the same amount, 19 meV.[34] The intensity of the lines due to decay of the exciton molecule is proportional to the square of the excitation intensity. In the exciton molecule the two electrons are apparently paired off in the same orbital state and do not contribute to the level splittings. The two holes in Γ_8 states couple to give states of total $J = 0$ and $J = 2$. Zeeman measurements confirm the degeneracies of the states. The Zeeman splittings have cubic anisotropy.

In GaP containing acceptors as well as nitrogen a characteristic wing appears in a 25-meV range below each of the two nitrogen bound exciton zero-phonon lines. The wing consists of many finely spaced lines modulated by an intensity variation of larger periodicity called undulations. Several explanations of the wing have been been proposed.[35] There is general agreement that the wing is due to decay of the exciton bound to the nitrogen trap but perturbed by the presence of a neutral acceptor. Street and Wiesner[36] believe that each of the sharp lines in the wing corresponds to a different nitrogen-acceptor separation. The origin of the undulatory modulation is not well understood.

As nitrogen concentration increases, a new series of luminescence peaks with intensity proportional to the square of the nitrogen concentration appears at lower energies and becomes stronger than the luminescence due to isolated nitrogen. This series is due to nitrogen pairs and consists of discrete groups of lines corresponding to various nitrogen-nitrogen separations.[37] Each group contains a number of zero-phonon lines and phonon replicas. The zero-phonon lines are derived from $J = 1$ and $J = 2$ states split in the lower symmetry of the pair. The nearest-neighbor group NN_1 occurs at lowest energy. The other groups NN_i form a series that converges to a higher energy limit at the position of the emission due to isolated nitrogen. An exciton is bound to the pair and the binding energy is larger the smaller the pair separation, amounting to 153 meV for NN_1. The lower solubility of Bi prevents the observation of Bi pairs. For the pairs NN_i with $i>3$, the difference between the pair emission energy and that of isolated nitrogen varies as the inverse cube of the pair separation R_i. Allen[38] has calculated a value of the coefficient of R_i^{-3} in good agreement with experiment assuming that the binding of an exciton to nitrogen N is perturbed by the strain field of N_i and that the energy shift due to strain is that of the band edges. Although the theory of isoelectronic impurities is in an early stage of development, it is clear that lattice relaxation plays an important part.[39] A calculation based on a square-well model of the short-range potential $U(r)$ and on the neglect of lattice strain effects leads to an erratic variation of the NN_i emission energy with increasing R_i rather than the smooth R_i^{-3} dependence observed.[40]

TRANSITION METAL IMPURITIES

Optical absorption due to transitions between levels of the d^n configurations of Fe(d^6), Co(d^7), and Ni(d^8) has been studied by Baranowski, Allen, and Pearson,[41] and by Loescher, Allen, and Pearson.[42] All show spectra characteristic of a tetrahedral environment and presumably substitute for gallium. Co is an acceptor with a level 0.4 eV above the band edge and has the d^7 configuration after accepting an electron. Two transitions from the ground state $^4A_2(^4F)$ to the states $^4T_1(^4F)$ and $^4T_1(^4P)$ occur with the large oscillator strengths 3.9×10^{-4} and 8.3×10^{-4}, respectively. The $e-t_2$ splitting parameter Δ is 5400 cm^{-1} and the Racah parameter $B=290$ cm^{-1}. For Fe the transition energy $^5E(^5D) \rightarrow {}^5_2T(^D)$ yields $\Delta \approx 3344$ cm^{-1}. Ni, like Co, is a deep acceptor. The observed transitions are $^3T_1(^3T) \rightarrow {}^3T_2(^3F)$, $^3T_1(^3F) \rightarrow {}^3T_1(^3P)$, and $^3T_1(^3F) \rightarrow {}^3A_2(^3F)$. The spectra are described by the parameters $B=310$ cm^{-1}, $\Delta=5600$ cm^{-1}. The small splitting of the $^3T_1(^3P)$ level by the spin-orbit interaction indicates a reduction of the spin-orbit coupling by a dynamic Jahn-Teller effect. The values of Δ are somewhat larger in GaP than in the II-VI compounds. B is considerably smaller than in the less covalent II-VI compounds. In the case of Co the value of B is smaller by a factor of 3 than in the free ion.

GALLIUM ARSENIDE

The band structure of GaAs is shown in Figure 8.3. The valence band is similar to that of Ge. The spin-orbit splitting is only slightly larger for GaAs. The band gap is direct. The simple conduction band minimum at Γ

makes GaAs an ideal case for application of the effective mass theory for shallow donors, and the theory describes the donors well.[43] Because of the direct band gap inter-band recombination is faster than in GaP. Recombination involving shallow impurities is similarly faster. The nearer the band edge is the impurity level, the more the impurity wavefunction resembles the band functions.

No bound states due to isoelectronic impurities have been observed in GaAs. On the basis of electronegativity differences N would be expected to bind an electron as in GaP. The smaller mass of an electron in the Γ valley may make such an electron too light to bind to N, however. The electronegativity difference between Bi and As is smaller than between Bi and P. This may explain the lack of observation of a bound state for holes associated with Bi in GaAs.

The simple group-VI and group-IV donors have very small ionization energies near 6 meV, as seen from Table 8.3. The large orbital radius of a donor implies that interactions between donors in the ground state become significant at concentrations greater than 10^{16} cm^{-3}. Concentrations less than 10^{14} cm^{-3} are necessary to avoid interactions between $2p$ excited states.[44] The delocalization of electrons at higher concentrations affects the optical spectra, so that spectra characteristic of isolated donors must be obtained with samples of very low concentration and high purity. In the lasers and incoherently emitting diodes made of GaAs the dopant concentration is higher. Many experiments have been carried out on high-purity samples containing residual donors of unknown identity. The absorption corresponding to transitions from the donor ground state to the excited states lies in the far infrared. These transitions can also be observed in the photoconductivity spectrum. In this simpler technique the incident infrared light excites the electron, or hole in the case of an acceptor, to an excited state from which it escapes to the band with absorption of a phonon. The effective mass theory gives for the ionization energy $E_D = 5.79$ meV. The central cell corrections are very small for the donors of Table 8.3 as a consequence of the large orbit. The electron spends very little time near the donor nucleus.

Table 8.3 Ionization energies of isolated donors and acceptors in GaAs (in meV)

As site donors		Ga site donors	
S	6.10	Si	5.854
Se	5.89	Ge	6.08
As site acceptors		Ga site acceptors	
C	26.7	Be	30
Si	35.2	Mg	30
Ge	41.2	Zn	31.4
Sn	171	Cd	35.4

The small ionization energies of donors and acceptors have the effect of crowding much of the extrinsic luminescence spectra into a narrow spectral region near the band-gap frequency. The weak binding is also responsible for much weaker phonon coupling, more like that shown in the upper part of Figure 3.4, than in the case of GaP, in which the situation is often that shown in the lower part of the figure. The zero-phonon line is uaually by far the strongest in the spectrum. The luminescence resulting from recombination of a free electron with a hole bound to an acceptor or of a free hole with an electron bound to a donor is recognized by the characteristic line shape, given in the former case at low temperature by[45]

$$I(x) = x^{1/2} \exp(-x) \qquad (8.6)$$

$$x = \frac{h\nu - E_g - E_A}{kT}$$

The maximum occurs at the transition energy $E_g - E_A + \frac{1}{2}kT$. The ionization energies of several acceptors have been determined from such spectra.[46,47] Excited states of acceptors have also been studied by means of the two-hole transitions accompanying decay of an exciton bound to a neutral donor.[47] The acceptor Sn has a much larger ionization energy than the other acceptors listed in Table 8.3. It is possible that the spectrum measured is not due to Sn_{As}, but to Sn associated with some intrinsic defect. However, no shallower level has been observed in GaAs containing Sn.[48] The EPR spectrum of the hole bound to Zn and Cd acceptors is observable when uniaxial stress is applied to split the Γ_8 ground state and produce narrower lines.[49]

Donor-acceptor pair spectra are not so useful in GaAs because of the lack of spectral detail. No resolved zero-phonon lines of close pairs occur. Even if an exciton could be bound to a pair with separation less than about 30 Å, the emission frequency would lie above E_g/h and would not be observed. The spectrum consists of a band in the region 1.475 to 1.49 eV, depending on the donor and acceptor.[50] It can be identified by changes in band shape and peak position with change of temperature or excitation intensity and from the time resolved spectra obtained with flash excitation. Because the peak of the spectrum corresponds to pairs with very large separation, these effects are less dramatic than in GaP.

SILICON

Silicon is the most-studied amphoteric impurity. Table 8.3 shows that it is a simple donor when it substitutes for Ga or a simple acceptor on an As site. Infrared emitters with external quantum efficiency (number of photons emitted per electron flowing into the diode) as high as 28% have been made from GaAs : Si, stimulating interest in the defect structure.[51] Electrical measurements indicate that at low Si concentration the crystal is n type, and the carrier concentration is proportional to Si concentration. Silicon largely substitutes on Ga sites. At a Si concentration of about 10^{18} cm^{-3} the carrier concentration begins to increase less rapidly with increasing silicon concentration, as the concentration of Si_{As} becomes comparable with that of Si_{Ga}. This behavior has been studied in detail through the local mode spectra.[52]

Because Si is electrically active, the crystal must be compensated to reduce the absorption due to intraband excitation of free carriers. This is accomplished either by diffusing an acceptor impurity such as Li or Cu into the crystal or by irradiation with electrons or neutrons. Both methods introduce other defects which may pair with Si or as isolated defects produce new spectra. These effects must be taken into account in analyzing experimental data.

As Si concentration increases from a low value, the first local mode absorption to appear[53,54] is a line at 383.7 cm^{-1}. The next is a line at 398.2 cm^{-1}. These are attributed to the local mode of T_2 symmetry of $^{28}Si_{Ga}$ and of $^{28}Si_{As}$, respectively. Gaur, Vetelino, and Mitra[55] have calculated the local mode frequencies of impurities in zincblende crystals assuming that the force constants are the same as for the prefect lattice. They find transition energies of 363 and 365 cm^{-1} for these modes. Lines due to $^{29}Si_{Ga}$ at 378.5 cm^{-1} and to $^{30}Si_{Ga}$ at 373.4 cm^{-1} have also been observed.[56] Three other lines occur at high Si concentration at 367, 393, and 464 cm^{-1} regardless of how compensation is effected. Their strengths are proportional to the product of the strengths of the lines due to Si_{Ga} and Si_{As}. They are due to Si_{Ga}—Si_{As} pairs. Such a nearest-neighbor pair has C_{3v} symmetry and has four infrared active local modes. Spitzer and Allred[54] have estimated the frequencies of these modes from the theory of Elliott and Pfeuty[57] for defects in silicon discussed in Chapter 7. The density of phonon states of GaAs is similar to that of Si except for a scale factor. The Ga and As atoms have nearly the same mass, and there is no gap of forbidden frequencies as, for example, in GaP. If the frequency scale of Elliott and Pfeuty is multiplied by $\omega_M(GaAs)/\omega_M(Si)$, where $\omega_M(GaAs)$ is ω_{lo} of Table 8.1, the predicted transition energies are 327, 369, 390, and 419 cm^{-1}. The line of lowest frequency is near ω_{lo} and is difficult to observe. The mode of highest frequency is one in which the two Si atoms vibrate against each other along the line connecting them. The frequency of this mode is very sensitive to the Si—Si force constant for this reason. In the other modes the Si—Si separation does not oscillate.

Other lines appear in the spectra, depending on the method of compensation.[52] Compensation with ^{6}Li produces six lines at 470, 480, 487,

374, 379, and 405 cm^{-1}. The first three lines shift to 438, 448, and 455 cm^{-1} when ^{7}Li replaces ^{6}Li. The other lines do not shift. The shifts are very nearly given by

$$\frac{\omega(^7\text{Li})}{\omega(^6\text{Li})} = \left(\frac{6}{7}\right)^{1/2} \tag{8.7}$$

This indicates that in these modes the light lithium atom vibrates almost alone. The three modes of lower frequency are mainly vibrations of the Si atom in a Si—Li pair. When Cu rather than Li is used for compenstation these lines shift to 374, 376, and 464 cm^{-1}. The structures of the Si—Li and Si—Cu pairs are unknown. Possibly the Si and Li or Cu atoms are on nearest Ga sites. The resulting orthorhombic symmetry would completely lift the threefold (T_2) degeneracy of the mode of each isolated impurity to give the observed number of lines. In crystals grown by the liquid encapsulation method with B_2O_3 as encapsulant boron contamination leads to the formation of B—Si_{Ga} pairs. Local mode spectra are produced. The structure of the pair is not yet known.[58]

If the Li diffusion takes place at a temperature less than 800 C, the concentrations of Si_{Ga}, Si_{As}, and Si_{Ga}—Si_{As} are independent of the diffusion temperature. When the diffusion temperature is higher, the Si_{As} concentration increases at the expense of Si_{Ga}, perhaps because Li donors are formed at the higher temperatures.[59] In crystals containing Zn acceptors the formation of Si_{As} acceptors and of Si—Si pairs is suppressed.[52] Three lines thought to be due to the Si vibrations of Si_{Ga}—Zn_{Ga} pairs appear at high Zn concentration. Similarly, the presence of Te donors suppresses Si_{Ga}.

LITHIUM

The small lithium atom diffuses rapidly interstitially in GaAs. When lithium is diffused into *p*-type GaAs containing a high concentration of acceptors, the lithium acts as a donor, compensating the acceptors to a high degree. Lithium may also compensate donors, acting in this case as an acceptor.[60] Interstitial lithium would be a simple donor as in silicon. Lithium associates with other impurities quite readily. Much evidence for association has been gathered in the form of local mode spectra, although definite models for the defects have not been established.

Isolated lithium at a tetrahedral interstitial or substitutional site may produce one threefold degenerate (T_2) local mode. When the lithium atom is paired with a much heavier impurity atom, this degeneracy may be lifted in the lower symmetry of the pair. In axial symmetry splitting into a doublet and a singlet would be expected. In orthorhombic symmetry three singlets would result. These modes might still represent nearly independent motion of the very light lithium atom. In addition other modes of the heavier impurity might occur at lower frequencies. In the experimental work reported in the literature, information about the symmetry of a pair has been deduced from counting the lines observed. Degeneracies of the modes have not yet been confirmed by stress studies.

In the case of Li—Mg complexes[61] a set of three local modes due mainly to the lithium vibration occur at 391.5, 404, and 419 cm^{-1} for ^{6}Li and 366.5, 377, and 391.5 cm^{-1} for ^{7}Li. There are also three modes at 318, 338, and 348.5 cm^{-1}, which are due almost entirely to vibration of Mg, because they shift in frequency only very slightly with a change of Li isotope. Magnesium is an acceptor when it substitutes for Ga. A band due to isolated Mg is observed at 331 cm^{-1} for ^{24}Mg. Because the Mg pair lines are near this value, the Mg atom in the pair is presumably also substitutional for Ga. Similarly, the Li pair lines are near those ascribed by Hayes[62] to isolated interstitial Li. The symmetry of the complex is lower than axial from the number of lines observed. Other complexes of Li with Mn,[63,64] Zn,[63] Cd,[64] or Te[65] have also been studied. In these cases the heavier member of the pair is heavier than Mg, and no pair bands due mainly to vibration of the heavier atom are observed.

TRANSITION METAL IONS

Manganese with the $(3d)^5(4s)^2$ configuration behaves as an acceptor and like Zn, $(3d)^{10}(4s)^2$, apparently substitutes for Ga. The EPR spectrum is that expected for the tetrahedral symmetry of this site.[66] The luminescence due to recombination of a free electron with the bound hole occurs at 1.409 eV.[67] The excited states have been observed in absorption.[68] Ionization energy is 0.111 eV. Copper also produces a deep acceptor level. From the luminescence spectrum the ionization energy is found to be 0.155 eV.[69] Copper readily associates with other defects to form complex centers. Two of these centers, thought to contain vacancies as well as copper, have been shown to have trigonal and orthorhombic symmetry by study of the splitting of luminescence lines under applied uniaxial stress.[70] Silver is also a deep acceptor; ionization energy is 0.239 eV.[71] The optical absorption spectra of Fe and Co are similar to the spectra seen in GaP. The crystal field splitting Δ is smaller in GaAs, however.[41]

OTHER CENTERS

In addition to the luminescence near the band gap frequency, other, generally broader bands often occur at lower frequencies. In crystals containing donors a band occurs near 1.2 eV with peak position at slightly higher energy for S, Se, or Te donors than for the Ga site donors Si, Ge, or Sn. The temperature dependence of peak position and band width are those expected for a transition between two localized states with strong vibrational coupling described by the configuration coordinate model of Chapter 3. There is a peak in the luminescence excitation spectrum at lower frequency than that for interband absorption, indicating that luminescence with a Stokes shift can be excited by pumping directly into the localized excited state.[72] Although neither the symmetry nor the structure of the defect center is known definitely, Williams[73] proposes a model consisting of a Ga vacancy associated with the donor, V_{Ga}—Ge_{Ga} or V_{Ga}—Se_{As}, for example, by analogy with the SA center in the II–VI compounds. Similar emission spectra near 1.37 eV with an excitation peak resolved from the band edge absorption occur in GaAs containing Zn or Cd. Hwang[74] attributes the spectra to V_{As}—Zn_{Ga} or V_{As}—Cd_{Ga} associates.

OTHER III–V COMPOUNDS

More is known about defects in GaP and GaAs than in any of the other III–V compounds. For the most part defect studies of the other compounds have been confined to electrical measurements of the effects of donor and acceptor dopants. Optical spectroscopic studies of InP and GaN[75] are in an early stage, but should, eventually, yield much new knowledge.

InP

The band structure of InP is very similar to that of GaAs. The InP band gap is slightly smaller and the spin-orbit splitting of the valence band is smaller by a factor of 3. When crystals are grown by the liquid encapsulation method with B_2O_3 as encapsulant, contamination by boron is apparently less than for GaAs or GaP. Boron local mode spectra are only observed in crystals containing Mn or Si, perhaps implying that these impurities react with B_2O_3 to give B contamination.[76] InP has been investigated much less extensively than GaAs, but impurity spectra in the two compounds seem to be quite similar.[77] The binding energy of an unidentified donor has been found[78] to be 7.7 meV, only about 30% larger than for donors in GaAs. From luminescence due to the recombination of a free electron with a bound hole the acceptor ionization energies 47.3, 56.3, and 95 meV have been measured for Zn_{In}, Cd_{In}, and Hg_{In}, respectively.[79] The luminescence spectra due to decay of an exciton bound to a neutral donor or to a neutral acceptor is also very similar in the two materials.

As in GaP, Bi is also an isoelectronic donor in InP.[80] The binding energy of an exciton to the impurity is 27 meV. In GaP the binding energy is 107 meV. In GaP the $J=1$ and $J=2$ states of the exciton are split apart by 2.8 meV. In InP this exchange splitting is not observed and must be less than 0.2 meV. The loosely bound electron has a very large orbit. The overlap of the electron and hole wave-functions is much smaller, and this accounts for the smaller exchange splitting. The Zeeman splitting of the exciton is described by the Hamiltonian

$$\mathcal{H} = g_e \beta \mathbf{H} \cdot \mathbf{j}' + g_h \beta \mathbf{H} \cdot \mathbf{j} + L\beta \left(j_x^3 H_x + j_y^3 H_y + j_z^3 H_z \right) \tag{8.8}$$

$$j' = 1/2, \qquad j = 3/2$$

$$g_e = 1.15, \qquad g_h = 0.72, \qquad L = -0.066$$

The first term describes the Zeeman effect of the electron; the second and third terms, the Zeeman effect of the hole.

The donor-acceptor luminescence band is similar to that in GaAs. No discrete pair lines occur, only a structureless band.[81] The nature of the spectrum is established by time resolved spectroscopic measurements.

InSb

Crystals of InSb are easily grown. The melting point and the vapor pressure at the melting point are low. The material has found use as an infrared detector because of the small direct band gap. Donor impurity levels are very shallow. From the data of Table 8.1, E_D should be less than 1 meV. Donor orbits are very large, and interactions between donors lead to metallic electronic properties even at low temperatures. In a magnetic field H, the conduction band is split into Landau levels of separation $h\nu_c$, where ν_c is the cyclotron frequency, $2\pi\nu_c = |e|H/m_e^*c$. When this splitting is large enough that $\frac{1}{2}h\nu_c$ is comparable with E_D, the donor wavefunction becomes compressed and distorted.[82] The distortion occurs because the compression is greater in the plane perpendicular to the magnetic field than along the field. The compression reduces the interaction between donors. Observable effects on the Hall coefficient occur in fields of the order of 10^4G.

Acceptor levels are deeper, typically about 9 meV. The energy level structures of Zn and Ge acceptors have been determined by Murzin, Demeshina, and Umarov[83] through far-infrared absorption spectroscopy. The assignment of energy levels to particular states is made by comparison with a calculation including the upper Γ_8 valance band. The spin-orbit splitting of the bands is so large that the Γ_7 band can be neglected. The energy level structures for Zn_{In} and Ge_{Sb} are very similar and are qualitatively similar to that of an acceptor in Ge. The ionization energies are 9.1 and 9.25 meV for Zn and Ge, respectively. From the temperature dependence of the Hall coefficient and of the resistivity Dashevskii et al.[84] find an activation energy $E_A = 9.5$ meV for Mn acceptors, which substitute for In. For Mn concentrations larger than about 10^{14} cm^{-3} interactions between acceptors become significant enough to affect the measurements. Reference 85 contains a summary of the effects of various other impurities on the electrical properties.

AlSb

Ahlburn and Ramdas[86] have studied the far-infrared absorption spectra of Se and Te donors in AlSb. The conduction band has minima along the $\langle 100 \rangle$ directions near the X points. Transitions are observed from the $1sA_1$ ground state to five excited states, the lowest of which is thought to be $1sT_2$. Level degeneracies were checked by measurements with uniaxial stress. The match between measured level positions and positions calculated by the simple effective mass theory is fair. For the intervals between excited states agreement is considerably worse than for donors in Si. Ionization energies are found by adding to the observed $1sA_1 \rightarrow 2p_\pm$ transition energy the calculated binding energy of the $2p_\pm$ state. The resulting ionization energies, which may not be accurately given by this method in this case, are 146.5 meV for Se and 71.3 meV for Te. The ionization energies of the $1sT_2$ state are 29.4 and 30.7 meV for Se and Te. Thus the central cell component of the binding is much stronger for Se than for Te, which should be more similar to Sb, coming from the same row of the periodic table. The position of the $1sT_2$ level is nearly the same for both donors, as is generally the case in other crystals. This analysis is based on the assumption that the conduction band has minima somewhere along the lines Δ, as in Si, but not at the zone edge X. If the minima were at the X points, the band would be similar to that of GaP. In this case there would be no excited 1s level to which a transition is allowed by symmetry.

Donor-acceptor pair spectra in AlSb contain discrete pair lines as in GaP. Type-I spectra, due to Se donors and unknown acceptors also on Sb sites, have lines corresponding to separation $R = 7.5$ Å (third shell) and larger.[87] Ionization energy of the acceptor is $E_A = 0.04$ meV.

GaSb

Although the band gap of GaSb is direct, the minima at the X and L points lie only 85 and 315 meV above the minimum at Γ and have an interesting effect on the donor states. The donor ionization energy expected from the data of Table 8.1 is very small, $E_D(\Gamma) \approx 2.7$ meV. This small value leads to large orbits, strong interaction between donors, and a merging of the donor level into an impurity band—an effective slight lowering of the minimum, even at small donor concentrations. However, the effective masses of electrons in the L and X minima are larger than for the Γ minimum. Impurity levels associated with these minima are deeper than those of the Γ minimum. For the S donor the ground state associated with the X minima is 390 meV below the X minima: $E_D(X)=390$ meV. Thus this level lies 75 meV below the Γ edge. The $E_D(X)$ for Se is apparently about 0.1 eV less, and for Te, about 0.2 eV less. The L ground state of Se lies just above the Γ edge, $E_D(L)\approx0.08$ eV. Probably $E_D(L)$ is greatest for S and least for Te.

It is possible to associate donor levels with particular minima because the minima shift in energy differently with a change of lattice constant, and the donor level shifts in the same way as the minimum with which it is associated. Analysis of the effect of hydrostatic pressure on the electrical properties and on the frequency of recombination radiation yielded these results.[88,89,90,91] Because other III–V compounds have conduction bands with similar properties, these phenomena may not be confined to GaSb. Reference 92 contains a compilation of impurity levels observed in GaSb.

REFERENCES

1. M. L. Cohen and T. K. Bergstresser, *Phys. Rev.* **141**, 789 (1966).
2. S. J. Bass and P. E. Oliver, *J. Crystal Growth* **3**, **4**, 286 (1968).
3. T. N. Morgan, *Phys. Rev. Lett.* **21**, 819 (1968).
4. T. N. Morgan, T. S. Plaskett, and G. D. Pettit, *Phys. Rev.* **180**, 845 (1969).
5. A. Onton, *Phys. Rev.* **186**, 786 (1969).
6. A. Onton, *Phys. Rev.* **B4**, 4449 (1971).
7. A. Onton and R. C. Taylor, *Phys. Rev.* **B1**, 2587 (1970).
8. R. S. Title, *Phys. Rev.* **154**, 668 (1967).
9. D. D. Manchon and P. J. Dean in *Proceedings of the Tenth International Conference on the Physics of Semiconductors*, S. P. Keller, J. C. Hensel, and F. Stern, Eds. (USAEC Division Technical Information, Oak Ridge, Tenn., 1970) p. 760.
10. P. J. Dean, R. A. Faulkner, and S. Kimura, *Phys. Rev.* **B2**, 4062 (1970).
11. F. Mehran, T. N. Morgan, R. S. Title, and S. E. Blum, *Phys. Rev.* **B6**, 3917 (1972).
12. P. J. Dean, C. J. Frosch, and C. H. Henry, *J. Appl. Phys.* **39**, 5631 (1968).
13. P. J. Dean, E. G. Schönherr, and R. B. Zetterstrom, *J. Appl. Phys.* **41**, 3475 (1970).
14. D. G. Thomas, M. Gershenzon, and F. A. Trumbore, *Phys. Rev.* **133A**, 269 (1964).
15. P. J. Dean and M. Ilegems, *J. Lumin.* **4**, 201 (1971).
16. P. J. Dean, C. H. Henry, and C. J. Frosch, *Phys. Rev.* **168**, 812 (1968).
17. J. D. Wiley and J. A. Seman, *Bell Syst. Tech. J.* **50**, 355 (1970).
18. L. Mehrkam and F. E. Williams, *Phys. Rev.* **B6**, 3753 (1972).
19. A. T. Vink, R. L. A. van der Heyden, and J. A. W. van der Does de Bye, *J. Lumin.* **8**, 105 (1973).

20. G. Munschy and B. Strebe, *Phys. Stat. Sol.* **59**, 525 (1973).

21. L. Patrick, *Phys. Rev.* **180**, 794 (1969).

22. T. N. Morgan and H. Maier, *Phys. Rev. Lett.* **27**, 1200 (1971).

23. P. J. Dean, *Bull. Acad. Sci. USSR (Physical Ser.)* **37**, 178 (1973); P. J. Dean in *Luminescence of Crystals, Molecules, and Solutions*, F. Williams, Ed. (Plenum, New York, 1973) p. 538.

24. C. H. Henry, J. J. Hopfield, and L. C. Luther, *Phys. Rev. Lett.* **17**, 1178 (1966).

25. C. H. Henry, P. J. Dean, and J. D. Cuthbert, *Phys. Rev.* **166**, 754 (1968).

26. R. N. Bhargava, *Phys. Rev.* **B2**, 387 (1970).

27. P. J. Dean and C. H. Henry, *Phys. Rev.* **176**, 928 (1968).

28. H. Kukimoto, C. H. Henry, and F. R. Merritt, *Phys. Rev.* **B7**, 2486 (1973).

29. A. S. Barker, Jr., R. Berman, and H. W. Verleur, *J. Phys. Chem. Solids* **34**, 123 (1973).

30. P. J. Dean, *Phys. Rev.* **B4**, 2596 (1971).

31. P. J. Dean, *J. Lumin.* **7**, 51 (1973).

32. P. J. Dean, J. D. Cuthbert, and R. T. Lynch, *Phys. Rev.* **179**, 754 (1969).

33. A. Onton and T. N. Morgan, *Phys. Rev.* **B1**, 2592 (1970).

34. J. L. Merz, R. A. Faulkner, and P. J. Dean, *Phys. Rev.* **188**, 1228 (1969).

35. P. J. Dean in *Progress in Solid State Chemistry*, Vol. 8, J. O. McCaldin and G. Somorjai, Eds. (Pergamon, Oxford, 1973) Ch. 1.

36. R. A. Street and P. J. Wiesner. *Phys. Rev. Lett.* **34**, 1569 (1975).

37. D. G. Thomas and J. J. Hopfield, *Phys. Rev.* **150**, 680 (1966).

38. J. W. Allen, *J. Phys.* **C1**, 1136 (1968).

39. A. Baldereschi and J. J. Hopfield, *Phys. Rev. Lett.* **28**, 171 (1972).

40. R. A. Faulkner, *Phys. Rev.* **175**, 991 (1968).

41. J. M. Baranowski, J. W. Allen, and G. L. Pearson, *Phys. Rev.* **160**, 627 (1967).

42. D. H. Loescher, J. W. Allen, and G. L. Pearson, *J. Phys. Soc. Japan* **21**, Suppl., 239 (1966).

43. G. E. Stillman, D. M. Larsen, C. M. Wolfe, and R. C. Brandt, *Solid State Commun.* **9**, 2245 (1971).

44. C. J. Summers, R. Dingle, and D. E. Hill, *Phys. Rev.* **B1**, 1603 (1970).

45. D. M. Eagles, *J. Phys. Chem. Solids* **16**, 76 (1960).

46. W. Shairer and W. Graman, *J. Phys. Chem. Solids* **30**, 2225 (1969).

47. A. M. White, P. J. Dean, D. J. Ashen, J. B. Mullin, M. Webb, B. Day, and P. D. Greene, *J. Phys.* **C6**, L243 (1973).

48. W. Schairer and E. Grobe, *Solid State Commun.* **8**, 2017 (1970).

49. R. S. Title, *IBM J. Res. Develop.* **7**, 68 (1963).

50. R. Dingle, *Phys. Rev.* **184**, 788 (1969).

51. P. J. Dean and A. A. Bergh, *Proc. IEEE* **60**, 156 (1972).

52. W. G. Spitzer in *Festkörperprobleme*, Vol. 11, O. Madelung, Ed., (Pergamon, London, 1971) p. 1.

53. O. G. Lorimor and W. G. Spitzer, *J. Appl. Phys.* **37**, 3687 (1966).

54. W. G. Spitzer and W. Allred, *J. Appl. Phys.* **39**, 4999 (1968).

55. S. P. Gaur, J. F. Vetelino, and S. S. Mitra, *J. Phys. Chem. Solids* **32**, 2737 (1971).

56. F. Thompson and R. C. Newman, *J. Phys.* **C5**, 1999 (1972).

57. R. J. Elliott and P. Pfeuty, *J. Phys. Chem. Solids* **28**, 1789 (1967).

58. S. R. Morrison and R. C. Newman, *J. Phys.* **C7**, 633 (1974).

59. W. G. Spitzer and W. Allred, *Appl. Phys. Lett.* **12**, 5 (1968).

60. C. S. Fuller and K. B. Wolfstirn, *J. Appl. Phys.* **34**, 1914 (1963).

61. P. C. Leung, L. H. Skolnik, W. P. Allred, and W. G. Spitzer, *J. Appl. Phys.* **43**, 4096 (1972).

62. W. Hayes, *Phys. Rev.* **138A**, 1227 (1965).

63. O. G. Lorimor and W. G. Spitzer, *J. Appl. Phys.* **38**, 3008 (1967).

64. R. S. Title, *J. Appl. Phys.* **40**, 4902 (1969).

65. O. G. Lorimor and W. G. Spitzer, *J. Appl. Phys.* **38**, 2713 (1967).

66. N. Almeleh and B. Goldstein, *Phys. Rev.* **128**, 1568 (1962).

67. T. C. Lee and W. W. Anderson, *Solid State Commun.* **2**, 265 (1964).

68. R. A. Chapman and W. G. Hutchinson, *Phys. Rev. Lett.* **18**, 443 (1967).

69. H. J. Queisser and C. S. Fuller, *J. Appl. Phys.* **37**, 4895 (1966).

70. E. F. Gross, V. I. Safarov, V. E. Sedov, and V. A. Marushchak, *Sov. Phys. Solid State* **11**, 277 (1969).

71. M. Blatte, W. Schairer, and F. Willman, *Solid State Commun.* **8**, 1265 (1970).

72. E. W. Williams and A. M. White, *Solid State Commun.* **9**, 279 (1971).

73. E. W. Williams, *Phys. Rev.* **168**, 922 (1968).

74. C. J. Hwang, *Phys. Rev.* **180**, 827 (1969).

75. M. Ilegems, R. Dingle, and R. A. Logan, *J. Appl. Phys.* **43**, 3797 (1972).

76. R. C. Newman, F. Thompson, J. B. Mullin, and B. W. Straughan, *Phys. Lett.* **33A**, 113 (1970).

77. A. M. White, P. J. Dean, L. L. Taylor, R. C. Clarke, D. J. Ashen, and J. B. Mullin, *J. Phys.* **C5**, 1727 (1972).

78. J. M. Chamberlain, H. B. Ergun, K. A. Gehring, and R. A. Stradling, *Solid State Commun.* **9**, 1563 (1971).

79. A. M. White, P. J. Dean, K. M. Fairhurst, W. Bardsley, E. W. Williams, and B. Day, *Solid State Commun.* **11**, 1099 (1972).

80. A. M. White, P. J. Dean, K. M. Fairhurst, W. Bardsley, and B. Day, *J. Phys.* **C7**, L35 (1974).

81. U. Heim, *Solid State Commun.* **7**, 445 (1969).

82. E. H. Putley in *Semiconductors and Semimetals*, Vol. 1, R. K. Willardson and A. C. Beer, Eds. (Academic, New York, 1966) Ch. 9.

83. V. N. Murzin, A. I. Demeshina, and L. M. Umarov, *Sov. Phys. Semicond.* **6**, 419 (1972).

84. M. Y. Dashevskii, V. S. Ivleva, L. Y. Krol, I. N. Kurilenko, L. B. Litvak-Gorskaya, R. S. Mitrofanova, and E. U. Fridlyand, *Sov. Phys. Semicond.* **5**, 757 (1971).

85. K. F. Hulme and J. B. Mullin, *Solid State Elect.* **5**, 211 (1962).

86. B. T. Ahlburn and A. K. Ramdas, *Phys. Rev.* **167**, 717 (1968).

87. M. R. Lorenz, T. N. Morgan, G. D. Detit, and W. J. Turner, *Phys. Rev.* **168**, 902 (1968).

88. B. B. Kosicki and W. Paul, *Phys. Rev. Lett.* **17**, 246 (1966).

89. B. B. Kosicki, W. Paul, A. J. Strauss, and G. W. Iseler, *Phys. Rev. Lett.* **17**, 1175 (1966).

90. B. B. Kosicki, A. Jayaraman, and W. Paul, *Phys. Rev.* **172**, 764 (1968).

91. A. Y. Vul, G. L. Bir, and Y. V. Shmartsev, *Sov. Phys. Semicond.* **4**, 2005 (1971).

92. I. I. Burdiyan, *Sov. Phys. Semicond.* **7**, 499 (1973).

Chapter *9*

II-VI Compounds

Of the twelve binary compounds formed by the metals Zn, Cd, and Hg of column II of the periodic table and the chalcogens O, S, Se, and Te of column VI, the seven compounds listed in Table 9.1 are now considered. Very little is known about defects in the Hg compounds; HgSe and HgTe are semimetals and are therefore very different from the other compounds. Because CdO has rock salt structure, it is excluded. The compounds listed have the zincblende or the wurtzite structure except at very high pressures, where some of them undergo a transition to the denser rock salt structure.

Although the main structures of these compounds are cubic zincblende and hexagonal wurtzite, shown in Figures 8.1 and 9.1, there are in addition polytypes, most often found in ZnS, which are derivatives of these two structures. For ZnS the wurtzite structure is more stable at temperatures above 1150 C and zincblende, at lower temperatures. A crystal at room temperature may contain regions of each structure in addition to regions (polytypes) transitional between the two structures. It is also possible to obtain crystals of pure zincblende structure or pure wurtzite structure, depending on growth conditions. ZnSe crystals are more often obtained with the zincblende structure. Whereas ZnTe and CdTe almost always crystallize in the zincblende structure, CdS and CdSe crystals are usually obtained with the wurtzite structure. A discussion of growth methods and the resulting structures of the crystals can be found in reference 1. ZnO has the wurtzite structure. The two structures are not so very different. In each case every atom has 4 nearest neighbors and 12 next nearest neighbors. In the wurtzite lattice the overall hexagonal structure destroys the crystallographic equivalence of some of these, dividing the 4 nearest neighbors into two sets of 1 and 3, for example.

The seven compounds have direct energy gaps. The band structure[2] of ZnSe is shown in Figure 9.2. Those of the other compounds with zinc-

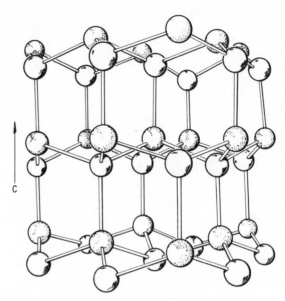

Figure 9.1 Wurzite lattice structure.

Table 9.1 Some Properties of II-VI Compounds; the Parameters Are the Same as for Table 8.1

	$E_g(4K)$ (eV)	f_i	R_{nn} (Å)	T_m (C)	$\hbar\omega_{lo}$ (cm^{-1}, meV)
ZnO[a]	3.44	0.616	1.99		580, 72.0
ZnS	3.80	0.623	2.36	1830	349, 43.3
ZnSe	2.80	0.630	2.45	1520	251, 31.1
ZnTe	2.38	0.609	2.63	1295	206, 25.6
CdS[a]	2.58	0.685	2.61	1475	304, 37.6
CdSe[a]	1.84	0.699	2.74	1239	217, 27.0
CdTe	1.60	0.717	2.78	1098	171, 21.2

	ε_s	ε_o	Δ_{so} (eV)	m_e^*/m
ZnO[a]	8.84($\parallel$), 8.47($\perp$)			0.38
ZnS	8.32	5.13	0.07	0.3
ZnSe	9.2	6.1	0.45	0.16
ZnTe	10.1	7.3	0.93	
CdS[a]	9.00($\parallel$), 8.37($\perp$)			0.180($\parallel$), 0.190($\perp$)
CdSe[a]	10.65($\parallel$), 9.70($\perp$)			0.13
CdTe	10.6	7.2	0.81	0.096

[a]Wurtzite structure.

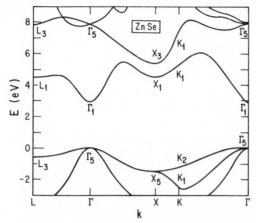

Figure 9.2 Band structure of ZnSe (from Cohen and Bergstresser, reference 2).

blende structure are similar.[2] The effective mass of an electron in the Γ minimum is larger than for the III-V compounds. Some discussion of the band structure of CdS has been given in Chapter 2. The band gap energies span the optical region from the near infrared to the near ultraviolet.

Alloy or mixed crystal formation is possible for many pairs of II-VI compounds. As in the case of III-V compounds, mixed crystals can be obtained if the difference in lattice constants of the two constituents is not too large. Compatible or miscible systems are ZnS-ZnSe, ZnS-CdS, ZnSe-CdSe, ZnSe-ZnTe, ZnTe-CdTe, CdTe-CdSe, and CdS-CdSe.

The difference between cation and anion is greater in II-VI compounds than in III-V compounds. The ionicities listed in Table 9.1 are much larger than those in Table 8.1. Electron transfer from column II atom to column VI atom is more nearly complete, so that the ionic picture of spherical distributions of charge $2|e|$ and $-2|e|$ around cation and anion, respectively, is closer to the truth for II-VI compounds. There are no known amphoteric impurities that occupy either cation or anion site, as there are in III-V compounds. At a defect center holes tend to be localized in orbitals of the more attractive anion and electrons, on cations. Although BeO has been omitted from Table 9.1, it is similar in defect structure to ZnO. The band gap is much larger, 11.2 eV, and the ionicity is similar, $f_i = 0.602$.

Magnetic resonance studies have been very useful in determining properties of defects in II-VI compounds. One reason for this is the low abundances of isotopes of the elements of the host compounds with nonzero nuclear magnetic moments. These are listed in Table 9.2. Except for ^{111}Cd and ^{113}Cd, all these abundances are, at most, a few percent. For this reason spectral lines are not greatly broadened by superhyperfine interactions, as in III-V compounds. But the narrow lines are accompanied by weak satellites due to these interactions, which provide much information about the structure of the defect center. Irradiation of a crystal with light of band-gap frequency at low temperature creates free electrons and holes that can be trapped at defect centers. Thus more than one charge state of a defect may be observed, and charge states may be created that would not occur in thermal equilibrium. Because only about half the number of possible charge states of a defect are paramagnetic, the probability of observing the defect by magnetic resonance is increased. Transfer of charge from one type of defect to another is also conveniently monitored in this way.

Table 9.2   Isotopes of Nonzero Nuclear Spin in II-VI Lattices

Isotope	Nuclear Spin I	Abundance
^{67}Zn	5/2	4.1%
^{111}Cd	1/2	12.8%
^{113}Cd	1/2	12.3%
^{17}O	5/2	0.037%
^{33}S	3/2	0.75%
^{77}Se	1/2	7.5%
^{123}Te	1/2	0.88%
^{125}Te	1/2	7.0%

Although the band gap energies of most of the compounds of Table 9.1 lie in the visible or near ultraviolet region, application in light-emitting devices has been impeded because of the difficulty of forming *p-n* junctions. ZnS powder phosphors are used in color television and in electroluminescent panels, where diode structures are not required. Commercially important photoconductors are based on CdS. It has not been possible to make CdS, CdSe, ZnSe, or ZnO *p* type with low resistivity at room temperature or to make ZnTe *n* type; however, CdTe has been made *n* type or *p* type. For *p*-type CdS, for example, shallow acceptors must be present and must not be entirely compensated by donors. When acceptors are introduced, however, they become compensated by donors that have not been intentionally added. This phenomenon is called self-compensation. The compensating donors might be native defects such as S vacancies or Cd interstitial atoms. Other possibilities are a donorlike complex of the acceptor impurity with a native defect or the acceptor impurity at a different type of site, where it functions as a donor. An example of this last possibility is an interstitial alkali atom (donor) that could compensate an alkali acceptor substitutional on a Cd site. When an acceptor impurity is compensated by a donor the energy of the system decreases by the amount $E_g - E_D - E_A$ if the electron on the donor is transferred to the acceptor. Self-compensation occurs if this energy is greater than the energy necessary to form the compensating donor defect.[3,4]

DONORS AND ACCEPTORS

The elements expected to be simple donors in the II-VI compounds are the halogens on chalcogen sites, the column III metals B, Al, Ga, In, and Tl on metal sites, and interstitial alkali atoms. The donors which have been studied and that bind electrons in shallow hydrogenlike states have ionization energies near the values predicted by the effective mass theory for such states. However, many of these impurities—F and Ga in ZnS, for example—bind the electron more tightly. More information is available about shallow donors in CdS than in other II-VI crystals. The ionization energies are near the hydrogenic value. In this respect CdS is similar to GaAs and unlike Si, Ge, and GaP. The depths of shallow-donor levels in ZnSe, CdSe, and CdTe also seem to be near the hydrogenic values.[5]

With acceptors the situation is similar. Of the column-V atoms on chalcogen sites or alkali atoms on metal sites some form shallow states, but others are deeper acceptors. The noble metals Cu, Ag, and Au on metal sites may also be acceptors.

Resolved discrete donor-acceptor pair lines are more difficult to observe in II-VI compounds than in GaP. They are often masked by the bound exciton luminescence, which is stronger in a crystal with a direct band gap. The lines are easily broadened by strain. Care must be taken to keep impurity concentrations low and otherwise to avoid introducing strains during crystal growth.[6] Discrete pair spectra have been observed in CdS,[6] CdSe,[7] ZnSe,[8] and CdTe.[7] In ZnO there are apparently no shallow acceptors, and luminescence near the band gap energy is due to recombination of free and bound excitons.

Native defects which act as donors or acceptors can often be created by annealing the crystal at high temperature in the vapor of one of the components to bring about a measureable deviation from perfect stoichiometry. Hall measurements at high temperature have shown in some cases whether the donor or acceptor is singly or doubly ionized at the temperature of interest.[9,10] Results of this sort of measurement are more easily interpreted than in cases in which measurements are made at low temperature and the high temperature composition is assumed to be frozen by quenching. The measurements do not determine whether the active species is a vacancy or an interstitial, however. Some donor ionization energies are given in Table 9.3.

Table 9.3 Ionization Energies E_D of Simple Donors in II-VI Compounds (in meV)

CdS	Ga	33.1	F	35.1
	In	33.8	Cl	32.7
	Li^a	28.6	Br	32.5
	Na^a	31.5	I	32.1
CdSe	unknown	19.5		
CdTe	Al	14		
	In	14		
ZnSe	Al	26.3	F	29.3
	Ga	27.9	Cl	26.9
	In	28.9		

^{a}Interstitial donor.

HALOGEN DONORS

Halogen atoms are incorporated into crystals grown by the chemical-vapor transport method in which a chemical reaction occurs in a higher temperature zone of a growth tube, and the reverse reaction takes place in a zone of slightly lower temperature. For example, in the case of ZnSe in the hotter zone, the ZnSe powder source reacts with I_2 gas,

$$ZnSe(s) + I_2(g) \rightarrow ZnI_2(g) + \tfrac{1}{2}Se_2(g) \qquad (9.1)$$

ZnI_2 is also a gas at the growth temperatures. In the cooler zone the reaction is reversed and the crystal, which will contain some iodine, grows.[11] Other incorporation methods are also used. The halogen substitutes for the chalcogen. Halogen donors have been studied in ZnO, ZnS, ZnSe, CdS, and in $ZnS_{1-x}Se_x$ mixed crystals.

A comprehensive optical spectroscopic study of donors in CdS has been made by Henry and co-workers.[12, 13] The bound exciton spectra are identified by the characteristic Zeeman splittings described in Chapter 2. From the two-electron transitions of the spectrum of the exciton bound to the neutral donor the positions of several excited states of the donor can be determined. Adding the observed transition energy of the $1s \rightarrow 2p_\pm$ transition to the calculated binding energy of the $2p_\pm$ state, 8.3 meV, gives the ionization energy E_D of the donor. For F, Cl, Br, and I donors the ionization energies are between 32 and 35 meV, the F level lying deepest. The dissociation energy Q of the exciton bound to the neutral donor is given by $Q = (0.23)E_D$.

In ZnSe the bound exciton spectra have also yielded information on the electronic states of the F and Cl donors.[14] The two-electron transitions shown in Figure 9.3 accompanying the 'I_2 line', the strong line due to decay of the exciton bound to the neutral donor, are identified by Zeeman splittings. Excitons may also be bound to the ionized donors. For ZnSe $Q = (0.15)E_D$.

The EPR spectrum of the donors Cl, Br, or I in ZnS and of Cl in CdS at concentrations sufficiently high that donor banding occurs consists of a single line with g tensor nearly identical with that of a free conduction electron.[15] Spectra of these donors have not been observed in the lower concentration ranges, where localization of the electron could lead to hyperfine structure from interaction of the electron with the magnetic moment of the donor nucleus, as in Si. In mixed crystals of composition $ZnS_{1-x}Se_x$ the g factor of the donor resonance varies smoothly with composition x from a value characteristic of ZnS to that for ZnSe.[16]

The fluorine donor in ZnS binds an electron more tightly than the other halogens.[17] The EPR spectrum consists of two strong lines due to the hyperfine splitting produced by the ^{19}F nucleus with nuclear spin $I = \frac{1}{2}$ and weaker satellite lines from the superhyperfine splittings from the 4 nearest-neighbor Zn nuclei and 12 next-nearest Zn nuclei. The resonance is due to an electron trapped largely on the 4 nearest Zn neighbors. The density of the extra electron on one of these sites is $|\phi(0)|^2 = 4 \times 10^{24} \text{ cm}^{-3}$, whereas the

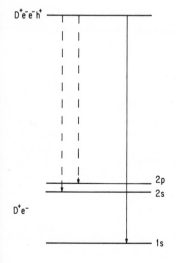

Figure 9.3 Lowest energy level of the exciton bound to a neutral donor and three lowest levels of the neutral donor on an expanded scale showing 2 two-electron transitions (dashed lines).

density on the central F site is half as large. This compares with a density of 0.4×10^{24} cm^{-3} for the electron at the site of the P donor in Si. The resonance is photosensitive and is enhanced by irradiation of the crystal with ultraviolet light. Electrons excited by the light are attracted by the net positive charge of F$^-$ on a S^{2-} site, but the electron becomes localized on the more positive Zn^{2+} neighbors. The g factor is 1.9964, negatively shifted from the free electron value 2.0023.

Fluorine in BeO also binds an electron tightly. At 4 K the electron occupies an orbital spread over the fluorine ion and two of the three Be neighbors in nonaxial positions.[18] At higher temperature the electron hops among all three nonaxial neighbors. An orbital on the axial neighbor apparently lies too high in energy to be occupied.

GROUP-III DONORS

The group III atoms Al, Ga, and In behave as donors, substituting on cation sites. In BeO substitutional B is a donor with a very strongly localized electron.[18, 19] In ZnS, Ga, In, and Tl are anomolous with more highly localized states, but Al is a shallow donor.

In CdS the ionization energies of the Ga and In donors determined from the bound exciton spectra are nearly identical and are similar to the ionization energies of the halogen donors, as can be seen from Table 9.3. Also in ZnSe the Al, Ga, and In donors are very similar to F and Cl donors in electronic properties.[13] The donor-acceptor pair spectra, in which the donor is Al, Ga, or In and the acceptor is Li, are of type I, indicating that donor and acceptor occupy the same type of site.[8]

The donors Al and In have the same ionization energy in CdTe.[20] Donor-acceptor pair spectra consist of broad bands due to recombination at distant pairs. No discrete lines due to closer pairs have been seen in most investigations. The pair band is identified by a shift of the band to shorter wavelengths with increasing excitation intensity.

When irradiated at low temperatures with blue light, ZnS containing Ga, In, or Tl impurities displays an EPR spectrum characteristic of the divalent ion that has a single unpaired electron tightly bound in an s orbital.[21] The resonance is isotropic and is described by the spin Hamiltonian

$$\mathcal{H} = g\beta \mathbf{H} \cdot \mathbf{S} + A\mathbf{I} \cdot \mathbf{S} + g_n \beta_n \mathbf{H} \cdot \mathbf{I} + \sum_i \mathbf{S} \cdot \mathbf{T}_i \cdot \mathbf{I}_i \qquad (9.2)$$

The electron spin $S = 1/2$ and I is the nuclear spin of Ga, In, or Tl. The superhyperfine interaction with the four sulfur neighbors, represented by the last term is observed for Ga. This establishes that the impurity is either substitutional for Zn or interstitial. The latter possibility can be discounted as very unlikely. The hyperfine splitting is large, and from it the electron density on the central atom can be calculated. These densities are more than an order of magnitude larger than for donors in Si. Almost as much electron density is localized on the four nearest S neighbors, largely in *p* orbitals with lobes pointing toward the central Ga.

The charge state of these substitutional impurities before irradiation is uncertain. Ga_{Zn}^{+} could trap a hole produced by the irradiation to become Ga_{Zn}^{2+}, or Ga_{Zn}^{3+} could trap an electron. Initially, both Ga^{+} and Ga^{3+} may be present, providing charge compensation. GaS contains equal numbers of trivalent and monovalent cations. Ga, In, and Tl in ZnS are thus quite different electronically from Al in ZnS[15] or Al, Ga, and In in ZnSe.

The boron donor in BeO binds an electron tightly in an antibonding orbital localized mainly on the boron ion and one of the three oxygen neighbors in nonaxial or basal sites. The electron is more tightly bound than in the case of the F donor in BeO and does not hop among the three equivalent neighbors.[18, 19]

ALKALI METAL ACCEPTORS AND DONORS

From Table 9.4 the only alkali metals shown to be shallow acceptors in II-VI compounds are Li and Na. In CdS and CdSe the other alkali metals K, Rb, and Cs are insoluble.[4] These atoms or their ions are larger than Cd and larger by an even greater margin than Zn, and so they may be insoluble in other II-VI compounds as well.

Table 9.4 Ionization Energies E_A of Simple Acceptors in II-VI Compounds (in eV)

CdS	Li	0.165	P	~1
	Na	0.169	As	~1
CdSe	Li	0.109		
	Na	0.109		
CdTe	Li	0.03		
	Na	0.03		
ZnO	Li	~1		
	Na	~1		
ZnSe	Li	0.114	P	~0.7
	Na	0.1		
ZnTe	Li	0.03	P	0.03

In ZnSe sharp line donor-acceptor pair spectra, where the donor is Al, Ga, or In and the acceptor is Li, have led to precise measurement of the acceptor ionization energy, $E_A(\text{Li}) = 114$ meV. The spectra are of type I, showing that the donor and the acceptor occupy the same type of substitutional site.[8] Small shifts occur in the pair lines when the Li is enriched in ^{6}Li, establishing conclusively that Li is the acceptor. Luminescence due to decay of an exciton bound to the neutral acceptor has also been observed. For Li the dissociation energy $Q = (0.083)E_A$. A bound exciton line thought to be due to Na has $Q = 8.5$ meV, leading to an estimated ionization energy for the Na acceptor, $E_A \approx 0.0085/0.083 = 0.1$ eV. In ZnSe the presence of K acceptors has been inferred from their compensation of Fe donors.[22]. In crystals containing K and Fe the EPR signal of substitutional Fe^{3+} with the $(3d)^5$ configuration is much enhanced compared with crystals containing only Fe, largely present in the form Fe^{2+}.

The alkali impurities Li and Na produce bound exciton and donor-acceptor pair spectra also in CdS and CdSe.[4] The donor that compensates the acceptor is probably interstitial Li or Na. In the case of Li in CdS Ploix and Dugue[23] have shown that as the Li concentration is varied, the intensity of the Li donor bound exciton line is proportional to the intensity of the Li acceptor bound exciton line. Assuming that the donor exciton spectra are due to decay of an exciton bound to the neutral donor, Henry, Nassau, and Shiever[4] estimate the ionization energies $E_D(\text{Li}) = 28.6$ meV and $E_D(\text{Na}) = 31.5$ meV in CdS from the measured value of Q and the ratio Q/E_D found for substitutional donors. The zero-phonon lines are very weak in the donor-acceptor pair spectra. From the position of the peak due to distant donor-acceptor pairs and these donor ionization energies the acceptor ionization energies $E_A(\text{Li}) = 165 \pm 6$ meV and $E_A(\text{Na}) = 169 \pm 6$ meV are obtained. These values are slightly less than 10 times the dissociation energy Q of the exciton bound to the neutral acceptor.

From electrical measurements ionization energies $E_A = 0.03$ eV have been associated with the Li acceptor in ZnTe and Li and Na acceptors in CdTe.[24] Luminescence spectra in ZnTe have been ascribed to radiative recombination of a free electron with the hole bound to the Li acceptor.[25]

In ZnO substitutional Li has a deep acceptor level about 1 eV above the valence band edge.[26,27,28] In the language of an ionic model the Li^+ ion substitutes for the Zn^{2+} ion. The closed electronic configuration of Li^+ is so stable that a hole is trapped on one of the four nearest neighbor O^{2-} ions rather than at the Li site. The unpaired electron produces a characteristic EPR spectrum, which is different depending whether the hole is trapped at the axial oxygen site or at one of the three nonaxial sites. These two kinds of center are shown in Figure 9.4. Both types of resonance are seen. The axial site is energetically favored by 15 meV compared with the

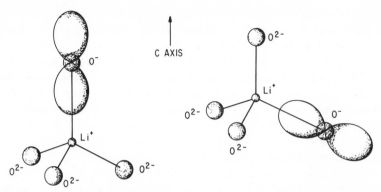

Figure 9.4 Lithium acceptor center in ZnO. The hole is trapped at an axial oxygen site (left) or at a basal oxygen site.

nonaxial sites, and this energy difference leads to a variation of the relative intensities of the two types of spectra with temperature. The greater stability of the hole at the axial site can be understood on the basis of a point-charge model of the lattice.

The axial center produces a resonance with axial symmetry, and the nonaxial centers lead to spectra with orthorhombic symmetry. The magnitudes of the resonance parameters are very nearly the same for both types of spectrum. Hyperfine interactions as a result of coupling of the electron with the nuclear moments of the four nearest metals ^{6}Li or ^{7}Li and three ^{67}Zn neighbors confirm the model. From the magnitudes of the components of the tensor characterizing the Li hyperfine interaction it can be inferred that the magnetic electron occupies a $2p$ orbital of O^- which is oriented along the Li-O axis, as shown in Figure 9.4. About 90% of the electron density is concentrated in this orbital. This Li-O separation is about 40% greater than the normal spacing because of the presence of the hole.

The magnetic properties, then, are those of an O^- ion in a strong axial field along the $Li^+ - O^-$ axis. In the case of the nonaxial centers, there is in addition the weak polar field along the c axis.[26] For the axial center the p_x and p_y orbitals are degenerate, lying below the p_z orbital by an energy ΔE. The z axis is taken to be parallel to the c axis. The five electrons of the $(2p)^5$ configuration of O^- fill the p_x and p_y orbitals, and half-fill the p_z orbital. These orbitals can be viewed as perturbed valence band states. The g tensor is then given by

$$g_{\parallel} = 2.0023, \qquad g_{\perp} = 2.0023 + 2|\lambda|/\Delta E \qquad (9.3)$$

The experimental values are $g_\parallel = 2.0028$, $g_\perp = 2.0053$. Inserting the spin-orbit parameter λ of a free O^- ion into Equation 9.3 yields $\Delta E = 1.4$ eV. An exactly similar Li center occurs in BeO.

The Li center produces a yellow luminescence peaked at 0.615 μm (2.02 eV). This band is generally only poorly resolved from a green band associated with residual Cu impurities, but can be excited separately in thermoluminescence.[27,28] The polarization of the yellow luminescence, and in particular the temperature dependence of the polarization, reflect the distribution of the holes among the axial and nonaxial centers. As the sample temperature rises, thermoluminescence occurs when electrons are released from shallow donor states and recombine with the holes at the Li acceptor centers. From the peak of the luminescence band the acceptor level is calculated to be $E_A = 1.3$ eV. If the high-energy limit 2.5 eV (the approximate position of a possible zero-phonon line) is used instead, the value $E_A = 0.8$ eV is obtained. Thus E_A is about 1 eV.

When Li is diffused into a relatively pure ZnO crystal, the conductivity decreases as the inverse square of the Li concentration in the range where the concentration of Li is much greater than that of residual donor impurities. The substitutional Li acceptor centers are compensated by Li donor centers, presumably interstitial Li.

Na forms a similar acceptor center in ZnO with associated EPR spectrum and luminescence band. Ionization energy is also about 1 eV. In the case of Na the nonaxial centers are more stable than the axial center, however. This stability cannot be explained on an electrostatic model. In BeO the Li acceptor is very similar to the ZnO center.[26] In BeO the axial center is more stable by 34 meV, compared with 14 meV for ZnO. In the cubic crystals MgO, CaO, and SrO a hole may also be trapped on one of the oxygen neighbors of a substitutional lithium ion.[30]

GROUP-V ACCEPTORS

The column V atoms P and As substitutional on anion sites are acceptors. Their electronic properties vary considerably from compound to compound. In ZnTe they are shallow acceptors, but in CdS and ZnSe they are deep acceptors.

Electrical measurements[24] indicate an acceptor level due to P in ZnTe with $E_A = 0.03$ eV. A luminescence band at 2.3 eV associated with P has been attributed to recombination of free electrons with holes bound to the acceptors.[25] The band shifts in the same way as the band gap with varying temperature. An absorption band with peak near 1.2 eV is due to excitation of an electron from the Γ_7 valence band to the P or As acceptor level. At higher temperatures some of the acceptors are ionized, and transitions from the Γ_7 band to empty states of the higher-lying Γ_8 valence band also contribute to the absorption.[31]

In CdS both P and As are apparently deep acceptors.[32] They produce broad luminescence bands, which have been ascribed to recombination of free electrons with holes bound to the acceptor. However, the spectral shapes of the bands are not simple. Other processes may contribute to this luminescence also. Photoconductivity occurs for wavelengths of the exciting light shorter than 0.80 or 0.85 μm for samples containing P or As, respectively. These thresholds have been interpreted as the minimum energy necessary to remove an electron from the ionized acceptor to the conduction band. Ionization energies for both acceptors are found to be near 1 eV.

Magnetic resonance measurements have shown that P and As in ZnSe, although they substitute for Se and are not associated with another defect, form trigonally distorted centers.[33] The distortion is a manifestation of the Jahn–Teller effect, which may be responsible for the larger ionization energy. It is not clear why this effect should occur in ZnSe but not in ZnTe.

The EPR spectra are easily identified by the hyperfine interaction with the ^{31}P$(I = 1/2)$ or ^{75}As$(I = 3/2)$ nuclei. Four equivalent centers occur corresponding to the different $\langle 111 \rangle$ symmetry axes. For one of these the spin Hamiltonian is

$$\mathcal{H} = \beta \mathbf{H} \cdot \mathbf{g} \cdot \mathbf{S} + \mathbf{I} \cdot \mathbf{A} \cdot \mathbf{S} + \mathbf{I}_{Zn} \cdot \mathbf{K}_{Zn} \cdot \mathbf{S} + \mathbf{S} \cdot \sum_{i=1}^{6} \mathbf{K}_{Se} \cdot \mathbf{I}_{Se} \qquad (9.4)$$

The electron spin $S = 1/2$. The tensors $\mathbf{g}$, $\mathbf{A}$, and $\mathbf{K}_{Zn}$ have $\langle 111 \rangle$ axial symmetry, $\mathbf{K}_{Se}$ represents a superhyperfine interaction with the six next nearest neighbor Se nuclei, and $\mathbf{K}_{Zn}$ is due to interaction with the nucleus of the nearest neighbor Zn atom on the symmetry axis. For As a term due to the As nuclear quadrupole moment must be added to (9.4). Analysis of the measured parameters shows that the paramagnetism is concentrated largely in a p orbital of P or As with lobes aligned along the $\langle 111 \rangle$ symmetry axis of the center.

When uniaxial stress is applied at 1.3 K along the [111] axis, the centers with this symmetry axis are lowered in energy with respect to the other centers, and a redistribution of population among the distortion directions occurs. Because the redistribution takes place at a low temperature where atomic motion is unlikely, the trigonal symmetry is shown to be due to a spontaneous distortion of the Jahn–Teller type, which removes the three-fold degeneracy associated with a hole in a p shell, rather than to association with another defect. The distortion is such that the P atom and three of the four nearest Zn neighbors tend toward a coplanar arrangement.

There are luminescence bands associated with P and As in ZnSe. In the case of P the luminescence has been studied in detail.[34] The luminescence band with peak at 1.9 eV is nearly Gaussian in shape with a width at half maximum $W(T)$ given by

$$W(T) = A + B \left(\coth \frac{\hbar \omega}{2kT} \right)^{\frac{1}{2}} \qquad (9.5)$$

This expression differs from Equation 3.45 because of the additional constant A. This constant may be due to another vibrational mode of frequency much larger than ω (and larger than $kT/\hbar$) associated with the excited state.

A complex consisting of a cluster of four phosphorus atoms has been identified in ZnSe with P concentration about 10^{18} cm^{-3}. The EPR spectra are characterized by axial symmetry about the $\langle 111 \rangle$ directions, g factors near two, a large hyperfine interaction with one phosphorus nucleus, and smaller hyperfine interactions with three other equivalent phosphorus nuclei.[35] The tensor characterizing the large hyperfine splitting has the same $\langle 111 \rangle$ axial symmetry as $\mathbf{g}$; whereas that of the smaller interaction has $\langle 110 \rangle$ axial symmetry. The associate is a cluster of four P atoms located at the points of a regular tetrahedron.

The paramagnetism is mainly concentrated in a p orbital on one phosphorus, and the configuration is probably p^5. It is not possible to infer from the data how this complex is incorporated in the ZnSe lattice. Such tetrahedra occur about each substitutional and interstitial site. All these sites could give rise to the observed hyperfine interaction with three equivalent selenium nuclei when the magnetic field is parallel to the $\langle 111 \rangle$ symmetry axis. It seems most likely that phosphorus would substitute for the four zincs about one of the two interstices, one phosphorus having approximately the $(3s)^2(3p)^5$ configuration and the others, $(3s)^2$. Thus the complex has one excess negative charge. The P_4O_6, P_4O_{10}, and P_4S_{10} molecules have the same structure as the complex and its immediate neighbors would have at this site. A compensating defect along the trigonal axis may make one phosphorus inequivalent to the other three.

VACANCIES

Both types of vacancy have been investigated in a number of II–VI compounds, largely by magnetic resonance studies of the ground electronic state. The metal vacancy is an acceptor center. It can trap a hole into the orbitals of the four surrounding chalcogen neighbors. Because of the degeneracy associated with the chalcogen p orbitals, a spontaneous distortion occurs, and the hole is localized on one of the four neighbors. In this way the metal vacancy is similar to the vacancy in silicon. The chalcogen vacancy is donorlike. The trapped electron is distributed equally over the s orbitals of the four metal neighbors, at least in the zincblende lattice, where the four neighbors are crystallographically equivalent. This vacancy is then similar to the halogen vacancy or $\mathscr{F}$ center in the alkali halides. Indeed, the method of Gourary and Adrian for the $\mathscr{F}$ center described in Chapter 1 has even been applied to calculate properties of the chalcogen vacancy.[36] Vacancies are thought to be important in self-compensation of donor and acceptor impurities and vacancies associate with certain impurities to form more complex defects. In ZnSe the disappearance of isolated metal vacancies and their association with donor impurities has been monitored.[37]

CHALCOGEN VACANCY

In ZnS this vacancy center produces a purple coloration of the crystal and a characteristic EPR spectrum.[38,39] Vacancies have been created by annealing crystals in liquid zinc, by bombardment with fast neutrons, or by crushing crystals to powder. The EPR spectrum consists of a strong isotropic line with g factor 2.0034 and weaker satellites. The satellites are due to hyperfine interaction of the magnetic electron with the nuclear moments of the four nearest neighbors, and the hyperfine interactions have $\langle 111 \rangle$ symmetry axes. The resonance is due to an electron trapped by the vacancy. The electron is localized largely in s orbitals on the four zinc neighbors. The ground state of the center is somewhat similar to that of the F_S center in ZnS. In the F_S center there is a larger electron density on the zinc neighbors, perhaps because the electron is repelled by the negative central fluorine ion.

The coloration is due to absorption bands peaked at 0.430 and 0.545 μm. These bands disappear and another appears at 0.355 μm when the crystal is heated to 300 C. Irradiation with ultraviolet light into the 0.355-μm band produces the EPR signal and the two visible absorption bands, which always have the same relative intensities. Both visible bands are bleached by irradiation into the 0.430-μm band, photoconductivity also is produced, and the EPR signal is reduced in strength. The two visible bands are, therefore, associated with the same defect as the resonance spectrum. The origin of the 0.355-μm band is unknown. The 0.430-μm light induces a transition to a state that lies near the conduction band edge. The electron easily escapes and produces photoconductivity, optical bleaching, and quenching of the EPR signal.

The oxygen vacancy V_O^- in ZnO and BeO is slightly different because of the trigonal symmetry of the lattice site.[40,41] In these crystals the EPR spectra show interactions also with the nearby Zn or Be neighbors. There is more electron density concentrated on each of the three basal neighbors than on the neighbor in the axial poisition.

A donor-acceptor discrete pair line optical spectrum in CdS has been attributed to S-vacancy double donors and Li acceptors.[42] In the excited state two electrons are bound to the donor ($V_S^{2+}e^-e^-$ or V_S), and a hole is bound to the acceptor. In the ground state one electron is bound to the donor. The Zeeman splittings are consistent with this number of particles: a single uncoupled hole in the excited state and a single electron in the ground state. The spectrum is produced by annealing the crystal in Cd vapor.

METAL VACANCY

The metal vacancy has been identified by magnetic resonance in ZnSe, CdS, ZnO, and BeO, the resonance being due to a hole trapped by the vacancy. In all cases the hole is trapped on one of the four chalcogen neighbors. In ZnSe the spectrum has axial symmetry about a $\langle 111 \rangle$ direction.[37] The ^{77}Se hyperfine interaction shows that the hole is trapped on a single Se neighbor, largely in a p orbital. Preferential alignment of the defects at low temperature after application of uniaxial stress has lowered the energy of one type of center shows that the trigonal symmetry is a result of a spontaneous Jahn–Teller distortion rather than association with another defect. Two absorption bands, at 0.885 and 0.468 μm, are associated with the vacancy. They have polarization properties consistent with the symmetry of the V_{Zn}^- center. When the electronic orientation is influenced by applied stress, the time constant characterizing the reorientation is found to be the same whether measured by the optical polarization or by the EPR signal intensities.

As sample temperature rises, the number of V_{Zn}^- centers decays with a decay time proportional to $\exp(E/kT)$. The energy $E = 1.26$ eV may be the activation energy for diffusion of the vacancy. As V_{Zn}^- centers disappear, other centers such as the SA center, a vacancy associated with a halogen donor, appear.

When the vacancy is produced by 1.5-MeV electron irradiation other centers which have been identified as the vacancy associated with an interstitial zinc atom nearby—a Frenkel pair—also occur.[43] The V_{Zn}^- ground state is perturbed only weakly by the interstitial. When the interstitial atom is along a $\langle 111 \rangle$ axis from the vacancy, the g tensor, and hyperfine interactions differ slightly from those of isolated V_{Zn}^-. However, in centers of one type the interstitial fixes the position of the symmetry axis

and prevents reorientation under applied uniaxial stress. In t¹
forms, V_{Zn}^-—Zn_i(II) and V_{Zn}^-—Zn_i(III), partial reorientation occ
V_{Zn}^-—Zn_i(II) the hole can hop between three of the four Se neighb
suggests that in this case also the perturbing atom is along
direction, but in a different type of site from that in the first complex. In
centers of type III the hole is free to hop only between two of the Se
neighbors, indicating an associated defect along a $\langle 100 \rangle$ direction, as
shown in Figure 9.5.

When the electron beam is incident along the $[1\bar{1}\bar{1}]$ direction, centers of
type I with symmetry axis $[1\bar{1}\bar{1}]$ are predominantly formed, because the
zinc atom knocked out to create the vacancy goes into an accessible
interstitial site. When a crystal with centers preferentially created in this
way is annealed, many of the interstitial atoms move to positions behind
the three nearest Se neighbors to form type-II centers shown in Figure 9.5.
For electrons incident in the $[1\bar{1}\bar{1}]$ direction the centers of type III which
are created also reflect the influence of the beam direction, the interstitial
atom being most often along the $[\bar{1}00]$, $[0\bar{1}0]$, or $[00\bar{1}]$ axis from the
vacancy.

In CdS the V_{Cd}^- center is similar to the vacancy in ZnSe.[44] The hole is
localized on one of the four sulfur neighbors. The axial neighbor is

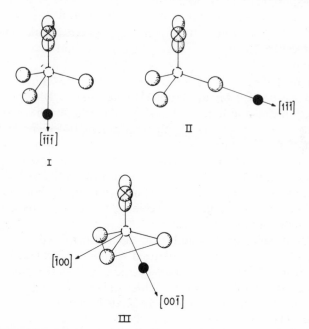

Figure 9.5 Interstitial-vacancy pairs in ZnSe. The open circles represent selenium atoms, and
the solid circles represent zinc atoms.

energetically favored. In BeO the situation is similar,[45] but in ZnO the nonaxial hole sites are more stable than the axial.[46] Ultraviolet illumination of ZnO containing zinc vacancies produces a resonance spectrum ascribed to the neutral vacancy, V_{Zn}. The corresponding spin Hamiltonian is

$$\mathcal{H} = \beta \mathbf{H} \cdot \mathbf{g} \cdot \mathbf{S} + DS_z^2 + E\left(S_x^2 - S_y^2\right) - (1/3)DS(S+1) \qquad (9.6)$$

The effective spin $S = 1$. The z axis, a principal axis of the g tensor, is along the line joining two of the three basal oxygen neighbors. Two holes are trapped on two of these oxygen neighbors.

SA CENTER

In ZnS or ZnSe containing chlorine, bromine, iodine, aluminum, or gallium, an associate of the donor impurity with a zinc vacancy may be formed. This SA defect is responsible for the well-known "self-activated" luminescence. The singly negative halogen ion replaces one of the four anions nearest the vacancy, or the tripositive group III ion substitutes for a nearest zinc neighbor. Such a complex has one excess negative charge with respect to the lattice and can compensate an ionized shallow donor—isolated substitutional Cl^- or Al^{3+}, for example. In this manner the donors are self-compensated and do not make the crystal n type conducting. When the *SA* center traps a hole it becomes paramagnetic. The structure of the *SA* center has been determined by optical and magnetic resonance experiments.

The optical properties of all the *SA* centers are quite similar.[47] The luminescence band is broad and of Gaussian shape. The emission occurs in the blue in ZnS and in the red in ZnSe, the peak for the halogen centers occuring at slightly higher energy than for group-III centers. The excitation spectrum of this luminescence has two peaks, one at bandgap energy and another at lower energy. The luminescence shifts to lower energies with decreasing temperature, opposite to the shift of the band gap. The functional forms of the dependence on temperature of the luminescence peak position and width are in good agreement with the predictions of the configuration coordinate model.

Koda and Shionoya[48] found that the self-activated luminescence in cubic ZnS is polarized under polarized excitation and that the polarized emission from the defect in its diamagnetic charge state is due to σ electric dipole oscillators oriented along the four $\langle 111 \rangle$ directions. From this they inferred that the chlorine atom is associated with another simple defect along the $\langle 111 \rangle$ axis.

The EPR spectra of the paramagnetic charge state of the *SA* center, produced by irradiation of the sample with ultraviolet light, have been

investigated by several workers.[47,49,50] At liquid helium temperatures all
the spectra display monoclinic symmetry with a {110} mirror plane, and
the hole is trapped on one sulfur or selenium atom, as shown by observation of a hyperfine interaction with a single ^{33}S nucleus. Models of these
two types of complex are shown in Figure 9.6. The presence of the halogen
or group-III impurity in the defect has also been established from hyperfine splittings.[51] In the case of the group-III SA centers it is obvious how
the low C_s symmetry of the defect arises. Of the four sulfur neighbors
around the vacancy the hole prefers the one farthest from the Al^{3+} ion.
For the halogen SA centers the three sulfurs are equivalent, however. But
because of the degeneracy associated with the three sulfur sites a spontaneous Jahn–Teller distortion of the complex occurs lowering the symmetry
from C_{3v} to C_s. With the distortion localization of the hole on a single
sulfur is now energetically more favorable. As the temperature rises, the
hole hops among the three equivalent distortions. When this hopping
becomes rapid enough, the lower symmetry is averaged out and a spectrum
with axial symmetry is observed. The SA center has also been seen in
hexagonal ZnS. Schneider, Dischler, and Räuber[52] have studied the SA
center in mixed $ZnS_{1-x}Se_x$ crystals. They find that when one of the sulfur
atoms in the complex is replaced by selenium, the hole becomes localized
on the selenium. When two sulfur atoms are replaced by selenium, the hole
is found in a molecular orbital localized on the two selenium atoms. This is
understandable, because the valence electrons of selenium are less tightly
bound than those of sulfur.

The photosensitivity of the EPR spectra has made it possible to show
that the same defect is responsible for the optical and magnetic properties.
As the temperature of the crystal rises from a low value, thermoluminescence occurs accompanied by a decrease in the EPR signal. Illumination
with 1.4 eV light quenches the EPR signal and the SA thermoluminescence.[53] Illuminating the crystal at low temperature with polarized light of

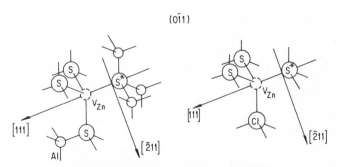

Figure 9.6 Two types of SA center in ZnS.

wavelength corresponding to the lower energy excitation band creates preferentially EPR spectra of SA centers of one orientation, dependent on the direction of polarization.[54]

The presence of the vacancy in the associate is difficult to prove, but it is indicated by several pieces of evidence. The optical polarization measurements show that the halogen atom is associated with another defect along the $\langle 111 \rangle$ axis. More SA centers are formed when the number of available zinc vacancies is increased by higher partial pressure of sulfur during preparation.[55] Molecular orbital models and crystal field calculations, which assume a vacancy present, are able adequately to describe the experimental results.[52, 53]

These optical and magnetic properties are described by the following model. The diamagnetic SA center has an occupied ground state above the valence band edge and an empty excited state slightly above the conduction band edge. Luminescence is produced in two ways. Band-gap excitation produces free holes and electrons, and some of the holes may be trapped at SA centers, subsequently to recombine with free electrons, producing luminescence. Or excitation into the lower energy band excites the center to the unrelaxed excited state. The lattice around the excited center relaxes lowering the excited state below the bottom of the conduction band, and luminescence occurs with the return to the ground state. The positions of these energy levels with respect to the band edges have been established. When electrons are released from traps to recombine with holes on SA centers, thermoluminescence and quenching of the EPR signal result. The polarization effect on the EPR occurs when diamagnetic centers of particular orientation are preferentially raised to the excited state from which ionization may occur before the state relaxes below the conduction band edge.

ISOELECTRONIC IMPURITIES

Three cases of isoelectronic impurities that form bound states in the II–VI compounds have been discovered. They are ZnTe:O, ZnS:Te, and CdS:Te. From the electronegativities listed in Table 2.4, O is an isoelectronic acceptor, and Te is an isoelectronic donor.

The luminescence spectrum of the oxygen isoelectronic acceptor is similar to those of isoelectronic impurities in the III–V compounds.[56] There are two zero-phonon lines caused by the exchange splitting of the excited state into two components corresponding to two values of the total angular momentum, $J = 1$ and $J = 2$. The electron in the Γ_1 minimum of the conduction band has $j = 1/2$, and the hole in the valence band has $j' = 3/2$.

The splitting is only 1.7 meV, with the $J = 1$ state lying above the $J = 2$ state. The oscillator strength of the lower-energy, forbidden transition, $J = 2 \rightarrow J = 0$, is weaker by a factor of 60 than that of the higher-energy transition, $J = 1 \rightarrow J = 0$. The strong phonon sidebands are replicas of the longitudinal optical Γ-point phonons. These are the phonons to which coupling would be expected if the coupling were due to the Γ_1 electron rather than the Γ_5 hole. This observation supports the idea that the electron is bound to the center by the short range potential $U(r)$, which leads to more rapid spatial variation in the electron wavefunction and resultant stronger coupling to phonons, and that the hole is bound in a larger orbit by Coulomb attraction to the electron; that is, it lends support to the view of O as an isoelectronic acceptor in ZnTe. The emission spectrum of ZnTe:O is mirrored in a similar absorption spectrum. Observation of the ^{18}O isotope effect in the emission spectrum confirms the presence of the oxygen atom. The Zeeman effect of the zero-phonon lines has shown that the symmetry of the center is that of the lattice site and that, therefore, there is no other defect nearby.[57] The binding energy, $E_g - h\nu$, is 0.4 eV; this is much larger than the values for N and Bi in GaP and Bi in InP.

Tellurium in ZnS and CdS produces a luminescence spectrum that is unlike those of the other isoelectronic impurities.[58,59,60] No zero-phonon lines are seen, only a broad band with half width given as a function of temperature by Equation 9.5. This is the spectral shape to be expected, as explained in Chapter 3, if coupling to phonons is strong. If the position of the zero-phonon transition is estimated to be at the high-energy extremity of the band, the binding energy is found to be 0.44 eV for ZnS and 0.25 eV for CdS. In CdS there is a small decrease in the lifetime associated with the luminescence spectrum between 10 and 20 K. This has been interpreted as implying that the upper member of the exchange-split excited state, from which a transition to the ground state is more strongly allowed, is becoming thermally populated.

High concentrations of Te are attainable in ZnS and CdS. For concentrations of 10^{19} cm^{-3} and higher, a second band appears at lower energy. This band is thought to be due to decay of an exciton trapped on a pair of substitutional Te atoms. The energy of binding to the pair is greater than for the single impurity, as in the case of nitrogen pairs in GaP.

GROUP-IV IMPURITIES

The elements from column IV of the periodic table are amphoteric impurities in the III–V compounds, substituting for cation or anion. There is no evidence that they occupy any site other than the substitutional metal site in II–VI crystals, however. They have been studied largely by magnetic resonance in ZnS,[61] ZnSe,[62,63] ZnTe,[64] ZnO,[65,66] CdS,[67] and CdSe.[67] The valence configuration is though to be $(ns)^2$ of the divalent ion. When electrons and holes are excited by illuminating the crystal with band gap light, the impurity can trap a hole.

The EPR spectrum of the center that has trapped a hole is easily identified by the large hyperfine interaction with the impurity nucleus; ^{29}Si, ^{73}Ge, ^{119}Sn, ^{117}Sn, and ^{207}Pb have nuclear spin $I = 1/2, 9/2, 1/2, 1/2$, and $1/2$, respectively. The spectrum is that of a $^2S_{1/2}$ state with g factor slightly greater than 2. In CdS and CdSe the g tensor is axial, but the anisotropy is small. The magnitude of the hyperfine interaction with the nuclei of the four chalcogen neighbors shows that the hole is largely confined to the $p\sigma$ orbitals of the chalcogens with less density in the s orbital of the impurity. This is evidence that the resonance is due to a trapped hole. A hole should be more easily localized on the negative anions than on the impurity.

Iida[68] and Watanabe[69] have constructed a model for the center based on molecular orbitals consisting of linear combinations of atomic orbitals of the impurity and the four neighbors. The s orbital of the impurity can combine with the a_1 ligand orbital of Equation 1.67. The shift of the g factor from the free electron value 2.0023 is due to admixture by the spin-orbit interaction of a ligand orbital into the ground state. The g shift on this model is proportional to the spin-orbit parameter of the chalcogen p orbitals and should increase with increasing atomic number of the chalcogen. This is found to be the case, as shown for Ge and Pb in Table 9.5.

Charge transfer effects involving these centers occur. The EPR signal is quenched by light of wavelength in a definite range. Luminescence has

Table 9.5 g Factors of Ge and Pb in ZnO, ZnS, ZnSe, and ZnTe

	ZnO	ZnS	ZnSe	ZnTe
Pb	2.013^a	2.024^b	2.0729^c	2.167^d
Ge		2.0086^e	2.0403^c	2.1375^b

^{a} Reference 65.
^{b} Reference 68.
^{c} Reference 62.
^{d} Reference 64.
^{e} Reference 61.

also been observed in some of these systems, the emission spectrum consisting of a broad band of Gaussian shape.[70] No definite assignment of the spectra to transitions of the center has yet been possible.

IRON GROUP TRANSITION METAL IONS

Transition metals of the iron group are common contaminants in the II-VI compounds. As far as is known, they always substitute on the metal site. Often more than one charge state of the impurity is stable under certain conditions. If the crystal is irradiated with band gap light at low temperature, holes or electrons excited by the light may be trapped by the impurities to create charge states that would not be stable at higher temperatures.[71] For example, iron and nickel have been observed as Fe^+, Fe^{2+}, Fe^{3+}, Ni^+, Ni^{2+}, and Ni^{3+}.

The tetrahedral site symmetry leads to smaller values of the cubic field parameter Δ than in octahedral coordination and to larger oscillator strengths for optical transitions because of the forced electric dipole mechanism. The Racah parameters which describe the interelectron electrostatic interaction are generally considerably reduced compared with values for free ions or for ions in octahedral coordination. The tetrahedral field splitting is generally smaller than the internal electrostatic splittings but much larger than spin orbit or Jahn–Teller splittings. To understand details of the spectra it is essential to consider Jahn–Teller coupling when orbitally degenerate states are in question, as will become evident from the discussion of the following sections. Analysis of the ligand superhyperfine structure in magnetic resonance spectra shows that the d electron states are highly localized on the impurity and that mixing with ligand orbitals is small.[72]

d^1

The ground state of Sc^{2+} substitutional on a Zn site in ZnS has been studied by EPR by Barksdale and Estle[73] and by Broser and Schulz.[74] The d state splits into a lower ^2E and an upper $^2T_2$ state in the tetrahedral symmetry of the site. As described in Chapter 3, the ground vibronic state that results from coupling of the ^2E electronic state with lattice vibrations of local E symmetry is a fourfold degenerate ^2E state. This vibronic ^2E state is split by random internal strains into two doublets, and each doublet produces at sufficiently low temperature an anisotropic EPR spectrum, identifiable by the $^{45}Sc(I = 7/2)$ nuclear hyperfine structure. At slightly higher temperatures the spectrum is isotropic, probably because of rapid thermally induced transitions between the two doublets which average out the anisotropy. The intensity of the resonance signal increases when the crystal is irradiated with ultraviolet light at low temperatures.

d^2

The Ti^{2+} and V^{3+} ions have the $(3d)^2$ configuration. The optical and magnetic resonance spectra have been studied in several II-VI crystals.[75-80] The energy levels are shown in Figure 1.6. The ground term is $^3A_2$, and strong transitions are seen in absorption to the two $^3T_1$ states. In crystals with the wurtzite structure the spectra are very similar to those in crystals with zincblende structure, because the axial field is small, and the terms are often labeled in this case also according to the labels appropriate to T_d symmetry. The crystal field parameter Δ ranges from 3070 cm^{-1} for Ti^{2+} in CdSe to 3750 cm^{-1} in ZnSe. The Racah parameter B is typically near 350 cm^{-1}, or half the free-ion value.

A poorly resolved splitting of the longer wavelength $^3A_2 \rightarrow ^3T_1$ absorption band in CdS:Ti^{2+} has been interpreted as a manifestation of a strong Jahn–Teller effect in the $^3T_1$ state.[78] The splitting of the $^3T_1$ state increases with increasing temperature in the expected way. The temperature dependence arises from the increase of vibrational amplitudes with temperature.

The ground state $^3A_2$ gives rise to a characteristic EPR spectrum described by the spin Hamiltonian with spin $S = 1$

$$\mathcal{H} = g\beta \mathbf{H} \cdot \mathbf{S} + A\mathbf{I} \cdot \mathbf{S} \tag{9.7}$$

in crystals with the zincblende structure. In addition to the expected transition $M = 0 \leftrightarrow M = \pm 1$, another, sharper line due to the double quantum transition $M = 1 \leftrightarrow M = -1$ is seen at higher microwave power levels. The g factor is slightly less than 2.0023 because of admixture by the spin-orbit interaction of higher-lying triplet states into $^3A_2$, as described in Chapter 1. In crystals with wurtzite structure the extra trigonal field splits the spin triplet into a doublet and a singlet.

d^4

The optical and EPR spectra of Cr^{2+} in cubic ZnS, ZnSe, ZnTe, and CdTe, and in hexagonal ZnS and CdS have been studied by Vallin and co-workers.[81-84] The spectra are qualitatively similar for all these host crystals. A Tanabe–Sugano diagram for the d^4 configuration in a tetrahedral field is considerably more complex than that for d^2 (Figure 1.6), because there are 10 free ion ^{2S+1}L terms. However, the near-infrared optical data are adequately explained by considering only the ^5D term. In a weak tetrahedral field this term splits into a lower orbital triplet $^5T_2$ and an upper orbital doublet ^5E. The triplet is split by a strong Jahn–Teller effect through coupling to a vibrational mode of E symmetry into a lower orbital singlet $^5B_2$ (of D_{2d}) and an upper doublet $^5E(D_{2d})$. These levels are shown in Figure 9.7. The ^5E state is apparently much more weakly coupled to vibrations and does not undergo a Jahn–Teller effect.

The EPR spectra of $^5B_2(D_{2d})$ clearly reveal the effect of the Jahn–Teller distortion.[82] Each of the three spectra has axial symmetry about a different $\langle 100 \rangle$ axis corresponding to the D_{2d} symmetry of the distortion. Uniaxial stress applied at low temperatures causes a redistribution of population among the three types of distortion and shows that the low symmetry is not due to association with another defect. The spin Hamiltonian is

$$\mathcal{H} = \beta g_\parallel S_z H_z + \beta g_\perp (H_x S_x + H_y S_y) + D S_z^2 + \frac{F}{180} \left[35 S_z^4 - 30 S(S+1) S_z^2 \right.$$

$$\left. + 25 S_z^2 \right] + \frac{a}{6} \left(S_x^4 + S_y^4 + S_z^4 \right) \tag{9.8}$$

In the hexagonal crystals the symmetry is lower, and an additional small term must be included. The spin $S = 2$. The parameter D varies greatly from crystal to crystal, being -2.48 cm^{-1} for ZnSe and $+2.30$ cm^{-1} for

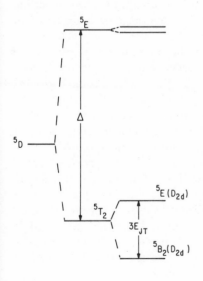

Figure 9.7 Lower energy levels of Cr^{2+} substitutional in II-VI compounds with zincblende structure.

ZnTe, for example. This variation of D can be understood if small admixtures of ligand orbitals with the d orbitals are considered. The spin-orbit interaction leads to mixing with excited states, and the ligand contribution to the interaction increases greatly in the order S, Se, Te.

The optical transitions are very similar in all crystals studied. The cubic field parameter Δ varies from 4070 cm^{-1} for CdS to 4650 cm^{-1} for ZnS. The transition $^5B_2(D_{2d}) \rightarrow {}^5E(D_{2d})$ seen in absorption gives the Jahn–Teller energy E_{JT}, which lies in the range 500 to 600 cm^{-1} for the different crystals. This is an order of magnitude larger than the vibrational energy of the E mode, the transverse acoustic mode at the L point, to which the d state couples.

d^5

In the d^5 configuration the d shell is half filled. The ground state of the free ion is ^6S, and for the ion in a tetrahedral field this is $^6A_1$. All excited states have lower spin, and transitions between them and the ground state are, therefore, weak. The excited states are numerous and crowded together in energy, making the optical spectra difficult to analyze. The theoretical interpretations are reviewed by Uehara[85] for Mn^{2+} in ZnS.

The ground state EPR spectrum is described by the spin Hamiltonian of Equation 9.8 with isotropic g tensor and $F=0$, $S=5/2$. Mn and Fe are very common contaminants in II-VI compounds. The Mn^{2+} spectrum is easily recognized by the ^{55}Mn (100% abundant, $I=5/2$) nuclear hyperfine splitting. Cr$^+$ and Fe^{3+} resonance spectra have also been observed in many II-VI crystals.[86] Several interesting complex defects in which Fe^{3+} is associated with another impurity have also been discovered.

Kravitz and Piper[87] have identified an associate of iron with six fluorine ions in CdTe. The resonance spectrum of Fe^{3+} with strong hyperfine interactions with six fluorines was studied both with EPR and ENDOR. The cluster is an $(FeF_6)^{3-}$ molecule ion with the fluorines along the six $\langle 100 \rangle$ directions. The complex does not bond appreciably with the host lattice, and it is not known at what site the cluster is incorporated.

Magnetic resonance spectra of associates of iron with copper, silver, or lithium have been observed in many of the II-VI compounds, including the Fe-Cu associate in ZnS, ZnSe, ZnTe, and CdTe, the Fe-Ag associate in ZnS, and the Fe-Li associate in ZnS and ZnO,[88] similar associates reported in CdS[89] containing iron and copper or iron and lithium and in CdS[90] intentionally doped only with iron. For Fe-Cu in ZnS, ZnSe, and ZnTe the defect with which the iron associates could be identified by hyperfine structure, but in the other cases this was not possible.

In crystals with cubic structure very anisotropic EPR spectra characteristic of Fe^{3+} in an electric field of C_s symmetry were observed. Resonances from both the lowest and next lowest doublets were seen. Analysis of the complex angular dependence of the spectra shows that they arise from twelve crystallographically equivalent iron defects. The reflection plane and the orientation of the principal axes of the g tensor of one of these are shown in Figure 9.8. In the case of Fe-Cu the ^{63}Cu and ^{65}Cu hyperfine splittings are seen and prove that one copper ion is associated with the iron in each center. It seems likely that both iron and copper would occupy cation sites. A Cu^+ ion on a nearby cation site provides charge compensation and would be diamagnetic, as observed. Fe-Ag and Fe-Li dopings each produce 12 spectra very similar to those of Fe-Cu. The same type of associate is probably formed in these cases also.

In the wurtzite lattices the 12 nearest cation sites corresponding to the first cation shell of the cubic structure are not all crystallographically equivalent. Consequently, there can be three different types of associate formed on these and the central cation site. All three types form in ZnO.

An Fe-As complex consisting of Fe^{3+} on a metal site and As^{3-} on a nearest anion site has been studied in ZnSe.[91] The g tensor of the resonance is axially symmetric about a $\langle 111 \rangle$ axis. The identity of the As atom is established from the hyperfine interaction with the ^{75}As nucleus (see Figure 5.3). Splittings due to interaction with the three nearby Se nuclei also occur. The trigonal field at the Fe site splits the spin sextet into three doublets, and the observed resonance is due to the lowest pair with $M = \pm 1/2$.

Iron associated with one or two selenium atoms has been observed by Schnieder, Dischler, and Räuber[16] in mixed $ZnS_{1-x}Se_x$ crystals with x

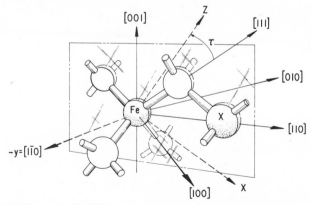

Figure 9.8 Fe-X associate in II-VI compounds with zincblende structure. X represents Cu, Ag, or Li. Axes of the g tensor and the mirror symmetry plane are shown. The angle τ depends on X and on the compound (from Holton, de Wit, Estle, Dischler, and Schneider, reference 88).

small. The Fe-Se center consists of Fe^{3+} on a zinc site with one neighboring sulfur replaced by a selenium and is similar to the Fe-As associate. The spectrum shows $\langle 111 \rangle$ axial symmetry and hyperfine interactions from ^{57}Fe and ^{77}Se nuclei. In the Fe-2Se associate two of the four nearest sulfurs are replaced by selenium. The spectrum has orthorhombic symmetry corresponding to the C_{2v} symmetry of the FeS_2Se_2 cluster. The rather large splitting of the Fe^{3+} ^6S state in this case probably is due to a difference in the covalent contributions to the Fe-Se and Fe-S interactions. Similar associates containing manganese instead of iron were also observed.

Cubic ZnS and βMnS have the same structure. Mn^{2+} pairs have been studied at high manganese concentrations where an observable number of pairs would be expected purely on a statistical basis if manganese substitutes for zinc in a random fashion.

The pairs are exchange coupled, probably by super-exchange through the sulfurs. The optical absorption spectra of the pairs have been studied by McClure.[92] Brummage et al.[93] have analyzed their contribution to the magnetic susceptibility. The most detailed information has been obtained by Röhrig and Räuber from EPR experiments.[94] In a pair the two spins 5/2 of the ^6S states are coupled to form two-particle states of total spin 0, 1, 2, 3, 4, and 5, the 0 state lying lowest. Resonances seen are due to the states of total spin 1, 2, and 3. Because these are excited states, the intensities of the resonance are strongly temperature dependent, and the energy separations of the states are extracted from the temperature dependence. The spectra are twelvefold and display C_{2v} symmetry with a $\langle 110 \rangle$ symmetry axis. There is a prominent hyperfine splitting from the nuclear moments of two equivalent Mn^{2+} nuclei. States 4 and 5 lie at much higher energy and are sparsely populated at the experimental temperatures. The exchange coupling is characterized by an isotropic exchange parameter $\mathcal{J} = 35.8$ cm^{-1} and much smaller anistropic interactions. The coupling is antiferromagnetic, that is, $\mathcal{J} > 0$, as in βMnS.

d^6

The d^6 configuration of Fe^{2+} has been studied optically in ZnS and CdTe in the infrared.[95-99] The configuration is similar to d^4. The levels of interest are ^5E and $^5T_2$, the ground state being ^5E for d^6. As in the case of d^4, the ^5E state is weakly coupled to lattice vibrations, but a Jahn–Teller effect occurs in the $^5T_2$ state. These states are separated by the single particle cubic field splitting Δ, which is 3400 cm^{-1} in ZnS and 2480 cm^{-1} in CdTe.

The ^5E state is unaffected by Jahn–Teller effects, and there is no first-order spin-orbit splitting. A small splitting does arise from admixture of excited states by the spin-orbit interaction, however. The result is that

^5E is split into five nearly equally spaced levels spanning only about 80 cm^{-1}. The ground state is a singlet, and for this reason no magnetic resonance can be observed. Transitions between these levels and between them and $^5T_2$ are observed optically. The $^5T_2$ state suffers a Jahn–Teller effect characterized by a smaller ratio of the Jahn–Teller splitting to the vibrational energy of the associated distortion mode than in the case of d^4. There is a dynamic Jahn–Teller effect, but the Jahn–Teller coupling is weaker than the spin-orbit interaction. There is not a simple reduction of the spin-orbit splitting, as when the Jahn–Teller coupling is stronger, but a mixing of vibrational states occurs. To explain the observed spectra it is necessary to consider coupling of the electrons with both acoustic phonons of energy 100 cm^{-1} and optical phonons of energy 300 cm^{-1}.

d^7

Co^{2+}, Fe$^+$, and Ni^{3+} have the $(3d)^7$ configuration. The lowest term of the free ion is ^4F. In T_d symmetry this splits into a $^4A_2$ ground term and two excited terms, $^4T_1(F)$ and $^4T_2$. The only other free ion term of spin $S = 3/2$ is ^4P which becomes $^4T_1(P)$ in the crystal. $^4T_1(P)$ and nearby terms coming from the free ion ^2G term produce an absorption band near 15,000 cm^{-1} for Co^{2+} in ZnO, ZnS, and CdS.[100] The weaker band due to $^4T_1(F)$ occurs near 6000 cm^{-1}, and the still-weaker $^4T_2$ band lies about 3000 cm^{-1} lower. Careful investigation of the $^4A_2 \rightarrow {}^4T_2$ and $^4A_2 \rightarrow {}^4T_1(F)$ bands in purely cubic ZnS shows that these absorption groups consist of a number of sharp lines.[101] When the lattice phonon sidebands are subtracted, the number is larger than four, the number of states into which the spin-orbit coupling would split each term.

The extra structure can be explained theoretically by considering coupling of the d electron states to vibrational modes of local E symmetry. The extra lines are due to higher excitations of the oscillator. However, because the electron-vibration interaction is comparable in strength with the spin-orbit interaction, a ladder of evenly spaced sidebands does not occur; rather, there is strong mutual repulsion of vibronic levels of the same symmetry and mixing of vibronic states. The theoretical treatment is similar to that of Ham and Slack[99] for Fe^{2+}. In addition to this Jahn–Teller coupling the electrons couple to lattice modes of A_1 symmetry to give the normal phonon replicas of the strongest line. Similar spectra have been recognized in hexagonal ZnS and in polytypes.[102] The $^4A_2$ and $^4T_2$ terms are separated in energy by Δ, which is 3850 cm^{-1} in cubic ZnS, 3720 cm^{-1} in hexagonal ZnS, 3300 cm^{-1} in CdS, and 3900 cm^{-1} in ZnO.

The ground state $^4A_2$, an orbital singlet and spin quartet, splits in a magnetic field according to the formula (1.54) for a Γ_8 state, or, equivalently, the splitting can be described by the spin Hamiltonian

$$\mathcal{K} = g\beta \mathbf{H}\cdot\mathbf{S} + u\beta\left\{ S_x^3 H_x + S_y^3 H_y + S_z^3 H_z - (1/5)(\mathbf{S}\cdot\mathbf{H})\left[3S(S+1)-1\right]\right\}$$

$$(9.9)$$

The g factor is slightly larger than 2. In hexagonal crystals the quartet is split into two doublets.[71,72,86,103,104]

d^8

The d^8 configuration has the same terms as d^2. In T_d symmetry the lowest are $^3T_1$, $^3T_2$, and $^3A_2$, which come from the free-ion term ^3F and are written in order of increasing energy, $^3T_1$ being the ground term. The spin-orbit interaction splits $^3T_1$ into four levels. The lowest is a singlet Γ_1; therefore, no magnetic resonance is expected. Similarly, $^3T_2$ is split into four states labeled Γ_4, Γ_3, Γ_5, and Γ_2. In cubic ZnS at sufficiently low temperature that only the lowest sublevel of $^3T_1$ is occupied, the only zero-phonon transition allowed in absorption is $\Gamma_1(^3T_1)\rightarrow\Gamma_5(^3T_2)$. In ZnO the additional trigonal field makes observation of other lines possible. The $^3T_2$ levels in ZnO are shown in Figure 3.7. Their positions are explained[105] by reduction of the spin-orbit splittings and trigonal field splittings by interaction with a vibrational mode of E symmetry. The situation is probably similar in ZnS. It is also necessary to consider the Jahn–Teller effect in explaining the structure of the absorption at higher energy to the $^3T_1(P)$ term coming from the free ion ^3P term.[106] The cubic field parameter Δ is 4260, 4130, 4110, and 4000 cm^{-1} for Ni^{2+} in cubic ZnS, hexagonal ZnS, ZnO, and CdS[100], respectively.

d^9

Cu^{2+} and Ni$^+$ with the $(3d)^9$ configuration have been studied in several II-VI compounds. The d^9 configuration, like d^1, has two terms $^2T_2$ and ^2E, separated by Δ, which is 6900 cm^{-1} for Cu^{2+} in ZnS. The ground $^2T_2$ term should be split by spin-orbit interaction into a lower Γ_7 doublet and an upper Γ_8 quartet; ^2E is not split. From luminescence spectra the $\Gamma_7-\Gamma_8$ splitting, $3\zeta_{3d}/2$, is found to be only 14 cm^{-1} for Cu^{2+} in cubic ZnS, nearly two orders of magnitude smaller than expected from the free ion spin-orbit parameter.[107] The g factor of the Γ_7 doublet is expected to be very near 2, but is found to be 0.71 for Cu^{2+} in ZnS;[108] 1.40 for Ni$^+$ in ZnS;[109] 1.437 for Ni$^+$ in ZnSe;[104] and $g_\parallel = 0.74$, $g_\perp = 1.531$ for Cu^{2+} in ZnO.[110]

These experimental results have been explained as a manifestation of Jahn–Teller coupling of the $^2T_2$ state with an E vibrational mode.[111] The

Jahn–Teller energy for Cu^{2+} in ZnS is $E_{JT}=900$ cm^{-1}, or about 3 times the vibrational energy of the E mode involved. For the hexagonal crystals the problem is somewhat more complex, but is, in general, similar.[112]

In ZnO:Cu there is also a high-energy luminescence band associated with copper.[113] The Zeeman effect of the sharp zero-phonon line shows that the ground state of the transition is Cu^{2+}_{Zn}. In the excited state the hole is probably localized on the oxygen neighbors and the copper d shell is filled: Cu^{+}_{Zn}.

Copper easily associates with other defects to form complexes. In CdS four resonances due to Cu^{2+} have been observed,[114,115] and in ZnS eight others have been reported,[109] excluding the Fe-Cu complex already described and isolated substitutional Cu^{2+}. The structures of these defects are unknown. One of the four resonances in CdS is probably due to isolated substitutional Cu^{2+}. Copper is a well-known activator of luminescence in ZnS, leading to bands in the blue, green, and red-orange spectral regions.[116] Urabe et al.[117] have established that the blue copper luminescence is due to well-localized electronic states fo a defect with C_{3v} symmetry. Interstitial Cu^{+} associated with Cu^{+} on a nearest Zn site has been proposed as a model for this center, but the idea is speculative.

RARE EARTH IMPURITIES

Rare earth ions are much larger than iron group transition metal ions and are less soluble in the II-VI compounds. They are incorporated as trivalent or divalent ions and may occupy metal sites or interstitial sites. Like the iron group impurities they readily form complex defects through association with other impurities. Many of the ions that have an odd number of $4f$ electrons have been studied by magnetic resonance, and the characteristic optical spectra consisting of groups of sharp lines have been produced in several II-VI compounds. The magnetic resonance spectra are not affected by irradiation of the crystal with light of band gap energy so that, in contrast to the iron group impurities, no charge transfer effects are observed.

In crystals with the cubic zincblende structure into which the rare earth is the only deliberately introduced impurity the trivalent ion substitutes for the cation. Some examples are Er^{3+} in ZnS, ZnSe, ZnTe, and CdTe; Nd^{3+} in CdTe; and Yb^{3+} in ZnSe, ZnTe, and CdTe.[118,119] The EPR spectrum is characterized by an isotropic g tensor, except in the case of a Γ_8 ground state, in which a cubic anisotropy occurs. The lowest $^{2S+1}L_J$ multiplet of the free ion splits in the tetrahedral field into a number of states of Γ_6, Γ_7, and Γ_8 type for an odd number of $4f$ electrons. The tetrahedral splitting is characterized by two parameters, as explained in Chapter 1. If one knows the ratio of these two parameters, he can determine which state is lowest, and, from the wavefunction of the state, tabulated by Lea, Leask, and Wolf,[120] calculate the g factor. It turns out that the ratio of the two tetrahedral field parameters is fairly accurately given by a simple point-charge model. Because, in general, a different state, with numerically different g factor, lies lowest for each type of site, it is possible from measurement of the g factor to decide what type of site the ion occupies. It is surprising that a point-charge model would be adequate, but it is theoretically justified by the small size of corrections due to overlap of the 4f orbitals with ligand orbitals.[121] In hexagonal crystals the spectrum has axial symmetry, but the "average g factor" ($(g_{\parallel}+2g_{\perp})/3$) lies close to the g factor measured in cubic crystals for the ion at a substitutional site. Other spectra in hexagonal crystals indicate that other types of site are also occupied.[118, 122, 123]

Kingsley and co-workers[119, 124, 125] discovered that rare earth luminescence in ZnS, CdS, and ZnSe is sensitized by copper or silver impurities. When Cu or Ag is present in the crystal, a new isotropic Er^{3+} resonance occurs with g factor characteristic of a rare earth in an interstitial site. The cubic anisotropy of the resonance indicates that the site symmetry is tetrahedral also. It is these ions, rather than the substitutional ones, that produce the observed optical spectra. This was proved by comparison of the ground state g factor measured by EPR with that obtained from an optical Zeeman experiment. The tetrahedral field parameters are slightly different, depending whether Cu or Ag is present, indicating that the rare earth ion is associated with Cu or Ag in such a way that tetrahedral symmetry is preserved.

Evidence that the rare earth ion in this complex is at an interstitial site also comes from analysis of super-hyperfine structure.[118, 126] In CdTe containing Yb and Cu, Ag, or Au, three types of EPR spectra occur: an isotropic resonance from unassociated substitutional rare earths, another isotropic resonance, and a resonance from the rare earth in a site with $\langle 111 \rangle$ axial symmetry. In CdTe the Yb^{3+} resonance of the second type shows a hyperfine splitting from interaction with six equivalent tellurium nuclei along the $\langle 100 \rangle$ axes. The isotropic nature of the resonance implies that the ion is at a site of tetrahedral symmetry. This site is the interstitial site with six nearest tellurium atoms along the $\langle 100 \rangle$ directions. The g factors are similar but quantitatively different depending whether the noble metal co-dopant is copper ($g=2.525$), silver ($g=2.511$), or gold ($g=2.501$). This implies that the noble metal is near the Yb^{3+}. The linewidth of the Yb—Cu resonance is 3 times greater than in other cases, probably because of an unresolved hyperfine interaction with the large copper nuclear moment. The model of this associate is shown in Figure 9.9. Four monovalent noble metal ions substitute for the four nearest cations. The g factors calculated from this model agree with the observed ones.

In CdTe the axial Yb^{3+} resonance also shows a hyperfine interaction with six tellurium nuclei. The six are not equivalent, but are divided into two groups of three equivalent atoms. Again the g factors are similar, but depend on the noble metal dopant. The average g factors are near the values for the $Yb_i^{3+}-4M_{Cd}^{+}$ defects. This resonance is probably due to Yb^{3+} in the same interstitial position, but with only three of the four nearest cadmiums replaced by Cu^+, Ag^+, or Au^+. A point-charge calculation based on this model predicts the observed average g factor and sense of anisotropy, $g_{\parallel} > g_{\perp}$.

By similar arguments Title and Mayo[127, 128] have identified associates of interstitial Er^{3+} or Yb^{3+} with one substitutional Li^+ in CdTe and of a substitutional Yb^{3+} or Er^{3+} with a P^{3-} ion substitutional for a nearest tellurium in ZnTe and CdTe. The phosphorus hyperfine interaction was observed in ZnTe.

Thulium–lithium associates have been observed by Masui[129] in ZnSe. The angular dependence of the optical Zeeman effect shows that the defect

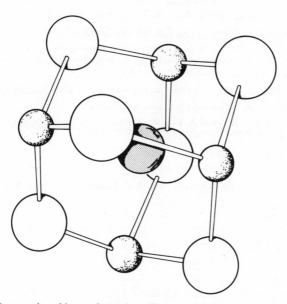

Figure 9.9 Rare earth–noble metal associate. The central dark circle represents an interstitial earth ion. The four small open circles represent substitutional noble metal ions, and the large circles are chalcogens.

has a $\langle 100 \rangle$ symmetry axis. The optical spectra were analyzed and found to be compatible only with the following model. The Tm^{3+} ion is located at the second type of interstitial site whose near neighbors are four seleniums and six zincs. Four of the zincs in a $\{100\}$ plane are replaced by Li^+.

REFERENCES

1. *Physics and Chemistry of II–VI Compounds*, M. Aven and J. S. Prener, Eds. (North-Holland, Amsterdam, 1967).

2. M. L. Cohen and T. K. Bergstresser, *Phys. Rev.* **141**, 789 (1966).

3. G. Mandel, *Phys. Rev.* **134A**, 1073 (1964).

4. C. H. Henry, K. Nassau, and J. W. Shiever, *Phys. Rev.* **B4**, 2453 (1971).

5. H. H. Woodbury and M. Aven, *Phys. Rev.* **B9**, 5195 (1974).

6. C. H. Henry, R. A. Faulkner, and K. Nassau, *Phys. Rev.* **183**, 798 (1969).

7. P. J. Dean in *Progress in Solid State Chemistry*, Vol. 8, J. O. McCaldin and G. Somorjai, Eds. (Pergamon, New York, 1973) Ch. 1.

8. J. L. Merz, K. Nassau, and J. W. Shiever, *Phys. Rev.* **B8**, 1444 (1973).

9. F. T. J. Smith, *Solid State Commun.* **8**, 263 (1970).

10. G. H. Hershman and F. A. Kröger, *J. Solid State Chem.* **2**, 483 (1970).

11. S. G. Parker and J. E. Pinnell, *Trans. Met. Soc. AIME* **245**, 451 (1969).

12. C. H. Henry and K. Nassau, *Phys. Rev.* **B2**, 997 (1970).

13. K. Nassau, C. H. Henry, and J. W. Shiever in *Proceedings of the Tenth International Conference on the Physics of Semiconductors*, S. P. Keller, J. C. Hensel, and F. Stern, Eds. (USAEC, Oak Ridge, Tenn., 1970) p. 629.

14. J. L. Merz, H. Kukimoto, K. Nassau, and J. W. Shiever, *Phys. Rev.* **B6**, 545 (1972).

15. K. A. Müller and J. Schneider, *Phys. Lett.* **4**, 288 (1963).

16. J. Schneider, B. Dischler, and A. Räuber, *J. Phys. Chem. Solids* **29**, 451 (1968).

17. P. H. Kasai, *J. Chem. Phys.* **43**, 4143 (1965).

18. O. F. Schirmer, K. A. Müller, and J. Schneider, *Phys. kondens. Mat.* **3**, 323 (1965).

19. A. R. Reinberg, *J. Chem. Phys.* **41**, 850 (1964).

20. N. V. Agrinskaya, E. N. Arkadeva, and O. A. Matveev, *Sov. Phys. Semicond.* **5**, 762 (1971).

21. A. Räuber and J. Schneider, *Phys. Stat. Sol.* **18**, 125 (1966).

22. J. Dieleman, J. W. De Jong, and T. Meijer, *J. Chem. Phys.* **45**, 3178 (1966).

23. J. L. Ploix and M. Dugue, *Phys. Stat. Sol.* **A18**, 323 (1973).

24. B. L. Crowder and W. N. Hammer, *Phys. Rev.* **150**, 541 (1966).

25. B. L. Crowder and G. D. Pettit, *Phys. Rev.* **178**, 1235 (1969).

26. O. F. Schirmer, *J. Phys. Chem. Solids* **29**, 1407 (1968).

27. O. F. Schirmer and D. Zwingel, *Solid State Commun.* **8**, 1559 (1970).

28. D. Zwingel, *J. Lumin.* **5**, 385 (1972).

29. D. Zwingel and F. Gärtner, *Solid State Commun.* **14**, 45 (1974).

30. O. F. Schirmer, *J. Phys. Chem. Solids* **32**, 499 (1971).
31. N. Watanabe and S. Usui, *Japanese J. Appl. Phys.* **5**, 569 (1966).
32. B. Tell, *J. Appl. Phys.* **41**, 3789 (1970).
33. R. K. Watts, W. C. Holton, and M. de Wit, *Phys. Rev.* **B3**, 404 (1971).
34. A. R. Reinberg, W. C. Holton, M. de Wit, and R. K. Watts, *Phys. Rev.* **B3**, 410 (1971).
35. R. K. Watts and W. C. Holton, *Phys. Rev.* **B2**, 4882 (1970).
36. D. Wruck, *Phys. Stat. Sol.* **B48**, 181 (1971).
37. G. D. Watkins, *Radiation Effects* **9**, 105 (1971).
38. J. Schneider and A. Räuber, *Solid State Commun.* **5**, 779 (1967).
39. K. Leutwein, A. Räuber, and J. Schneider, *Solid State Commun.* **5**, 783 (1967).
40. J. M. Smith and W. E. Vehse, *Phys. Lett.* **31A**, 147 (1970).
41. A. K. Garrison and R. C. DuVarney, *Phys. Rev.* **B7**, 4689 (1973).
42. C. H. Henry, K. Nassau, and J. W. Shiever, *Phys. Rev. Lett.* **24**, 820 (1970).
43. G. D. Watkins, *Phys. Rev. Lett.* **33**, 223 (1974).
44. A. L. Taylor, G. Filipovich, and G. K. Lindeberg, *Solid State Commun.* **9**, 945 (1971).
45. A. Herve and B. Maffeo, *Phys. Lett.* **32A**, 247 (1970).
46. D. Galland and A. Herve, *Phys. Lett.* **33A**, 1 (1970).
47. W. C. Holton, M. de Wit, and T. L. Estle in *International Symposium on Luminescence*, N. Riehl and D. Kallmann, Eds. (Karl Thiemig, Munich, 1966) p. 454.
48. T. Koda and S. Shionoya, *Phys. Rev.* **136A**, 541 (1964).
49. A. Räuber, J. Schneider, and F. Matossi, *Z. Naturf.* **17A**, 654 (1962).
50. J. Schneider, W. C. Holton, T. L. Estle, and A. Räuber, *Phys. Lett.* **5**, 312 (1963).
51. J. Schneider, A. Räuber, B. Dischler, T. L. Estle, and W. C. Holton, *J. Chem. Phys.* **42**, 1839 (1965).
52. J. Schneider, B. Dischler, and A. Räuber, *J. Phys. Chem. Solids* **31**, 337 (1970).
53. J. Schneider in *II-VI Semiconducting Compounds*, D. G. Thomas, Ed. (Benjamin, New York, 1967). p. 40.
54. R. K. Watts, unpublished data.
55. J. S. Prener and D. J. Weil, *J. Electrochem. Soc.* **106**, 409 (1959).
56. J. J. Hopfield, D. G. Thomas, and R. T. Lynch, *Phys. Rev. Lett.* **17**, 312 (1966).
57. J. L. Merz, *Phys. Rev.* **176**, 961 (1968).
58. T. Fukushima and S. Shionoya, *Japanese J. Appl. Phys.* **12**, 549 (1973).
59. J. D. Cuthbert and D. G. Thomas, *J. Appl. Phys.* **39**, 1573 (1968).
60. D. M. Roessler, *J. Appl. Phys.* **41**, 4589 (1970).
61. K. Sugibuchi and Y. Mita, *Phys. Rev.* **147**, 355 (1966); ibid., **153**, 404 (1967).
62. W. C. Holton and R. K. Watts, *J. Chem. Phys.* **51**, 1615 (1969).
63. K. Suto and M. Aoki, *J. Phys. Soc. Japan* **26**, 287 (1969).
64. K. Suto and M. Aoki, *J. Phys. Soc. Japan* **22**, 1307 (1967); ibid. **24**, 955 (1968).
65. G. Born, A. Hofstaetter, and A. Scharmann, *Phys. Lett.* **36A**, 447 (1971).
66. A. Hausmann and P. Schreiber, *Z. Phys.* **245**, 184 (1971).
67. R. Böttcher and J. Dziesiaty, *Phys. Stat. Sol.* **31**, K71 (1969).
68. T. Iida, *J. Phys. Chem. Solids* **33**, 1423 (1972).
69. H. Watanabe, *Phys. Rev.* **149**, 402 (1966).

70. Y. Mita, *J. Phys. Soc. Japan* **20**, 1822 (1965).

71. W. C. Holton, J. Schneider, and T. L. Estle, *Phys. Rev.* **133A**, 1638 (1964).

72. T. L. Estle and W. C. Holton, *Phys. Rev.* **150**, 159 (1966).

73. A. O. Barksdale and T. L. Estle, *Phys. Lett.* **42A**, 426, 71973).

74. I. Broser and M. Schulz, *J. Phys.* **C7**, L147 (1974).

75. R. K. Watts, *Phys. Lett.* **27A**, 469 (1968).

76. P. A. Slodowy and J. M. Baranowski, *Phys. Stat. Sol.* **49B**, 499 (1972).

77. J. Schneider and A. Räuber, *Phys. Lett.* **21**, 380 (1966).

78. R. Boyn and G. Ruszczynski, *Phys. Stat. Sol.* **48B**, 643 (1971).

79. R. Böttcher and J. Dziesiaty, *Phys. Stat. Sol.* **57B**, 617 (1973).

80. E. M. Wray and J. W. Allen, *J. Phys.* **C4**, 512 (1971).

81. J. T. Vallin and G. D. Watkins, *Solid State Commun.* **9**, 953 (1971).

82. J. T. Vallin and G. D. Watkins, *Phys. Rev.* **B9**, 2051 (1974).

83. J. T. Vallin, G. A. Slack, S. Roberts, and A. E. Hughes, *Phys. Rev.* **B2**, 4313 (1970).

84. B. Nygren, J. T. Vallin, and G. A. Slack, *Solid State Commun.* **11**, 35 (1972).

85. Y. Uehara, *Bull. St. Marianna University School of Medicine* **3**, 57 (1974).

86. R. S. Title in *Physics and Chemistry of II-VI Compounds*, M. Aven and J. S. Prener, Eds. (North Holland, Amsterdam, 1967) Ch. 6.

87. L. C. Kravitz and W. W. Piper, *Phys. Rev.* **146**, 322 (1966).

88. W. C. Holton, M. de Wit, T. L. Estle, B. Dischler, and J. Schneider, *Phys. Rev.* **169**, 359 (1968).

89. R. C. Kemp, *J. Phys.* **C2**, 1416 (1969).

90. K. Morigaki and T. Hoshina, *J. Phys. Soc. Japan* **23**, 318 (1967).

91. R. K. Watts, *Phys. Rev.* **B2**, 1239 (1970).

92. D. S. McClure, *J. Chem. Phys.* **39**, 2850 (1963).

93. W. H. Brumage, C. R. Yarger, and C. C. Lin, *Phys. Rev.* **133A**, 765 (1964).

94. R. Rohrig and A. Räuber, paper presented at meeting of German Physical Society, Freudenstadt, April 13, 1972.

95. G. A. Slack, F. S. Ham, and R. M. Chrenko, *Phys. Rev.* **152**, 376 (1966).

96. G. A. Slack, S. Roberts, and F. S. Ham, *Phys. Rev.* **155**, 170 (1967).

97. G. A. Slack and B. M. O'Meara, *Phys. Rev.* **163**, 335 (1967).

98. G. A. Slack, S. Roberts, and J. T. Vallin, *Phys. Rev.* **187**, 511 (1969).

99. F. S. Ham and G. A. Slack, *Phys. Rev.* **B4**, 777 (1971).

100. H. A. Weakliem, *J. Chem. Phys.* **36**, 2117 (1962).

101. P. Koidl, O. F. Schrmer, and U. Kaufmann, *Phys. Rev.* **B8**, 4926 (1973).

102. P. Kodl and A. Räuber, *J. Phys. Chem. Solids* **35**, 1061 (1974).

103. P. Van Engelen, W. Boon, and J. Dieleman, *J. Phys. Chem. Solids* **33**, 1041 (1972).

104. R. K. Watts, *Phys. Rev.* **188**, 568 (1969).

105. U. Kaufmann, P. Koidl, and O. F. Schirmer, *J. Phys.* **C6**, 310 (1973).

106. U. Kaufmann and P. Koidl, *J. Phys.* **C7**, 791 (1974).

107. I. Broser and M. Maier, *J. Phys. Soc. Japan* **21** Suppl., 254 (1966).

108. M. de Wit, *Phys. Rev.* **177**, 441 (1969).

109. W. C. Holton, M. de Wit, R. K. Watts, T. L. Estle, and J. Schneider, *J. Phys. Chem. Solids* **30**, 963 (1969).

110. R. E. Dietz, H. Kamimura, M. D. Sturge, and A. Yariv, *Phys. Rev.* **132**, 1559 (1963).

111. B. Clerjaud and A. Gelineau, *Phys. Rev.* **B9**, 2832 (1974).

112. T. Yamaguchi and H. Kamimura, *J. Phys. Soc. Japan* **33**, 953 (1972).

113. R. Dingle, *Phys. Rev. Lett.* **23**, 579 (1969).

114. M. Schulz, *Solid State Commun.* **11**, 1161 (1972).

115. K. Morigaki in *II-VI Semiconducting Compounds*, D. G. Thomas, Ed. (Benjamin, New York, 1967) p. 1348.

116. S. Shionoya in *II-VI Semiconducting Compounds*, D. G. Thomas, Ed. (Benjamin, New York, 1967) p. 1.

117. K. Urabe, S. Shionoya, and A. Suzuki, *J. Phys. Soc. Japan* **25**, 1611 (1968).

118. R. K. Watts and W. C. Holton, *Phys. Rev.* **173**, 417 (1968).

119. J. D. Kingsley and M. Aven, *Phys. Rev.* **155**, 235 (1967).

120. K. R. Lea, M. J. M. Leask, and W. P. Wolf, *J. Phys. Chem. Solids* **23**, 1381 (1962).

121. J. E. Lowther, *Phys. Stat. Sol.* **50B**, 287 (1972).

122. J. E. Nicholls, *J. Phys.* **C7**, L88 (1974).

123. J. E. Lowther, *Phys. Stat. Sol.* **48B**, K107 (1971).

124. J. D. Kingsley, J. S. Prener, and M. Aven, *Phys. Rev. Lett.* **14**, 136 (1965).

125. J. D. Kingsley in *Optical Properties of Ions in Crystals*, H. M. Crosswhite and H. W. Moos, Eds. (Wiley-Interscience, New York, 1967) p. 151.

126. R. K. Watts and W. C. Holton, *Phys. Lett.* **24A**, 365 (1967).

127. R. S. Title and J. W. Mayo, *Bull. Am. Phys. Soc.* **12**, 41 (1967).

128. R. S. Title, *Bull. Am. Phys. Soc.* **11**, 14 (1966).

129. H. Masui, *J. Phys. Chem. Solids* **33**, 1129 (1972).

Chapter *10*

Alkali Halides

Defects in alkali halide crystals have been studied for many decades. More is known about point defects—their structures, interactions, and effects on crystal properties—in these materials than in any others. The crystals are optically transparent over a very wide spectral range. From Table 10.1 the band gaps are generally very large and the maximum phonon frequencies range from small to medium. The melting points are fairly low, and crystals of high quality and purity are relatively easily grown. On the other hand, the melting points are high enough that a wide temperature range is available for study of defect phenomena. Some of these compounds are hygroscopic and thus must be kept in a dry atmosphere for most experiments. The simple crystal structure provides an environment of high symmetry for impurities and native defects, making defect structure determination easier experimentally and simplifying theoretical calculations. Often defect-related phenomena have been studied in alkali halides with the hope that they could first be understood in these materials and that this understanding would lead to insight into defect properties of more complicated ionic crystals.

Most of the alkali halides have the rock salt structure shown in the upper part of Figure 10.1. Each ion is octahedrally coordinated by six nearest-neighbor ions of opposite charge. The Bravais lattice is face centered cubic. The first Brillouin zone of the reciprocal lattice has the same shape as that for the zincblende lattice shown in Figure 7.2. CsCl, CsBr, and CsI have the structure shown at the bottom of the figure. Each ion is surrounded by eight nearest-neighbor ions of opposite charge. The lattice is simple cubic with two ions per unit cell—one at a corner and one at the center of the cube. It can also be viewed as two interpenetrating simple cubic lattices, one of positive ions and the other of negative ions.

Table 10.1 Some Properties of Alkali Halides. Most of the Band Gaps Listed Are the Low-Temperature Values; m_e^p Is the Polaron Mass of the Electron; Other Parameters Are as in Tables 8.1 and 9.1

	E_g (eV)	f_i	R_{nn} (Å)	T_m (C)	$\hbar\omega_{lo}$ (cm^{-1}, meV)	ε_s	ε_o	m_e^*/m	m_e^p/m
LiF	13.7	0.915	2.01	845	660, 81.8	9.04	1.92		
LiCl	9.4	0.903	2.57	605	419, 51.9	11.86	2.79	~0.5	
LiBr	7.6	0.899	2.75	550	351, 43.5	13.33	3.22		
LiI	6.1	0.890	3.00	449		11.0	3.80		
NaF	11.5	0.946	2.32	993	420, 52.1	5.07	1.74		
NaCl	8.75	0.935	2.82	801	260, 32.2	5.89	2.35	~0.5	
NaBr	7.1	0.934	2.99	747	209, 25.9	6.40	2.64		
NaI	5.9	0.927	3.24	661	178, 22.1	7.28	3.08		
KF	10.8	0.955	2.67	858	334, 41.4	6.05	1.85		
KCl	8.7	0.953	3.15	770	210, 26	4.81	2.20	0.43	0.92
KBr	7.4	0.952	3.30	734	163, 20.2	4.90	2.39	0.37	0.71
KI	6.34	0.950	3.53	681	141, 17.5	5.09	2.68	0.33	0.54
RbF	10.3	0.960	2.82	795	289, 35.8	5.91	1.93		
RbCl	8.5	0.955	3.29	718	173, 21.4	4.92	2.91	0.43	1.04
RbBr	7.2	0.957	3.45	693	126, 15.6	5.0	2.33		
RbI	6.3	0.951	3.67	647	104, 12.9	4.94	2.61	0.37	0.72
CsF	9.8		3.00	682	245, 30.4	8.08	2.2		
CsCl^a	8.3		3.57	645	161, 20	6.95	2.67		
CsBr^a	7.3		3.72	636	113, 14	6.66	2.83		
CsI^a	6.1		3.96	626	91, 11	6.59	3.09	0.42	0.96

^{a} CsCl structure.

The band structures of the alkali halides are not as well known as those of the more covalent compounds already discussed. Optical data are more difficult to obtain because of the large band gaps. Figure 10.2 shows as a typical example the band structure of KCl calculated by Fong and Cohen[1] by the empirical pseudopotential method. Besides the large energy range on the vertical scale, two striking features are immediately obvious. The valence bands are nearly flat, that is, nearly independent of wave vector, and the conduction bands are not so flat. The valence bands are flat because the electrons in these states are tightly bound to the ions and are not so much affected by the rest of the crystal. The available valence electrons are the $(3s)^2(3p)^5$ electrons of chlorine. With the valence bands full of electrons the charge distribution is that of large, nearly spherical Cl^- ions and smaller spherical K^+ ions. The upper Γ_{4u} valence band is formed largely from Cl^- $3p$ states. When electron spin is taken into account, this band splits into a lower Γ_{6u} and an upper Γ_{8u} component at

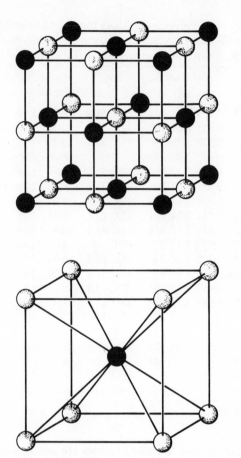

Figure 10.1   Rock salt (above) and CsCl lattice structures.

the Γ point, just as in the semiconductors. This splitting is not shown in the figure. It is given by the Cl spin orbit splitting, 0.1 eV. For F, Br, and I this splitting is 0.03, 0.4, and 0.9 eV, respectively.[2] The band gap is direct. The lowest conduction band is of the $s$like Γ_1 type. The conduction bands depend more strongly on wave vector, and a conduction electron is not so well localized on one type of ion. Calculation of the conduction bands is complicated by the necessity of including the $3d$ states of potassium.

The ionicities listed in the table are all near one, and the static dielectric constant is much larger than the optical dielectric constant. We expect strong coupling of electrons to the ionic motion. Polaron effects are important for the electron, as can be seen from comparison of the polaron mass of an electron with the bare electron effective mass in Table 10.1.

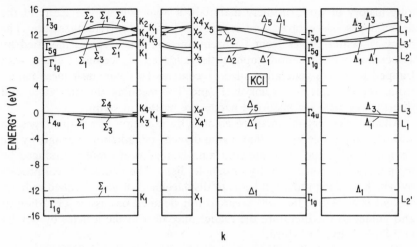

Figure 10.2 Band structure of KCl (from Fong and Cohen, reference 1).

The large enhancement is due to the large value of the polaron coupling constant α, which is about 3 for alkali halides.[3] The effect is much more dramatic for the hole. The hole is self-trapped into a bond between two halogen ions and thus has infinite mass. The "free" hole and "free" exciton with an electron bound to the hole are localized point defects! At higher temperatures the hole may hop through the lattice in random fashion.

Because the alkali halides are the simplest ionic crystals, many attempts have been made to understand why a particular compound has the rock salt or CsCl structure.[4,5] The cohesive energy—the work required to separate the solid into isolated atoms at O K—can be rather accurately calculated. It consists of a large Coulomb interaction of the ions (Madelung term), a small repulsive term due to overlap of the electron clouds, a smaller van der Waals term, and a very small zero-point lattice vibrational energy. The CsCl structure has a slightly larger Madelung constant, but the repulsive interactions of nearest and next nearest neighbors favor the rock salt structure, by larger amounts for the lighter alkali atoms than the heavier ones. The difference in cohesive energies of the two structures is very small—it is of the same order as the uncertainty in the calculations. Many alkali halides with rock salt structure undergo a phase transition to the CsCl structure at high pressures, and CsCl changes to the rock salt structure at 450 C.

A wide variety of intrinsic defects and impurities have been studied in alkali halides. Many of these lead to absorption bands that color the crystal and for this reason have been called color centers. Because of the long history of work on these materials, a nomenclature for many of the defect types has evolved and is still used in the current literature. The name $\mathscr{F}$ center comes from Farbzentrum and an $\mathscr{F}$ center modified by association with a cation impurity is called an $\mathscr{F}_A$ center. Centers with trapped holes are generally called $\mathscr{V}$ centers. To distinquish these names from, for example, the symbols F and V_F, meaning fluorine atom and fluorine vacancy, the historical names are written in script.

Alkali halide crystals may be colored by radiation of various types. In addition to particle beams that create defects, or radiation damage, by the obvious mechanism of elastic collision, damage is efficiently produced by high-energy photons, even by ultraviolet light. The photochemical process by which ionizing radiation leads to displacement of ions is called radiolysis and is an active field of research. For detailed discussion of radiolysis and guides to the literature the reader may turn to the articles by Sonder and Sibley[6] and Crawford.[7]

There are no shallow donors or acceptors in the alkali halides, although effective masslike spatially extensive states have been invoked to explain excited levels of certain centers including excitons and the $\mathscr{F}$ center. However, defects can act as donors or acceptors, and their charge states can be changed in various ways, but the ground state of the center generally lies far from the band edge. The $\mathscr{F}$ center has been called the most common donor in the alkali halides.[3] The ionization energy of the $\mathscr{F}$ center electron is a few electron volts.

HALOGEN VACANCY

The isolated halogen vacancy can have three charge states. The vacancy with no trapped electron is positively charged with respect to the lattice and is called the $\mathscr{F}^+$ center. The vacancy which has trapped one or two electrons is called the $\mathscr{F}$ center or the $\mathscr{F}^-$ center, respectively. The $\mathscr{F}$ center has been studied for many decades and is without doubt the most thoroughly investigated point defect. Indeed a whole book has been written on this center.[8] There are still open questions about the $\mathscr{F}$ center, however, and it will attract the attention of researchers for many more years.

In Chapter 1 the simple point-ion theory of Gourary and Adrian[9] for the electronic states of the $\mathscr{F}$ center was described. This theory gives a fairly good description of the ground state wavefunction. This function, from Equation 1.89, written to display the dependence on the electron coordinate $\mathbf{r}$, has the form

$$\Psi(\mathbf{r}) = N \left[\phi(\mathbf{r}) - \sum_n \langle n|\phi\rangle \psi_n(\mathbf{r}) \right]$$

It can be shown that the electron density at nucleus j at position $\mathbf{r}_j$, $|\Psi(\mathbf{r}_j)|^2$, is given approximately by $A|\phi(\mathbf{r}_j)|^2$, where A is a proportionality or "amplification" factor which can be calculated if the core functions ψ_n are known. ENDOR measurements that map $|\Psi(\mathbf{r})|^2$, and thus the envelope function $|\phi(\mathbf{r})|^2$, are in agreement with theory for distances r through the first few shells of neighbors around the vacancy, but for larger r the actual density is much larger than theoretical values.[10] The shape of $|\phi(\mathbf{r})|^2$ is roughly that of a hydrogenic $1s$ function with a maximum at the origin and no nodes.

The excited states of the electron are considerably more complex than indicated in Chapter 1. On the high-energy side of the main absorption or $\mathcal{F}$ band is a weak asymmetrical band called the $\mathcal{K}$ band. In the spectra of Li and Na halides it is not seen and may lie under the $\mathcal{F}$ band; it is best resolved from the $\mathcal{F}$ band in RbCl. Spinolo and Smith[11] have shown that this absorption is probably due to transitions to $3p$, $4p$, and other higher-lying levels of $p$like states. They converge to a limit that is supposedly the bottom of the conduction band. However, in some alkali halides there are also other weaker absorption bands of the $\mathcal{F}$ center, called the $\mathcal{L}_1$, $\mathcal{L}_2$, and $\mathcal{L}_3$ bands, which lie at even higher energies than this limit. Excitation at low temperature into these states, and into the high-energy side of the $\mathcal{K}$ band releases electrons from the $\mathcal{F}$ centers, producing photoconductivity. At room temperature, absorption into the $\mathcal{F}$ band also leads to release of electrons by a thermally activated process with activation energy in the range 0.1 to 0.2 eV, measured from the position of the relaxed excited state.[12] The nature of the excited states responsible for the $\mathcal{L}$ band absorptions is unknown.

The $\mathcal{F}$ band absorption and the corresponding emission band are broad and differ greatly in transiton energy. For example, at very low temperature in KCl the peak absorption occurs at 2.31 eV and the width of the band at half maximum intensity is 0.16 eV. The corresponding emission band peaks at 1.22 eV and has a width 0.26 eV. The temperature dependence of the position of band peak and of the width are those expected from the configuration coordinate model. The large shift between absorption and emission indicates very strong coupling of the electron to lattice vibrations. The Huang–Rhys factor for the excited state is typically $S \approx 20$, and for the ground state S is several times larger.[13] From analysis of the spectra on the basis of the configuration coordinate model also comes the effective frequency of the vibration associated with the configuration coordinate. This is not the same for the two states, the effective frequency in the excited state being the larger. The effective frequencies are less than the maximum lattice phonon frequency. The effective frequency represents an average over those vibrational modes that contribute to the motion represented by the configuration coordinate. Henry[14] has shown that this distribution of frequencies for the ground state is the same as that observed directly by Raman scattering[15] in crystals containing $\mathscr{F}$ centers. In the absence of $\mathscr{F}$ centers or other defects there is no first-order Raman scattering because of the high crystal symmetry. The configuration coordinate model is too simple for the $\mathscr{F}$ center, however. The difference between the peak energies of emission and absorption bands is in some cases 30% different from the expected value $S_g \hbar \omega_g + S_s \hbar \omega_s$.

Three complicating features of the absorption spectrum of the $\mathscr{F}$ center are spin-orbit coupling in the excited states, splitting and mixing of electronic states by interaction with vibrations, and the presence of states of even parity to which transitions from the ground state are not normally allowed. Because the absorption band is so broad, these splittings do not lead to well-resolved structure in the spectrum and are difficult to study. Let us first consider the spin-orbit interaction and splitting by interaction with vibrations. An atomic p state could be split by spin-orbit coupling into two sublevels, $^2P_{3/2}$ and $^2P_{1/2}$. Magneto-optic measurements interpreted on the basis of an atomic model imply that $^2P_{3/2}$ lies below $^2P_{1/2}$ in contrast with the case of an alkali atom. The inversion, as explained by Smith,[16] arises because the interaction occurs near the nuclei of the host lattice rather than near a single central nucleus. The magnitude of the spin-orbit splitting thus increases with increasing atomic number of the host lattice atoms. In the Cs halides the $\mathscr{F}$ band shows two or three poorly resolved peaks in contrast to the featureless band seen in other crystals. A p state can also be split by interaction with vibrations of E_g and T_{2g} local symmetry (Jahn–Teller effect). The interaction with modes of A_{1g} symmetry leads to broadening. Henry, Schnatterly, and Slichter[17] have shown that it is possible, by measuring the changes in the moments of the absorption band due to application of external perturbations such as a magnetic field or a uniaxial stress, to extract these splittings, even though the band width is very large. In CsF and CsCl the three-peaked structure of the $\mathscr{F}$ band is due to a large spin-orbit interaction combined with vibrational splitting[18] of $^2P_{3/2}$. There is no such structure in the emission band, however.

Analysis of the change in the shape of the $\mathscr{F}$ absorption due to application of an external electric field has revealed the presence of a $2s$ like state just above the $2p$ level.[19] The $2s$-$2p$ separation ranges from 0.07 eV in KBr to 0.17 eV in NaCl. The electric field mixes the $2s$ and $2p$ states, making $1s \rightarrow 2p$ absorption slightly weaker and the $1s \rightarrow 2s$ transition slightly allowed. The p and d states are split by the field. For a field along the [001] direction the local symmetry is reduced from O_h to C_{4v}, and the correspondences between irreducible representations are shown in Table 10.2. For light polarized along the z direction of the field, transitions are possible from the ground state to sa_1 states, to pa_1 states and to da_1 states. In KCl with perpencidular polarization (x or y) a transition to another state above the 2s state has also been observed and has been attributed to a *de* state. [20] Even without the electric field a *d* state would, in general, be split into two components (e_g and t_{2g}). The s and p electronic states can also be mixed by vibrations of local T_{1u} symmetry. The problem of the excited states becomes very complex if several closely spaced electronic states interacting with many types of vibrational modes must be considered. The observed levels probably belong to rather complicated vibronic states. Perregaux and Ascarelli[21] have pointed out that the spacing in energy of the $\mathcal{F}$, $\mathcal{K}$, and $\mathcal{L}$ levels seen in absorption is that expected for rotational levels of a molecule.

As the absorption spectrum contains information about the excited states of the $\mathcal{F}$ center in a nonequilibrium situation, so the emission spectrum tells something about these states after the surrounding ions have adjusted their positions. The absorption in the $\mathcal{F}$ band is characterized by an oscillator strength near 1, the same value as for the $1s \rightarrow 2p$ transition of a free atom. Although the radiative quantum efficiency of the emission is near 1 also, the observed lifetime, $\sim 10^{-6}$ sec, is about two orders of magnitude longer than expected for $2p \rightarrow 1s$ emission.[13] Fowler[22] has attributed the long lifetime to the more diffuse nature of the relaxed excited state compared with the unrelaxed state. Another possibility is that the emission comes largely from the 2s state which may lie just below 2p in the relaxed configuration. The two states would be mixed by T_{1u} vibrations. Experimental results on the Stark effect in the emission spectrum have been interpreted on the basis of strong mixing of 2s and 2p states by vibrations.[23,24] Ham[25] has considered theoretically the mixing of the 2s and 2p states by T_{1u} vibrations, neglecting the effect of coupling of the 2p state to E_g and T_{2g} modes. He finds that the lowest vibronic excited state has almost entirely either 2s or 2p character, with negligible mixing. A theoretical treatment of Wang, Matsuura, Wong, and Inoue[26] based on the different approach of polaron theory, on the other hand, predicts that the lowest vibronic state is a mixture with 70% 2s character and 30% 2p character and that it is more diffuse spatially than the unrelaxed state.

Because the lifetime of the relaxed excited state is rather long, it has been possible to maintain by strong optical pumping an adequate population in the state long enough to observe EPR in this state and to measure in absorption transitions from it to higher-lying states. Mollenauer and co-workers[27] have been able to observe EPR and ENDOR spectra of the relaxed excited state in KI. The resonance is isotropic and the g factor is only slightly less than that of the ground state. Their mapping of the wave function of the state shows an electron density with the general shape of a 2p function with negligible admixture of 2s. However, the measured g factor suggests an s state. The composition of the lowest relaxed excited state is, then still an open question. Kondo and Kanzaki[28] have observed absorption from the lowest relaxed excited state to several slightly higher-lying states. They see no continuum absorption rising with increasing transition energy which might be expected as a result of a transition to the conduction band edge.

The halogen vacancy can also bind two electrons to form the $\mathscr{F}^-$ center. This has a very broad asymmetrical absorption band which partially overlaps on its high-energy side the $\mathscr{F}$ band.[29] Absorption in the $\mathscr{F}^-$ band releases an electron to the conduction band, and this electron may be trapped at a vacancy $\mathscr{F}^+$ to form an $\mathscr{F}$ center or at an $\mathscr{F}$ center to form another $\mathscr{F}^-$ center. The positive vacancy $\mathscr{F}^+$ is observable through its perturbation on the spectra of an exciton trapped near it.[30,31]

HALOGEN VACANCY ASSOCIATED WITH FOREIGN ALKALI ION

This center has been studied in KCl, KBr, RbCl, and RbBr containing alkali metal impurities.[32] The defect, called the $\mathscr{F}_A$ center, consists of an $\mathscr{F}$ center, with one of the six alkali ions neighboring the vacancy replaced by a foreign alkali ion: Na^+ or Li^+ in KBr, for example. The foreign ion reduces the local symmetry to C_{4v}. The absorption spectrum is very similar to that of the $\mathscr{F}$ center except for the appearance of two bands separated by about 0.2 eV in place of the $\mathscr{F}$ band. One band is polarized parallel to the axis of the center and the other perpendicular, as is expected for transitions to a p-state split into two components in the lower symmetry (see Table 10.2).

At moderate temperatures the $\mathscr{F}_A$ center, like the $\mathscr{F}$ center, can be ionized by absorption of light in these bands. Because of the polarization property it is possible to align the centers with polarized light. The possibility of optical alignment is a very convenient property of many color centers with symmetry lower than that of a lattice site. Once a preferential alignment is obtained, the axis of the transition dipole responsible for optical absorption can be determined by taking the absorption

Table 10.2 Reduction of Symmetry from O_h to C_{4v}; Some Basis Functions for C_{4v} Are Given in Parentheses

	O_h	C_{4v}
s	a_{1g}	a_1 (z or a constant)
p	t_{1u}	a_1 (z or a constant) $+ e(x,y)$
d	e_g	a_1 (z or a constant) $+ b_1(x^2 - y^2)$
	t_{2g}	$b_2(xy) + e(x,y)$

spectrum with light of various polarizations. Before alignment there is an equal number of centers with dipole axis along several equivalent directions—the three $\langle 100 \rangle$ directions, for example—and the spectrum is independent of the polarization of the incident light. Changes in the alignment brought about by, for example, a thermally activated hop of the vacancy to one of the other equivalent positions around the impurity or by dissociation of the complex and reformation after diffusion of the vacancy through the lattice can also be measured. For an $\mathscr{F}_A$ center in the ground state the activation energy for the former process is about 1 eV, whereas for the excited state the corresponding activation energy is 10 times smaller.[32] The model of the $\mathscr{F}_A$ center has been confirmed by ENDOR measurements. Preferential alignment of the center axis can be monitored in the ENDOR spectra.[10,33] In this way the resonance spectra are shown to be due to the same defect as the optical spectra.

Associated with the two main absorption bands is a single emission band. The splitting of the p state is very small in the relaxed excited configuration. For most $\mathscr{F}_A$ centers, labeled type I by Luty,[32] the emission band occurs near the same transition energy as that of the corresponding $\mathscr{F}$ center, and the luminescence decay time is also similar to that of $\mathscr{F}$ center emission. In KCl:Li, RbCl:Li, and RbBr:Li, on the other hand, the Stokes shift is much larger and the decay time much shorter; these $\mathscr{F}_A$ centers are labeled type II. The different properties of the type II $\mathscr{F}_A$ center, in which the impurity ion is small, are thought to be connected with the greater ease of movement of the vacancy around the impurity in this case.

The $\mathscr{F}_A$ center can also bind a second electron to form an $\mathscr{F}_A^-$ center. Like the $\mathscr{F}^-$ center this also has a very broad absorption band which consists of two overlapping bands of different polarization.

Other complexes of unknown structure, called $\mathscr{Z}$ centers, are formed from divalent impurities such as Ca^{2+}, Sr^{2+}, or Ba^{2+}. The divalent impurity requires charge compensation. Vacancies are thought to be involved in the complex. The optical and magnetic resonance data are reviewed by Radhakrishna and Chowdari.[34]

HALOGEN VACANCY CLUSTERS

The complex formed by two $\mathscr{F}$ centers on nearest anion neighbor positions, two halogen vacancies and two trapped electrons, is called the $\mathscr{M}$ center. The two vacancies lie on a $\langle 110 \rangle$ axis in crystals with rock salt structure and on a $\langle 100 \rangle$ axis in those with the CsCl structure. $\mathscr{M}$ centers can be formed from $\mathscr{F}$ centers by optical excitation in the $\mathscr{F}$ band to ionize the center, forming $\mathscr{F}^+$ and a free electron. The liberated electron may be trapped by an $\mathscr{F}$ center to form an $\mathscr{F}^-$ center. An $\mathscr{F}^+$ center may diffuse toward the $\mathfrak{F}^-$ center to form the $\mathfrak{M}$ center. Two other charge states of the defect, the $\mathfrak{M}$ center which has trapped a third electron ($\mathfrak{M}^-$) and the singly ionized center ($\mathfrak{M}^+$) have also been observed.[35,36]

There are several absorption bands associated with the $\mathfrak{M}$ center. The most prominent bands are labeled $\mathfrak{M}_1$, $\mathfrak{M}_2$, and $\mathfrak{M}_2'$. The $\mathfrak{M}_2$ and $\mathfrak{M}_2'$ bands occur in the same spectral region as the $\mathfrak{F}$ band and are difficult to study because a crystal containing $\mathfrak{M}$ centers generally contains even more $\mathfrak{F}$ centers. The $\mathfrak{M}_1$ band lies at lower energy. For an $\mathfrak{M}$ center with axis [110] the $\mathfrak{M}_1$, $\mathfrak{M}_2$, and $\mathfrak{M}'_2$ bands are polarized in the [110], [1$\bar{1}$0], and [100] directions, respectively.[37] These polarization properties suggest the model of the center. The ground state is not paramegnetic.

If the $\mathfrak{F}$ center is similar to an hydrogen atom, we might expect the $\mathfrak{M}$ center to be roughly approximated by an hydrogen molecule. In the ground state of the free hydrogen molecule, both electrons are paired with opposite spin in the $1s\sigma_g$ orbital $\psi_{1s}^{(a)} + \psi_{1s}^{(b)}$, where $\psi_{1s}^{(a)}$ means the $1s$ orbital localized on proton a, to form a $^1\Sigma_g^+$ state. In the molecular notation the superscript + indicates parity on reflection of the molecule in a plane containing the axis. This state has no paramagnetism. In the first excited state with spin 1, $^1\Sigma_u^+$, one electron has been promoted to a $2p\sigma$ orbital, and in the $^1\Pi_u$ state it has been promoted to a $2p\pi$ orbital. These two excited states are analogous, except for the extra s electron, to the excited p states of the $\mathfrak{F}_A$ center; that is, one has the lobes of the p function along the axis of the center ($p\sigma$) and the other, perpendicular ($p\pi$). The levels are shown in Figure 10.3 as obtained[38] for an hydrogen molecule in a continuous medium with the dielectric properties of KCl. These are the levels that would be seen in absorption— that is, before lattice relaxation occurs. Also shown are some of the levels in the true D_{2h} symmetry of the center as calculated by Evarestov[39] using a point-charge model for the KCl crystal environment. Agreement with experimental spectra is rather good. The degeneracy of the $p\pi$ orbital is lifted in D_{2h} symmetry. Associated with these three absorption bands is a single polarized emission band[40] corresponding to the transition $^1B_{1u} \rightarrow {}^1A_g$. The Stokes shift is much smaller than in the case of the $\mathfrak{F}$ center, and the lifetime of the state is much shorter.

In the triplet state shown in Figure 10.3, $^3\Sigma_u^+$, one electron is in the $1s\sigma_g$ orbital and the other is in the $1s\sigma_u$ orbital $\psi_{1s}^{(a)} - \psi_{1s}^{(b)}$. In a free molecule this state is unstable against dissociation,[41] but in the confined $\mathfrak{M}$ center it is stable. It can be populated by irradiation in the $\mathfrak{F}$ band at lower temperatures. The lifetime of this state is very long—50 sec in KCl—because a radiative transition to the ground state is forbidden, and nonradiative decay by multiphonon emission is unlikely because the separation from the ground singlet is probably many times the energy of an optical phonon. When the triplet state is populated, new absorption bands appear corresponding to transitions from the triplet to other triplet excited states.[42,43]

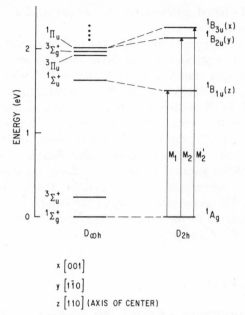

x [001]

y [1$\bar{1}$0]

z [110] (AXIS OF CENTER)

Figure 10.3 Energy levels of the $\mathfrak{M}$ center in KCl. The center is approximated by a free molecule at left.

EPR and ENDOR spectroscopy of the triplet state confirm the model of the $\mathfrak{M}$ center.[44] The new optical spectra and the magnetic resonance signal decay with the same long time constant of the metastable triplet. The spin Hamiltonian is

$$\mathcal{H} = g\beta \mathbf{H} \cdot \mathbf{S} + D S_z^2 + E \left(S_y^2 - S_x^2 \right) \qquad (10.1)$$

The axes are those indicated in Figure 10.3. The spin $S = 1$ and the g factor is slightly less than 2. As with the optical spectra, there is some experimental difficulty because of overlap of the resonance spectra with the $\mathfrak{F}$ center line.

The cluster of three nearest $\mathfrak{F}$ centers is called an $\mathfrak{R}$ center. In the rock salt structure the vacancies form a triangle in a {111} plane. There are several absorption bands, the most prominent ones being the $\mathfrak{R}_1$ and $\mathfrak{R}_2$ bands which lie between the $\mathfrak{F}$ and $\mathfrak{M}_1$ bands. The other bands are not resolved from those of other halogen vacancy centers usually also present in the crystal. Silsbee[45] has shown that the lowest states can be understood on the basis of the states of three $\mathfrak{F}$ centers that are brought together and allowed to interact. The three $1s$ ground states lead to a lower ^2E and a slightly higher $^4A_1$ state of the cluster, which has C_{3v} symmetry. Similarly, a large number of excited states should be derived from the p states of the isolated $\mathfrak{F}$ centers.

The $^4A_1$ state is thought to lie only a fraction of an electron volt above ^2E. EPR of this metastable state has been observed, and the spectrum is described by the spin Hamiltonian of Equation 10.1 with $S = 3/2$, $E = 0$, and the z axis along a $\langle 111 \rangle$ direction.[46] The metastable state is populated, as in the case of the $\mathfrak{M}$ center, by optical excitation through higher excited states. Changes in the optical spectrum accompany the population of $^4A_1$ and decay, with the EPR signal, with a time constant of 15 sec.

In addition to the $\mathfrak{M}$ and $\mathfrak{R}$ centers and their several charge states other, more complex vacancy clusters have been discussed in the literature.[47] For larger aggregates models are more difficult to establish, not only because of the greater complexity of the defects but also because of interference produced in the experimental spectra by the simpler aggregates which are invariably present also.

SELF-TRAPPED HOLE

In most of the alkali halides a hole in the valence band becomes trapped into a highly localized state on two nearest-neighbor halogen ions. The only crystals in which this self-trapped hole or $\mathcal{V}_k$ center has not yet been identified are LiBr, LiI, and CsF. The $\mathcal{V}_k$ center is created by X-irradiation of the crystal at low temperature to produce electrons and holes. The electrons can be trapped at cation vacancies or at some impurity electron trap, such as substitutional Ag^+, Tl^+, or Pb^{2+}, incorporated for this purpose.[48,49] The holes are trapped on a pair of halogen neighbors to form molecule ions X_2^-, where X stands for a halogen atom. These molecule ions are oriented along $\langle 110 \rangle$ directions in NaCl-type crystals and along $\langle 100 \rangle$ directions in CsI. This model for the $\mathcal{V}_k$ center is well established as a result of much work on the associated optical and magnetic resonance spectra.[50] In particular, it is certain that the localization of the hole is not due to a nearby impurity, vacancy, or native interstitial.

The hole is so well localized that a description of the electronic states in terms of an isolated X_2^- molecule ion explains the observed properties well. For Cl_2^-, for example, the single particle states formed from the $3p$ states of each chlorine atom are σ_g, π_u, π_g, and σ_u, as shown on the left in Figure 10.4.[51] When these are filled with the available eleven electrons, an unpaired electron is left in σ_u in the ground state. In the excited states one of the lower electrons is excited to σ_u, or equivalently, the hole in σ_u is excited to π_g, π_u, or σ_g. If we describe the states in this way, concentrating on the hole rather than the electrons, it is easy to see that the many

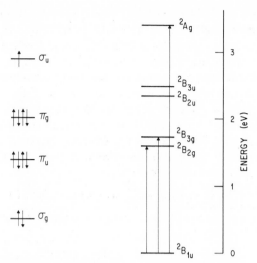

Figure 10.4 Energy levels of the self-trapped hole in KCl.

electron (or single hole) states are also a ground state $^2\Sigma_u^+$ and excited states $^2\Pi_g$, $^2\Pi_u$, and $^2\Sigma_g^+$. In a free molecule the Π states are split into two sublevels by the spin-orbit interaction, which is large for the heavier halogens. In the D_{2h} symmetry of a rock-salt crystalline environment such a splitting would result even in the absence of spin-orbit interaction, so that the observed splitting is due to these two effects combined. The molecular ground state $^2\Sigma_u^+$ becomes $^2B_{1u}$ in the crystal, and the other experimentally determined levels are shown on the right in Figure 10.4 for the $\mathcal{V}_k$ center in KCl.[52] The positions are those that would be observed in an absorption measurement. The arrows indicate electric dipole allowed transitions.

The EPR spectrum of the $\mathcal{V}_k$ center is characterized by a strong hyperfine interaction with the two equivalent halogen nuclei.[53,54] This is quite different from $\mathcal{F}$ center resonance spectra where such interactions are weak and hyperfine structure is so poorly resolved that it can only be studied by ENDOR. The hole of the $\mathcal{V}_k$ center is much more localized than the $\mathcal{F}$ center electron. The spin Hamiltonian is given by

$$\mathcal{H} = \beta \left(g_x H_x S_x + g_y H_y S_y + g_z H_z S_z \right)$$

$$+ \sum_{i=1}^{2} \left(A_x^{(i)} S_x I_x^{(i)} + A_y^{(i)} S_y I_y^{(i)} + A_z^{(i)} S_z I_z^{(i)} \right) \tag{10.2}$$

The axes x, y, z are the same as for the $\mathcal{M}$ center, Figure 10.3. The axis of the X_2^- molecule ion is z or [110]. The sum is over the two halogen nuclei. Typical g factors are those of the $\mathcal{V}_k$ center in KCl given in Table 10.3; g_x and g_y are very nearly equal. For a free molecule ion, which has $D_{\infty h}$ axial symmetry, they would be exactly equal. Analysis of the halogen hyperfine structure shows that the hole orbital is not entirely made up of atomic p orbitals, but has a small s component as well. The shifts of the g factors from the free electron value are determined by admixture of $^2B_{3u}$ and $^2B_{2u}$ into the ground state by the spin-orbit interaction. The positions of these levels can be inferred from the g shifts, whereas the positions of the other levels are determined by optical absorption. A fine collection of data on $\mathcal{V}_k$ centers can be found in reference 52.

Table 10.3   Some Molecule Ion Centers in KCl

Description	Name	g_x	g_y	g_z	$h\nu_{uv}$(eV)	$h\nu_{ir}$(eV)
$(Cl_2)^{-a}$	$\mathcal{V}_k$	2.0445	2.0424	2.00145	3.38	1.73, 1.59
$(Cl_2)^- V_K{}^a$	$\mathcal{V}_F$	2.0439	2.0414	2.0015	3.38	1.65
$(Cl_2)^- Na_K{}^b$	$\mathcal{V}_{kA}$(Na)	2.0394	2.0426	2.0015	3.49	
$Na_K(Cl_2)^- Na_K{}^b$	$\mathcal{V}_{kAA}$(Na)	2.0364	2.0409	2.0015		
$(BrCl)^{-a}$		2.1329	2.1350	1.9839	3.25	1.63
$(BrCl)^- V_K{}^a$		2.177	2.188	1.9702	3.37	1.36
$(Cl_2)^{-c}$	$\mathcal{H}$	2.0221	2.0227	2.0018		

[a] Reference 58.
[b] Reference 52.
[c] Reference 72.

Jette, Gilbert, and Das[55] have calculated wave functions and energy levels for $\mathcal{V}_k$ centers in several chlorides and fluorides in the following way. Initially the lattice is imagined to be perfect and the hole is not localized. The lower energy equilibrium configuration is approached in a thought experiment of several steps. First, the hole is localized by forming a wave packet from valence band states. This requires expenditure of energy about equal to half the valence band width.[47] Next, the surrounding ions are allowed to relax in response to the electric field of the localized hole, but the interionic separation of X_2^- is kept fixed. This polarization energy gain nearly cancels the localization energy. Finally, the X_2^- molecule ion is allowed to relax to an equilibrium separation. The energy gain in this last step, the "bonding energy," is the dominant contribution and is obtained from the potential energy curve (energy versus separation) of the free molecule ion. This discussion shows why no corresponding electron localization occurs in the perfect lattice. The conduction band width is much greater than the valence band width, and the halogen molecule ion is much more stable than the alkali molecule ion.

Jette, Gilbert, and Das find energy levels in good agreement with experimental optical spectra for LiF and NaF and fair agreement for the chlorides. The internuclear separation of X_2^- is found to be 10% larger than for the free molecule ion but smaller than the normal crystal spacing. The two halogen ions are drawn together by the hole. The displacements of other neighbor ions, obtained in calculating the polarization energy, are in good agreement with ENDOR data.[56]

The optical and magnetic resonance spectra have been correlated by bleaching with polarized light.[48] During such bleaching, centers with a particular orientation are destroyed on being raised to an excited state and are re-formed with other orientations. In most crystals optically induced orientation can be destroyed by a thermally activated process, the activation energy being 0.5 eV in KCl.[52] The geometric nature of the jumps by which reorientation and motion of the center through the lattice proceed has been studied in several alkali halides.[50] The high-energy optical absorption band has polarization parallel to the z axis. The lower-energy bands have mixed polarization, the parallel component becoming stronger with increasing halogen atomic weight.[52,57] The separation of the $^2B_{2g}$ and $^2B_{3g}$ bands also increases in this fashion, because the magnitude of the spin-orbit couplings for F, Cl, Br, and I stand in the ratios $1:2:9:19$.[52]

Mixed molecule ions $(XY)^-$, in which X and Y are halogen atoms of different types, also occur and have properties similar to those of the V_k center.[58] In an alkali halide MX containing a small concentration of Y, holes may be trapped to form $(XY)^-$ if Y has a smaller electronegativity than X. For example, $(BrCl)^-$ can exist in KCl, but $(BrCl)^-$ has not been observed in KBr. The mixed-molecule ion is an isoelectronic trap. Molecule ions Y_2^- in an MX crystal have also been observed. Examples are I_2^- in KCl[59] and Br_2^- in NaCl.[52] In Table 10.3 some experimental data for $(BrCl)^-$ in KCl are shown.[52] The g factors are more different from those of Cl_2^- in KCl than the difference for Cl_2^- from one chloride to another. This is another indication of the strong localization of the hole.

The V_k center is an efficient electron trap, and a V_k center with a bound electron is called a self-trapped exciton. Recombination of the electron and hole is accompanied by a characteristic luminescence consisting of two bands. In some crystals a third, much weaker band is also present.[60] In a cystal in which the V_k centers have been preferentially aligned, the luminescence bands resulting from recombination of free electrons with the trapped holes are observed to be polarized parallel to the alignment axis (higher-energy band) or perpendicular to it (lower-energy band).[61] The lifetime of the higher-energy luminescence is short, about 10^{-8} sec, and that of the lower energy luminescence is longer by several orders of magnitude. The higher-energy band is ascribed to an electric dipole allowed singlet→singlet transition, and the other, to a triplet→ singlet transition made weakly allowed by spin-orbit interaction. The multiplicity of the triplet has been confirmed by excited state EPR experiments.

The model for the self-trapped exciton consists in a hole tightly bound in the σ_u orbital of X_2^- with an electron in a somewhat more diffuse orbital called "$1s\sigma_g$" by Williams and Kabler.[62] The electron orbital outside the region of X_2^- is thought to have an envelope with the general contour of a $1s$ state. The triplet state, for example, is $(\sigma_u)(1s\sigma_g)^3\Sigma_u^+$. In the ground state to which the luminescence transition goes there is no hole, but two electrons paired in σ_u of Figure 10.4. The long-lived triplet state has been studied by EPR in KBr and CsBr.[63,64] The spin Hamiltonian is

$$\mathcal{K} = \beta \mathbf{H} \cdot \mathbf{g} \cdot \mathbf{S} + D\left(S_z^2 - 2/3\right) + E\left(S_x^2 - S_y^2\right) + \sum_{i=1}^{2} \mathbf{S} \cdot \mathbf{A}^{(i)} \cdot \mathbf{I}^{(i)} \quad (10.3)$$

with spin $S = 1$. In CsBr the symmetry is higher than in KBr and $E = 0$, $g_x = g_y$. The hyperfine coupling tensor $\mathbf{A}$ is axial and is due to interaction with the two halogen nuclei of X_2^-. The similarity of the parameters to those of the V_k center shows that the hole is little affected by the presence of the electron.

Optical absorption spectra have also been observed that correspond to transitions from the $(\sigma_u)(1s\sigma_g)^3\Sigma_u^+$ excited state to higher excited triplets. The absorption spectra decay with time constant equal to the lifetime of $(\sigma_u)(1s\sigma_g)^3\Sigma_u^+$. This identifies the initial state of the transitions.[62] These transitions are thought to be of two types: excitations of the hole only or excitations of the electron only. Two examples of the first type are

$$(\sigma_u)(1s\sigma_g)^3\Sigma_u^+ \rightarrow (\sigma_g)(1s\sigma_g)^3\Sigma_g^+$$

$$(\sigma_u)(1s\sigma_g)^3\Sigma_u^+ \rightarrow (\pi_g)(1s\sigma_g)^3\Pi_g$$

These spectra are very similar to the corresponding absorption spectra of the V_k center itself. The transitions of the second type comprise two converging series of the form

$$(\sigma_u)(1s\sigma_g)^3\Sigma_u^+ \rightarrow (\sigma_u)(np\sigma_u)^3\Sigma_g^+$$

$$(\sigma_u)(1s\sigma_g)^3\Sigma_u^+ \rightarrow (\sigma_u)(np\pi_u)^3\Pi_g$$

where $n \geqslant 2$. In all these expressions the hole orbital is written in the first parentheses and the electron orbital in the second.

The system $(XY)^-$ plus an electron is an exciton bound to an isoelectronic trap. The properties of such a center are similar to those of the self-trapped exciton.[65,66] The exciton is not free to migrate away from the impurity, however. In an optical absorption experiment in which bound excitons are created at iodine impurities in KCl, lines ascribed to a Rydberg series of effective masslike states of the electron in the Coulomb field of the hole have been seen by Baldini and Teegarden.[67]

One reason for the great interest in the self-trapped exciton is the role it plays in radiation damage. In competition with the radiative recombination process just described is a damage-creating process in which the energy released on recombination is not radiated, but rather is used to create lattice defects. It is not difficult to see intuitively how such a process might occur. With the exciton present the halogen nuclei of X_2^- are drawn together. On recombination of the electron and the hole they spring apart and may impart enough kinetic energy to one of the members of the pair, or to another halide ion in the direct line of fire, along the $\langle 110 \rangle$ axis to knock it into an interstitial site.[66,68] It has been shown that within 11×10^{-12} sec after exciton formation an $\mathscr{F}$ center in its ground state and an interstitial neutral chlorine atom appear as products of such a radiolysis process.[70] This damage is of the Frenkel type. In contrast the disorder produced by random thermal motion of the lattice of a pure alkali halide crystal is predominantly of the Schottky type—equal numbers of anion vacancies and cation vacancies. Damage creation by large atomic displacements induced by electron-hole recombination may be a rather general phenomenon. Although it has been largely studied in alkali halides, there is evidence for such processes in III-V compounds as well.

CATION VACANCY

The cation vacancy may trap a hole into an orbital localized on two of the nearest-neighbor halogens. This center is called $\mathscr{V}_F$ to indicate that it is the antimorph of the $\mathscr{F}$ center. The $\mathscr{V}_F$ center has been identified in LiF, KCl, and NaCl by EPR.[71,52] It is not at all similar to the $\mathscr{F}$ center, of course, but has in common with the cation vacancy centers in II–VI compounds and the vacancy in silicon an asymmetrical charge distribution and a lower symmetry than the normal lattice site. From Figure 10.5 it looks like a $\mathscr{V}_k$ center adjacent to a cation vacancy. The only difference, except for the vacancy, is a slight bend of the X_2^- molecule toward the vacancy. At temperatures above 95 K the Cl_2^- molecule in KCl can reorient by jumping

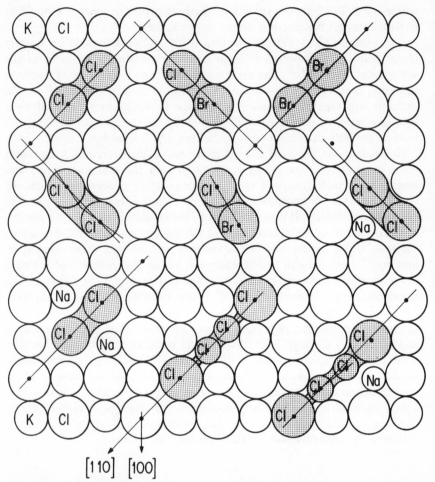

$[\overline{1}10]$ $[100]$

Figure 10.5 Various hole centers in KCl. From left to right, top to bottom the centers are the self-trapped hole or V_k center, a hole trapped at an isoelectronic Br impurity, a hole trapped on two Br impurities, a hole trapped at a cation vacancy or V_F center, a hole trapped at a Br impurity adjacent to a cation vacancy, a V_{kA} center, a V_{kAA} center, an $\mathcal{H}$ center, and an $\mathcal{H}_A$ center, (from D. Schoemaker, reference 52).

from one to another of the 12 equivalent $\langle 110 \rangle$ positions around the vacancy. At higher temperatures the $Cl_2^- - V_K$ complex migrates through the crystal as a unit and can be trapped at a bromine impurity to form the similar $(ClBr)^- - V_K$ complex.[12] Irradiation into the higher-energy absorption band of $(ClBr)^- - V_K$ results in the decomposition of this center and the formation of $Cl_2^- - V_K$.

OTHER HALOGEN MOLECULE ION CENTERS

Other centers similar to the $\mathcal{V}_F$ center are formed by trapping of the hole on two halogen ions beside a foreign alkali ion.[52] This is called a $\mathcal{V}_{kA}$ center. It has the same structure as the $\mathcal{V}_F$ center except that a foreign alkali ion occupies the position of the vacancy. In the $\mathcal{V}_{kAA}$ center there is a foreign alkali ion on each side of the X_2^- molecule ion, as shown in Figure 10.5.

Another common hole center is the $\mathcal{H}$ center, an X_2^- molecule ion on a single halogen site in a split interstitial arrangement. In KCl the X_2^- axis lies along a $\langle 110 \rangle$ direction,[72] but in LiF the axis is $\langle 111 \rangle$.[73,74] The heteronuclear analogues $(FX)^-$ are also oriented along $\langle 111 \rangle$ directions.[75]

In KCl the orientation of the $\mathcal{H}$ center is the same as that of the self-trapped hole. But in the $\mathcal{H}$ center there is considerable hole density on the two halogen ions along the Cl_2^- axis adjacent to the central pair. The $\mathcal{H}$ center with a foreign alkali ion on a nearest cation site is called a $\mathcal{V}_1$ or $\mathcal{H}_A$ center.[76-79] In KCl the Na $\mathcal{H}_A$ center is only slightly different, because of a small bending of the molecule ion, from the $\mathcal{H}$ center. But the Li $\mathcal{H}_A$ center has a quite different geometrical structure which can be considerably altered by application of uniaxial stress.[80] The $\mathcal{V}_t$ center in LiF consists of a hole localized on three adjacent fluorine ions that form a triangle in a $\{100\}$ plane.[81] Two cation vacancies and an anion vacancy are thought to be associated with the triangle. The hole orbital in the ground state is made of π orbitals on the three anions in contrast with the centers already discussed.

HYDROGEN CENTERS

Hydrogen can occupy halogen sites or interstitial sites in the alkali halides. The substitutional negative hydrogen ion, H^-, is called a $\mathcal{U}$ center. For example, KH and KBr form a solid solution and the lattice constant decreases linearly with increasing KH content.[82,83] The single absorption band of the $\mathcal{U}$ center, called the $\mathcal{U}$ band, lies at higher energy that the $\mathcal{F}$ band.[84] There is apparently no luminescence band associated with the absorption band.

The $\mathcal{U}$ center has been treated theoretically by Gourary,[85] among others. He calculates the energy of the excited state responsible for the absorption band on a point charge model similar to that used for the $\mathcal{F}$ center. The ground state is s^2, a singlet. In the singlet excited state one electron is in the relatively compact s orbital and the other is in a more extended p orbital. This contrasts with the heavy metal ions with s^2 ground-state configuration and compact sp excited states.

Irradiation in the $\mathcal{U}$ band at low temperature converts the $\mathcal{U}$ center into an interstitial neutral hydrogen atom or $\mathcal{U}_2$ center, leaving an $\mathcal{F}$ center in the substitutional site. At higher temperatures the interstitial hydrogen diffuses away to form H_2 molecules in the crystal. These mobile molecules can be trapped by $\mathcal{M}$ centers to form H^-—H^- nearest-neighbor substitutional pairs.[86,87] The $\mathcal{U}_2$ center has an optical absorption band at slightly longer wavelength than the $\mathcal{U}$ band, and also has a characteristic EPR spectrum with a hyperfine splitting nearly identical with that of a free hydrogen atom.[84] Irradiation in the $\mathcal{F}$ band frees the electron from the $\mathcal{F}$ center. It is trapped on the hydrogen atom, producing an interstitial H^- ion, or $\mathcal{U}_1$ center. An assocated $\mathcal{U}_1$ absorption band lies at longer wavelength than the $\mathcal{U}_2$ band.

A local vibrational mode of the $\mathcal{U}$ center occurs at more than twice the freguency of the longitudinal optical phonon. This mode has been studied for H^- and D^- in many alkali halides. A list of the observed frequencies is given in reference 88. Second harmonic local vibrations are seen by Raman scattering.[89] In crystals containing foreign alkali ions the triply degenerate mode of the isolated $\mathcal{U}$ center is split into a doublet and a singlet when the foreign cation is at a nearest-neighbor position.[90] When pairing with a nearest-neighbor halide impurity in KCl occurs, the symmetry is lower (C_{2v}) and three lines are seen.[91] The local mode of the $\mathcal{U}_1$ center occurs at lower frequency. In addition to the single absorption line due to a relatively isolated $\mathcal{U}_1$ center, other lines are seen which have been attributed to the $\mathcal{U}_1$ center perturbed by a nearby vacancy. [92]

HEAVY METAL IMPURITIES

The substitutional ions Tl^{2+}, Cd^+, and Pb^{3+} and neutral atoms Ag and Cu have a single s electron outside closed shells. These impurities are easily recognized by the characteristic EPR spectrum with large impurity nuclear hyperfine interaction.[49,93-96] Similar centers occur in the II–VI compounds and have been described in Chapter 9. Four optical absorption bands are associated with Tl^{2+} in KCl [97] and one with Ag in KCl.[96]

Four absorption bands have been identified with the neutral Tl atom in KCl, KBr, and KI. Two of these are in the near infrared and are thought to correspond to the $^2P_{1/2} \rightarrow {}^2P_{3/2}$ transition of the $(6s)^2(6p)$ configuration of the free atom.[98,99,100] Optical spectra assigned to Cu^+ and Ag^+ have also been studied. Although the ground state has a simple closed-shell d^{10} configuration, the open d shell and extra s or p electron in the excited configurations make analysis difficult. [101,102]

In some crystals Cu^+ and Ag^+ occupy positions displaced slightly from the substitutional site. The displacement is in a $\langle 111 \rangle$ direction for Cu^+ in the potassium and rubidium halides and in a $\langle 110 \rangle$ direction for Ag^+ in the rubidium halides. The off-center position is apparently favored when the noble metal replaces an alkali ion of larger size. A dipole moment is associated with the off-center ion and the paraelectric properties afford means of studying these defects.[103,104]

Ag^{2+} with the d^9 configuration is produced in KCl or NaCl when Ag^+ traps a hole. The lower-lying e state of the hole in the d shell is split by interaction with a distortion of E_g symmetry. The ground state observed by magnetic resonance shows a tetragonal distortion.[96] Hyperfine interactions with four equivalent nearby chlorine nuclei show that the hole occupies the orbital $|x^2 - y^2\rangle$, where the z axis is taken as the $\langle 001 \rangle$ axis of the distortion. This orbital represents a "cloverleaf" charge distribution in the $x - y$ plane, and admixture of σ orbitals on the four nearest-neighbor chlorines in the plane accounts for the chlorine hyperfine interaction. Several absorption bands are associated with the center. Their polarization properties allow correlation with the EPR spectra, but the exact nature of the excited states involved is not known.

In the ground state the positive ions Ga^+, In^+, Tl^+, Ge^{2+}, Sn^{2+}, Pb^{2+}, Sb^{3+}, and Bi^{3+}, the neutral atoms Zn, Cd, and Hg, and the negative ions Cu^-, Ag^-, and Au^- have two electrons in an s orbital. Most of these have been observed to substitute for the cation and to have an s^2 configuration in the crystal also. The ground state is not paramagnetic, but there are optical spectra that have been studied extensively, especially in the case of Tl^+.[105-110]

If the ground state of the ion in the crystal corresponds to $s^2\,{}^1S_0$ of the free ion, the lowest excited states should be derived from the sp configuration. These are ${}^3P_0$, ${}^3P_1$, ${}^3P_2$, and ${}^1P_1$ for the free ion. Only the transition ${}^1S_0 \rightarrow {}^1P_1$ is allowed in the limit of small spin-orbit interaction. However, the spin-orbit interaction is not small compared with the inter-electron electrostatic interaction (intermediate coupling) in these heavy metals, and ${}^1S_0 \rightarrow {}^3P_1$ is also allowed because of mixing of ${}^1P_1$ and ${}^3P_1$. The three absorption bands, called A, B and C in order of increasing energy, were first assigned to transitions to the states ${}^3P_1$, ${}^3P_2$, and ${}^1P_1$, respectively. The band B is very weak compared with A and C.

In the crystal the states are more complex because of admixture with ligand orbitals and interaction with lattice vibrations. The ratio of the oscillator strengths of the A and C bands is well explained by a molecular orbital model for the center in which the s and p atomic orbitals are replaced by a_{1g} and t_{1u} molecular orbitals; it is not explained by the simpler atomic model.[111,112] Although a configuration coordinate model involving a single distortion of A_{1g} symmetry explains some features of the spectra,[113] it seems essential to consider the Jahn–Teller effect to explain the structure observed in many cases in the A and C bands, the temperature dependence of this structure, and the polarization effects in the emission spectra. Toyozawa and Inoue[114] have explained the structure in the absorption spectra by considering coupling to distortions of T_{2g} as well as A_{1g} symmetry. The partially polarized emission observed under polarized excitation indicates a tetragonal distortion.[110,115,116] The monovalent impurities are probably not associated with another defect. The distortion is probably due to the Jahn–Teller effect. Much work has been done to deduce the shapes of the complicated potential energy surfaces in multidimensional distortion space, which could explain the observations.[110,117,118,119]

At higher impurity concentration, absorption bands due to paired ions appear. These can be recognized by the square-law dependence of absorption coefficient on concentration,[105] From polarization studies the Au^- pairs in KI are oriented along $\langle 110 \rangle$ directions, as expected for nearest-neighbor pairs.[120] The Tl^+ pairs may apparently be oriented along $\langle 110 \rangle$ or $\langle 100 \rangle$ directions.[121,122,123]

NaI and CsI containing Tl^+ are especially useful as scintillation detectors of radiation or particles. The high atomic numbers of Cs and I lead to high photoelectric absorption coefficients, of use in X-ray and γ ray detection. The incident radiation or particle beam creates electrons and holes and results in Tl^+ emission, perhaps by the following processes.[97] The Tl^+ traps a free electron to form neutral Tl. Recombination of this electron with a nearby self-trapped hole leaves Tl^+ in an excited state and leads to luminescence, or Tl^+ may trap an electron to form excited Tl^+.

CHALCOGEN IMPURITIES

Oxygen, sulfur, and selenium form many different complex centers in the alkali halides. Some of these are paramagnetic in the ground state, and definite models have been established by EPR studies. There are many others than those discussed in this section which remain to be investigated.

The diatomic molecule ions O_2^-, S_2^-, Se_2^-, and $(SeS)^-$ have been identified in several alkali halides with rock salt structure. The molecule substitutes for an anion, and the molecular axis is oriented in a $\langle 110 \rangle$ direction. The electronic states are rather compact spatially, as in the case of the halogen molecule ion. Magnetic resonance, Raman spectra, optical absorption and emission spectra have been studied.

The electronic states of the molecule ion are those that result from filling the one-electron orbitals of Figure 10.4 with the nine available p electrons.

This leaves a hole in the π_g molecular orbital to give a $^2\Pi_g$ ground state.[124] This is split by about 200 cm^{-1} in the case of O_2^-, into two components in the lower symmetry of the crystal. The resonance of the ground Kramers doublet is characterized by an orthorhombic g tensor and a hyperfine interaction with the two nuclei of the S_2^- molecule.[125,126] For O_2^- hyperfine splittings from interaction with four of the six neighboring cations are also seen.[127] The lobes of the p orbitals on the two atoms of the molecule point in a $\langle 110 \rangle$ direction for O_2^- in potassium and rubidium halides and for S_2^- in KCl, RbBr, and RbI, but for O_2^- in sodium halides and S_2^- in NaI, KBr, and KI they point in a $\langle 100 \rangle$ direction. In all cases they are perpendicular to the molecular axis, of course, since the molecular orbital is a π type. The molecular axis is able to flip from one equivalent orientation to another. Reorientation effects have been studied by EPR and optically.[128,129]

Several other sulfur and selenium complexes have been observed by Schneider, Dischler, and Räuber[133] in KCl, KBr, and KI. The triangular molecule ions S_3^- and Se_3^- are identified by a strong hyperfine interaction with a single chalcogen nucleus and a weaker interaction with two equivalent ones. The symmetry of the center is that of an isosceles triangle lying in a $\{100\}$ plane. The C_2 axis is a $\langle 110 \rangle$ direction. ENDOR spectra reveal hyperfine interactions with neighboring lattice ions and show that the triatomic molecule substitutes for two halogen ions and one alkali ion.[134]

When crystals containing these centers are heated in air, new centers are formed as sulfur or selenium reacts with dissolved oxygen. These new centers are thought to be $(SO_2)^-$ or $(SeO_2)^-$ molecule ions substitutional for an anion. At low temperatures the orientation of the molecule is fixed. At higher temperatures reorientation occurs in contrast with S_3^- and Se_3^-, which do not reorient. The resonance parameters of the $(SO_2)^-$ and $(SeO_2)^-$ molecule ions trapped in the crystal are similar to those of the free isoelectronic ClO_2 moledule. Another center consisting of four sulfur or selenium atoms lying in a $\{100\}$ plane has also been identified.[133] It is not known how this center fits into the lattice.

The emission spectra of the molecule ion have a striking appearance at low temperature.[130] The electronic transition $^2\Pi_u \rightarrow {}^2\Pi_g$ is strongly coupled to the intramolecular stretching mode and weakly coupled to lattice phonons. The result is a progression of lines due to transitions from the lowest (intramolecular) vibrational level of the $^2\Pi_u$ well to many of the vibrational levels of the ground well. Each line has a weak wing on the low-energy side corresponding to emission of phonons.[131,132] The lines are not evenly spaced because many of the terminal levels are high enough in the well that the parabolic approximation for the shape of the well as a function of internuclear distance is very poor. The position ν of a line corresponding to a transition from the zeroth vibrational level of the excited well to the level with vibrational quantum number v of the lower well is given approximately by

$$\nu = \nu_0 - v\omega + v^2 a - v^3 b$$

For O_2^- the energy $\hbar\omega$ is very near 1100 cm^{-1} for all host crystals studied. For S_2^- $\hbar\omega$ is very near 600 cm^{-1}, and for Se_2^- $\hbar\omega$ is 330 cm^{-1}. The intramolecular vibration is insensitive to the crystal environment. The vibrational frequency in the excited well is smaller than in the lower well.

Application of uniaxial stress raises the orientational degeneracy of the molecule and leads to splitting of the sharp emission lines. From the polarization of the split lines Ikezawa and Rolfe[129] have shown that the transition dipole moment lies along the $\langle 110 \rangle$ molecular axis for O_2^- in KCl, RbCl, RbBr, and RbI; for S_2^- in KI and KBr; and for Se_2^- in KI. But for O_2^- in NaCl, KBr, and KI the transition dipole moment lies along a $\langle 111 \rangle$ axis, according to these authors.

The $(OH)^-$ molecule ion can be incorporated substitutionally at halogen sites, and high concentrations are obtainable. A vibrational absorption band occurs at 2.7 μm and a strong electronic absorption at 0.204 μm in KCl. The molecule has a dipole moment and is oriented with the internuclear axis along a $\langle 100 \rangle$ direction. The molecules can be preferentially oriented along a particular $\langle 100 \rangle$ axis with an electric field. This leads to strong polarization of the ultraviolet absorption band from which Kuhn and Lüty[135] deduced the $\langle 100 \rangle$ orientation of the dipole. The molecule ion reorients at low temperatures by a quantum mechanical tunneling through the energy barrier separating the orientations of minimum energy.[136] Absorption in the ultraviolet band produces luminescence bands, which at low temperature correspond to transitions from the lowest vibrational level of the excited potential well to the first few vibrational levels of the lower well.[137] Because the quantum of the internal molecular vibration of the molecule is so large—0.4 eV for the molecule ion trapped in the solid as well as for the free molecule—several discrete bands result. The width of the bands relative to the spacing is larger than for the O_2^- spectra.

Similarly, trapped $(SH)^-$ and $(SeH)^-$ molecule ions have infrared vibrational absorption bands and electronic absorption in the ultraviolet.[138,139] Absorption in the ultraviolet band leads to dissociation, and substitutional S^- or Se^- is one of the products. Substitutional O^- can be formed in other ways.[140,141] These negative ions are paramagnetic. The free ion has the p^5 configuration—a single hole in the p shell. The EPR spectra are characterized by an axial g tensor.[140,141,142] For O^- the symmetry axis along which the lobes of the hole orbital extend is $\langle 100 \rangle$. For S^- and Se^- in KCl and KBr the axis is $\langle 111 \rangle$. An S^- spectrum of another sort has also been observed in KCl, KBr, KI, NaBr, and RbBr. This spectrum has orthorhombic symmetry with the lobes of the hole orbital lying along a $\langle 110 \rangle$ axis.[143] At least one of the two types of S^- center ($\langle 111 \rangle$ and $\langle 110 \rangle$) must be a more complex center involving association of the sulfur atom with another defect. An $(HSH)^-$ center with $\langle 111 \rangle$ symmetry axis has been identified in KCl and KBr by the hyperfine structure due to the two protons.[142]

For the O^- center in KI association with another defect does not appear to be the cause of the axial symmetry, because application of uniaxial stress at 1 K leads to redistribution of population among the six $\langle 100 \rangle$ orientations.[144] The proof is not conclusive, however, because some molecule ions reorient under applied stress at low temperature. Hyperfine splitting due to a single potassium nucleus indicates that the oxygen ion is displaced toward one of the alkali neighbors. Although the threefold degeneracy of the p state could be lifted by a tetragonal Jahn–Teller distortion, interaction with a symmetric distortion would lead to hyperfine interaction with an even number of K^+ nuclei, rather than the single nucleus.

Many other molecule ions have been incorporated in alkali halide crystals. Some examples are N_2^-, $(BO_3)^{3-}$, $(PO_4)^{3-}$, and $(CN)^-$.[145-149] We do not discuss other molecule ions; as isolated impurities and as components of more complex defects, they form a large class of impurity centers. The properties of the optical spectra of O_2^-, S_2^-, $(NO_2)^-$, and $(PO_2)^-$ in alkali halides are described and compared in detail by Rebane and Rebane.[150]

TRANSITION METAL IMPURITIES

Ions of the transition series are rarely found in a monovalent charge state in crystals. In alkali halides charge compensation in the form of another defect is required for electrical neutrality. The compensator may be near the metal impurity or distant from it. In most cases it is apparently nearby, and the result is a defect center of low symmetry. The charge compensator is thought to be the cation vacancy, although this is difficult to prove experimentally.

Bron and co-workers[151,152,153] have studied the optical spectra of Sm^{2+}, Eu^{2+}, and Yb^{2+} in several alkali halides with rock salt structure. The spectra consist of broad bands due to transitions of the type $(4f)^n \rightarrow (4f)^{n-1}(5d)$ and, in the case of Sm^{2+}, of weaker sharp lines due to transitions between states of the $(4f)^6$ configuration as well. From the polarization properties of the sharp lines the local symmetry is found to be that which would result from the presence of a charge compensator along a

$\langle 110 \rangle$ axis. A point charge model of the crystal field that allows for the presence of a charge compensating nearest cation vacancy is able to predict the observed crystal field splittings fairly well if the vacancy and the rare earth ion are allowed to relax toward each other slightly. Magnetic resonance studies of Eu^{2+} also indicate this sort of local symmetry.[154,155]

Superimposed on the broad absorption and emission bands are series of sharp lines. There are, in general, several series associated with each broad band. The spacing of the lines of a series is nearly independent of the identity of the rare earth ion and decreases with increasing atomic weight of the host lattice ions, the dependence on the mass of the nearest-neighbor anion being stronger than on the cation mass. These lines are due to strong coupling to vibrational pseudolocalized or resonance modes with frequencies less than the maximum phonon frequency of the perfect lattice. The width of the broad background band is due to coupling to lattice vibrations of a broad frequency range.

An exception is Eu^{2+} in NaCl, KCl, and RbCl. Here one of the quanta of localized vibration, measured in absorption spectra, is 800 cm^{-1}. This is much larger than the maximum phonon energy. Bron and Wagner[156] explain this anomaly in the following way. Some states of the $(4f)^7$ and $(4f)^6(5d)$ configurations lie in the same spectral range, and two of these may be nearly degenerate. These two excited states have two configuration coordinate curves that, because of strong vibrational coupling, interact and "repel" each other. Because of the interaction, the upper well has increased curvature and hence larger vibrational quanta associated with it. In absorption, transitions to this upper well have the large 800 cm^{-1} separation.

Sharp line optical spectra of Pr^{3+} in NaCl, KCl, RbCl, and KBr show that the trivalent ion can also be incorporated.[157] The degeneracy of each $^{2S+1}L_J$ multiplet is completely lifted because of association with charge compensating defects, thought to be two cation vacancies. Broad absorption bands produced by gamma irradiation of KCl containing Sm^{2+} have been attributed to Sm^+.[158]

Mn^{2+} forms several different types of center which have been studied by EPR.[159] Two of the centers have the symmetry expected from association of Mn^{2+} with a charge compensating vacancy at a nearest cation site along the $\langle 110 \rangle$ axis or at a next-nearest site along a $\langle 100 \rangle$ axis. At higher temperatures the vacancy hops among the equivalent sites about the Mn^{2+} impurity. This motion has been detected as a broadening in the EPR spectra and also by dielectric loss measurements.[160] The lifetime τ of the vacancy at a nearest cation site in NaCl is given by

$$\tau^{-1} = (1.2 \times 10^{14}) \exp \frac{-0.63\ eV}{kT}$$

At higher temperatures the vacancy dissociates completely from the impurity, and at 450 C only the spectrum of isolated Mn^{2+} remains.

The complex formed by a transition metal ion M^{n+} and six octahedrally coordinated cyanide ligands $[M^{n+}(CN)_6^-]^{(6-n)-}$ can fit into the rock salt lattice of many alkali halides, replacing, for example, the unit $[K^+Cl_6^-]^{5-}$ in KCl. Depending on the value of n, charge compensation by other defects may be necessary. In the complex the d electrons of the transition metal experience a strong ligand field. In a strong octahedral field the ground states for d^5, d^6, and d^7 metal configurations are $^2T_2$, $^1A_1$, and ^2E, respectively. Ligand-metal charge transfer transitions are seen in the optical spectra as well as transitions involving only the d electrons.[161] The molecular orbital theory for these complexes has been given by Alexander and Gray,[162] and the vibrational spectra[163] have been treated by Hawkins and co-workers.[164]

Irradiation with γ rays or heat treatment produces complexes of lower symmetry.[165,166] EPR studies in the cases in which the transition metal M^{n+} is Fe^+, Co^{2+}, Ni^{3+}, or Rh^{2+} in chlorides show hyperfine interactions with one or two chlorine nuclei. The results have been interpreted in terms of replacement of one or two $(CN)^-$ ligands by Cl^-.

REFERENCES

1. C. Y. Fong and M. L. Cohen, *Phys. Rev.* **185**, 1168 (1969).

2. N. F. Mott and R. W. Gurney, *Electronic Processes in Ionic Crystals* (Dover, New York, 1964) p. 95.

3. F. C. Brown in *Point Defects in Solids*, Vol. 1, J. H. Crawford, Jr. and L. M. Slifkin, Eds. (Plenum, New York, 1972) Ch. 8.

4. M. P. Tossi in *Solid State Physics*, Vol. 16, F. Seitz and D. Turnbull, Eds. (Academic, New York, 1964) Ch. 1.

5. A. Ghosh, A. K. Sarkar, and A. N. Basu, *J. Phys.* **C8**, 1332 (1975).

6. E. Sonder and W. A. Sibley in *Point Defects in Solids*, Vol 1, J. H. Crawford, Jr. and L. M. Slifkin, Eds. (Plenum, New York, 1972) Ch.4.

7. J. H. Crawford, Jr., *Adv. Phys.* **17**, 93 (1968).

8. J. J. Markham, *F Centers in Alkali Halides* (Academic, New York, 1966).

9. B. S. Gourary and F. J. Adrian, *Phys. Rev.* **105**, 1180 (1957).

10. H. Seidel and H. C. Wolf in *Physics of Color Centers*, W. B. Fowler, Ed. (Academic, New York, 1968) Ch.8

11. G. Spinolo and D. Y. Smith, *Phys. Rev.* **140A**, 2117 (1965); Smith and Spinolo, ibid., 2121 (1965).

12. R. K. Swank and F. C. Brown, *Phys. Rev.* **130**, 34 (1963).

13. W. Gebhart and H. Kunhert, *Phys. Stat. Sol.* **14**, 157 (1966).

14. C. H. Henry, *Phys. Rev.* **152**, 699 (1966).

15. J. M. Worlock and S. P. S. Porto, *Phys. Rev. Lett.* **15**, 697 (1965).

16. D. Y. Smith, *Phys. Rev.* **137A**, 574 (1965).

17. C. H. Henry, S. E. Schnatterly, and C. P. Slichter, *Phys. Rev.* **137A**, 583 (1965); C. H. Henry and C. P. Slichter in *Physics of Color Centers*, W. B. Fowler, Ed. (Academic, New York, 1968) Ch. 6.

18. P. R. Moran, *Phys. Rev.* **137A**, 1016 (1965).

19. U. M. Grassano, G. Margaritondo, and R. Rosei, *Phys. Rev.* **B2**, 3319 (1970).

20. M. Bonciani, U. M. Grassano, and R. Rosei, *Phys. Rev.* **B8**. 5855 (1973).

21. A. Perregaux and G. Ascarelli, *Phys. Rev.* **B10**, 1683 (1974).

22. W. B. Fowler, *Phys. Rev.* **135A**, 1725 (1964).

23. L. D. Bogan and D. B. Fitchen, *Phys. Rev.* **B1**, 4122 (1970).

24. L. F. Stiles, Jr., M. P. Fontana, and D. B. Fitchen, *Phys. Rev.* **B2**, 2077 (1970).

25. F. S. Ham, *Phys. Rev.* **B8**, 2926 (1973).

26. S. Wang, M. Matsuura, C. C. Wong, and M. Inoue, *Phys. Rev.* **B7**, 1695 (1973).

27. L. F. Mollenauer and S. Pan, *Phys. Rev.* **B6**, 772 (1972). L. F. Mollenauer and G. Baldacchini, *Phys. Rev. Lett.* **29**, 465 (1972).

28. Y. Kondo and H. Kanzaki, *Phys. Rev. Lett.* **34**, 664 (1975).

29. R. S. Crandall and M. Mikkor, *Phys. Rev.* **138A**, 1247 (1965).

30. F. Bassani and N. Inchauspé, *Phys. Rev.* **105**, 819 (1957).

31. T. Timusk, *J. Phys. Chem. Solids* **26**, 849 (1965).

32. F. Lüty in *Physics of Color Centers*, W. B. Fowler, Ed. (Academic, New York, 1968) Ch. 2.

33. R. L. Mieher, *Phys. Rev. Lett.* **8**, 362 (1962).

34. S. Radhakrishna and B. V. R. Chowdari, *Phys. Stat. Sol.* **14**, 11 (1972).

35. M. Hirai, M. Ikezawa, and M. Ueta, *J. Phys. Soc. Japan* **17**, 1483 (1962).

36. I. Schneider and H. Rabin, *Phys. Rev.* **140A**, 1983 (1965).

37. F. Okamoto, *Phys. Rev.* **124**, 1090 (1961).

38. R. Herman, M. C. Wallis, and R. F. Wallis, *Phys. Rev.* **103**, 87 (1956).

39. R. A. Evarestov, *Opt. Spectr.* **16**, 198 (1964).

40. C. J. Delbecq, *Z. Phys.* **171**, 560 (1963).

41. R. S. Mullikan, *Rev. Mod. Phys.* **4**, 1 (1932).

42. M. Ikezawa, *J. Phys. Soc. Japan* **19**, 529 (1964).

43. I. Schneider and M. E. Caspari, *Phys. Rev.* **133A**, 1193 (1964).

44. H. Seidel, *Phys. Lett.* **7**, 27 (1963).

45. R. H. Silsbee, *Phys. Rev.* **138A**, 180 (1965).

46. H. Seidel, M. Schwoerer, and D. Schmid, *Z. Phys.* **182**, 398 (1965).

47. W. B. Fowler in *Physics of Color Centers*, W. B. Fowler, Ed. (Academic, New York, 1968) Ch. 2.

48. C. J. Delbecq, B. Smaller, and P. H. Yuster, *Phys. Rev.* **111**, 1235 (1958).

49. D. Schoemaker and J. L. Kolopus, *Solid State Commun.* **8**, 435 (1970).

50. M. N. Kabler in *Point Defects in Solids,* J. H. Crawford and L. M. Slifkin, Eds. (Plenum, New York, 1972) Ch. 6.

51. M. H. Cohen, *Phys. Rev.* **101**, 1432 (1956).

52. D. Schoemaker, *Phys. Rev.* **B7**, 786 (1973).

53. T. G. Castner and W. Känzig, *J. Phys. Chem. Solids* **3**, 178 (1957).

54. C. P. Slichter, *Principles of Magnetic Resonance* (Harper and Row, New York, 1963) p. 195.

55. A. N. Jette, T. L. Gilbert, and T. P. Das, *Phys. Rev.* **184**, 884 (1969).

56. D. F. Daly and R. L. Mieher, *Phys. Rev. Lett.* **19**, 637 (1967).

57. C. J. Delbecq, W. Hayes, and P. H. Yuster, *Phys. Rev.* **121**, 1043 (1961).

58. C. J. Delbecq, D. Schoemaker, and P. H. Yuster, *Phys. Rev* **B3**, 473 (1971); *Phys. Rev.* **B7**, 3933 (1973); *Phys. Rev.* **B9**, 1913 (1974).

59. D. Schoemaker, *Phys. Rev.* **174**, 1060 (1968).

60. R. B. Murray and F. J. Keller, *Phys. Rev.* **153**, 993 (1967).

61. M. N. Kabler, *Phys. Rev.* **136A**, 1296 (1964).

62. R. T. Williams and M. N. Kabler, *Phys. Rev.* **B9**, 1897 (1974).

63. A. Wasiela, G. Ascarelli, and Y. Merle d'Aubigné, *Phys. Rev. Lett.* **31**, 993 (1973).

64. M. J. Marrone, F. W. Patten, and M. N. Kabler, *Phys. Rev. Lett.* **31**, 467 (1973).

65. L. S. Goldberg, *Phys. Rev.* **168**, 989 (1968).

66. T. Kamejima, S. Shionoya, and A. Fukuda, *J. Phys. Soc. Japan* **32**, 729 (1972).

67. G. Baldini and K. Teegarden, *J. Phys. Chem. Solids* **27**, 943 (1966).

68. D. Pooley, *Proc. Phys. Soc.* **87**, 245 (1966).

69. F. J. Keller and F. W. Patten, *Solid State Commun.* **7**, 1603 (1969).

70. J. N. Bradford, R. T. Williams, W. L. Faust, *Phys. Rev. Lett.* **35**, 300 (1975).

71. W. Känzig, *J. Phys. Chem. Solids* **17**, 80 (1960).

72. D. Schoemaker and J. L. Kolopus, *Phys. Rev.* **B2**, 1148 (1970).

73. Y. H. Chu and R. L. Mieher, *Phys. Rev.* **188**, 1311 (1969).

74. S. Susman, *Phys. Stat. Sol.* **37**, 561 (1970).

75. D. Schoemaker, *Phys. Rev.* **149**, 693 (1966).

76. M. L. Dakss and R. L. Mieher, *Phys. Rev.* **187**, 1053 (1969).

77. C. J. Delbecq, E. Hutchinson, D. Schoemaker, E. L. Yasaitis, and P. H. Yuster, *Phys. Rev.* **187**, 1103 (1969).

78. I. L. Bass and R. L. Mieher, *Phys. Rev.* **175**, 421 (1968).

79. W. J. Plant and R. L. Mieher, *Phys. Rev.* **B7**, 4793 (1973).

80. D. Schoemaker, *Phys. Rev.* **B9**, 1804 (1974).

81. M. H. Cohen, W. Känzig, and T. O. Woodruff, *J. Phys. Chem. Solids* **11**, 120 (1959).

82. R. Hilsch and R. W. Pohl, *Trans. Faraday Soc.* **34**, 883 (1938).

83. N. F. Mott and R. W. Gurney, *Electronic Processes in Ionic Crystals* (Dover, New York, 1964) p. 147.

84. C. J. Delbecq, B. Smaller, and P. H. Yuster, *Phys. Rev.* **104**, 599 (1956).

85. B. S. Gourary, *Phys. Rev.* **112**, 337 (1958).

86. M. de Souza and F. Lüty, *Phys. Rev.* **B8**, 5866 (1973).

87. I. Schneider, *Solid State Commun.* **12**, 161 (1973).

88. M. V. Klein in *Physics of Color Centers*, W. B. Fowler, Ed., (Academic, New York, 1968) Ch. 7.

89. G. P. Montgomery, Jr., W. R. Fenner, M. V. Klein, and T. Timusk, *Phys. Rev.* **B5**, 3343 (1972).

90. W. Barth and B. Fritz, *Phys. Stat. Sol.* **19**, 515 (1967).

91. D. N. Mirlin and I. I. Reshina, *Sov. Phys. Solid State* **8**, 116 (1966).

92. B. Fritz, *J. Phys. Chem. Solids* **23**, 375 (1962).

93. W. Dreybrodt and D. Silber, *Phys. Stat. Sol.* **20**, 337 (1967).

94. R. A. Zhitnikov, V. B. Koltzov, and N. I. Melnikov, *Phys. Stat. Sol.* **26**, 371 (1968).

95. Y. Toyotomi and R. Onaka, *J. Phys. Soc. Japan* **34**, 623 (1973).

96. C. J. Delbecq, W. Hayes, M. C. M. O'Brien, and P. H. Yuster, *Proc. Roy. Soc.* **A271**, 243 (1963).

97. C. J. Delbecq, A. K. Ghosh, and P. H. Yuster, *Phys. Rev.* **151**, 599 (1966).

98. C. J. Delbecq, A. K. Ghosh, and P. H. Yuster, *Phys. Rev.* **154**, 797 (1967).

99. R. S. Knox, *Phys. Rev.* **154**, 799 (1967).

100. K. Cho, *Solid State Commun.* **13**, 439 (1973).

101. R. S. Knox, *J. Phys. Soc. Japan* **18**, Suppl. 2, 268 (1963).

102. S. Nagasaka, M. Ikezawa, and M. Ueta, *J. Phys. Soc. Japan* **20**, 1540 (1965).

103. M. S. Li, M. de Souza, and F. Lüty, *Phys. Rev.* **B7**, 4677 (1973).

104. S. Kapphan and F. Lüty, *Phys. Rev.* **B6**, 1537 (1972).

105. P. H. Yuster and C. J. Delbecq, *J. Chem. Phys.* **21**, 892 (1953).

106. W. Kleeman, *Z. Phys.* **234**, 362 (1970).

107. S. G. Zazubovich, N. E. Lushchik, and C. B. Lushchik, *Opt. Spectr.* **15**, 203 (1964).

108. S. B. S. Sastry, V. Viswanathan, and C. Ramasastry, *J. Phys. Soc. Japan* **35**, 508 (1973).

109. A. Fukuda, K. Inohara, and R. Onaka, *J. Phys. Soc. Japan* **19**, 1274 (1964).

110. A. Fukuda, S. Makishima, T. Mabuchi, and R. Onaka, *J. Phys. Chem. Solids* **28**, 1763 (1967).

111. S. Sugano, *J. Chem. Phys.* **36**, 122 (1962).

112. T. Mabuchi, A. Fukuda, and R. Onaka, *Sci. Light* **15**, 79 (1966).

113. F. E. Williams and P. D. Johnson, *Phys. Rev.* **113**, 97 (1959).

114. Y. Toyozawa and M. Inoue, *J. Phys. Soc. Japan* **21**, 1663 (1966).

115. C. C. Klick and W. D. Compton, *J. Phys. Chem. Solids* **7**, 170 (1958).

116. R. Edgerton, *Phys. Rev.* **138A**, 85 (1965).

117. H. Kamimura and S. Sugano, *J. Phys. Soc. Japan* **14**, 1612 (1959).

118. A. Fukuda, *Phys. Rev.* **B1**, 4161 (1970).

119. A. Ranfagni and G. Viliani, *Phys. Rev.* **B9**, 4448 (1974).

120. A. Yoshikawa, H. Takezoe, and R. Onaka, *J. Phys. Soc. Japan* **33**, 1632 (1972).

121. Y. Uchida and E. Matsui, *J. Phys. Soc. Japan* **20**, 874 (1965).

122. E. Matsui, *J. Phys. Soc. Japan* **22**, 819 (1967).

123. G. K. Herb, M. P. Fontana, and W. J. Van Sciver, *Phys. Rev.* **168**, 1000 (1968).

124. H. R. Zeller and W. Känzig, *Helv. Phys. Act.* **40**, 845 (1967).

125. L. E. Vannotti and J. R. Morton, *Phys. Rev.* **161**, 282 (1967).

126. L. E. Vannotti and J. R. Morton, *J. Chem. Phys.* **47**, 4210 (1967).

127. W. Känzig and M. H. Cohen, *Phys. Rev. Lett.* **3**, 509 (1959).

128. R. H. Silsbee, *J. Phys. Chem. Solids* **28**, 2525 (1967).

129. L. A. Rebane, A. B. Treshchalov, and T. Y. Khaldre, *Sov. Phys. Solid State* **16**, 1460 (1975).

130. M. Ikezawa and J. Rolfe, *J. Chem. Phys.* **58**, 2024 (1973).

131. K. K. Rebane, A. I. Laisaar, L. A. Rebane, and O. I. Sild, *Bull. Acad. Sci. USSR* **31**, 2053 (1968).

132. L. A. Rebane, O. I. Sild, and T. Y. Khaldre, *Bull. Acad. Sci. USSR* **35**, 1276 (1971).

133. J. Schneider, B. Dischler, and A. Räuber, *Phys. Stat. Sol.* **13**, 141 (1966).

134. J. Suwalski and H. Seidel, *Phys. Stat. Sol.* **13**, 159 (1966).

135. U. Kuhn and F. Lüty, *Solid State Commun.* **2**, 281 (1964).

136. H. B. Shore and L. M. Sander, *Phys. Rev.* **B6**, 1551 (1972).

137. D. A. Patterson and M. N. Kabler, *Solid State Commun.* **4**, 75 (1965).

138. F. Fischer and H. Gründig, *Phys. Lett.* **13**, 113 (1964).

139. F. Fischer, *Phys. Lett.* **16**, 246 (1965).

140. W. Sander, *Z. Phys.* **169**, 353 (1962).

141. J. R. Brailsford, J. R. Morton, and L. E. Vannotti, *J. Chem. Phys.* **49**, 2237 (1968).

142. A. Hausmann, *Z. Phys.* **192**, 313 (1966).

143. L. E. Vannotti and J. R. Morton, *Phys. Rev.* **174**, 448 (1968).

144. M. de Wit, R. K. Watts, and W. C. Holton, *Bull. Am. Phys. Soc.* **14**, 354 (1969).

145. J. R. Brailsford, J. R. Morton, and L. E. Vannotti, *J. Chem. Phys.* **50**, 1051 (1969).

146. R. H. Silsbee and I. Bojko, *J. Phys. Chem. Solids* **34**, 1971 (1973).

147. A. Diaz-Góngora and F. Lüty, *Solid State Commun.* **14**, 923 (1974).

148. Y. P. Tsyashchenko and V. M. Zaporozhets, *Sov. Phys. Solid State* **16**, 1594 (1975).

149. S. C. Jain, A. V. R. Warrier, and H. K. Sehgal, *J. Phys.* **C6**, 189 (1973).

150. K. K. Rebane and L. A. Rebane in *Optical Properties of Ions in Solids*, B. Di Bartolo, Ed. (Plenum, New York 1975) Ch. 7.

151. W. E. Bron and W. R. Heller, *Phys. Rev.* **136A**, 1433 (1964).

152. M. Wagner and W. E. Bron, *Phys. Rev.* **139A**, 223 (1965).

153. W. E. Bron and M. Wagner, *Phys. Rev.* **139A**, 233 (1965).

154. R. Röhrig, *Phys. Lett.* **16**, 20 (1965).

155. G. Aguilar S., E. Muñoz P., H. Murrieta S., L. A. Boatner, and R. W. Reynolds, *J. Chem. Phys.* **60**, 4665 (1974).

156. W. E. Bron and M. Wagner, *Phys. Rev.* **145**, 689 (1966).

157. S. Radhakrishna and B. D. Sharma, *Phys. Rev.* **B9**, 2073 (1974).

158. F. K. Fong, J. A. Cape, and E. Y. Wong, *Phys. Rev.* **151**, 299 (1966).

159. G. D. Watkins, *Phys. Rev.* **113**, 79 (1959).

160. G. D. Watkins, *Phys. Rev.* **113**, 91 (1959).

161. S. C. Jain, A. V. R. Warrier, and H. K. Sehgal, *J. Phys.* **C6**, 193 (1973).

162. J. J. Alexander and H. B. Gray, *J. Am. Chem. Soc.* **90**, 4260 (1968).

163. L. H. Jones, *J. Chem. Phys.* **36**, 1209 (1962).

164. N. J. Hawkins, H. C. Mattraw, W. W. Sabol, and D. R. Carpenter, *J. Chem. Phys.* **23**, 2422 (1955).

165. R. P. A. Muniz, N. V. Vugman, and J. Danon, *J. Chem. Phys.* **54**, 1284 (1971).

166. S. C. Jain, K. V. Reddy, and T. R. Reddy, *J. Chem. Phys.* **62**, 4366 (1975).

Index